Environmental Impacts of Hydraulic Fracturing

Environmental Impacts of Hydraulic Fracturing

Frank R. Spellman

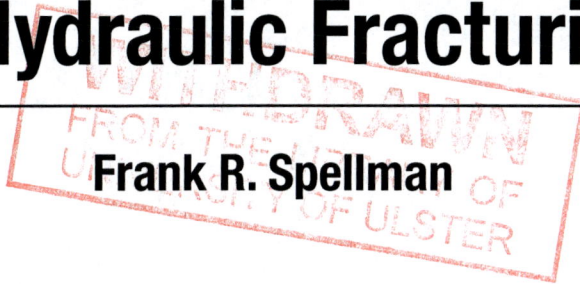

CRC Press is an imprint of the
Taylor & Francis Group, an **informa** business

CRC Press
Taylor & Francis Group
6000 Broken Sound Parkway NW, Suite 300
Boca Raton, FL 33487-2742

Printed in the United States of America on acid-free paper
Version Date: 20120501

International Standard Book Number: 978-1-4665-1467-6 (Hardback)

Library of Congress Cataloging-in-Publication Data

Spellman, Frank R.
Environmental impacts of hydraulic fracturing / Frank R. Spellman.
p. cm.
Summary: "Hydraulic fracturing is a well stimulation process used to maximize the extraction of hard-to-reach underground resources. While hydraulic fracturing allows access to previously unobtainable energy supplies, the technology can cause serious damage to environmental media. Public concerns have focused on the impacts of the process used during natural gas production from shale and coalbed methane formations. This book discusses the pros and cons of hydraulic fracturing in a balanced manner. It emphasizes the need for the mining of natural gas to provide future clean energy supplies but also stresses the need for close monitoring of the process, chemicals, and fate of waste products"-- Provided by publisher.
Includes bibliographical references and index.
ISBN 978-1-4665-1467-6 (hardback)
1. Hydraulic fracturing--Environmental aspects. I. Title.

TD195.G3S75 2012
333.8'2314--dc23 2012015793

Visit the Taylor & Francis Web site at
http://www.taylorandfrancis.com

and the CRC Press Web site at
http://www.crcpress.com

For

Nancy Lutje

Suzanne Wilson

Nobody realizes that some people expend tremendous energy merely to be normal.

—Albert Camus (1913–1960)

Contents

Preface xiii
Author xv
Acronyms and Abbreviations xvii

1. **Introduction** 1
   1.1 Shalenanza 1
   1.2 Other Side of the Coin 2
   1.3 Earthquakes 4
      1.3.1 What Causes Earthquakes? 4
      1.3.2 Seismology 5
      1.3.3 Earthquake Magnitude and Intensity 6
   1.4 Internal Structure of Earth 7
   1.5 Key Definitions 8
      1.5.1 Fracking Terminology 9
   1.6 Purpose of Text 17
   1.7 Thought-Provoking Discussion Questions 17
   References and Recommended Reading 18

2. **Shale Gas** 19
   2.1 Importance of Shale Gas 19
   2.2 Versatility of Natural Gas 22
   2.3 Alternative or Renewable Energy 25
   2.4 The 411 on Natural Gas 27
   2.5 The 411 on Processing and Refining Natural Gas 29
      2.5.1 Natural Gas Refining: The Process 29
   2.6 Thought-Provoking Discussion Questions 32
   References and Recommended Reading 33

3. **Shale Gas Geology** 35
   3.1 Introduction 35
   3.2 Sedimentation 35
   3.3 Types of Sedimentary Rocks 36
      3.3.1 Clastic Sedimentary Rocks 37
      3.3.2 Recoverable Resources 40
      3.3.3 Determining Grade of Oil Shale 41
      3.3.4 Origin of Organic Matter 42
      3.3.5 Thermal Maturity of Organic Matter 43
      3.3.6 Classification of Oil Shale 43
      3.3.7 Evaluation of Oil Shale Resources 45
      3.3.8 U.S. Oil Shale Deposits 46

3.4 Chemical and Organic Sedimentary Rocks ... 52
3.4.1 Chemical Sedimentary Rocks ... 52
3.4.2 Biochemical Sedimentary Rocks ... 53
3.5 Physical Characteristics of Sedimentary Rocks ... 54
3.6 Sedimentary Rock Facies ... 55
3.7 Thought-Provoking Discussion Questions ... 56
References and Recommended Reading ... 56

4. Shale Gas Plays ... 61
4.1 Unconventional Natural Gas Resources ... 61
4.1.1 Natural Gas Measurement ... 63
4.1.2 Natural Gas Calculation ... 63
4.1.3 U.S. Natural Gas Reserves ... 64
4.2 Distribution of Unconventional Natural Shale Gas ... 64
4.3 Fast Forward to the Future ... 66
4.4 Thought-Provoking Discussion Questions ... 67
References and Recommended Reading ... 67

5. Shale Gas Sources ... 69
5.1 Shale Gas in the United States ... 69
5.2 Geologic Time Scale ... 70
5.3 Active U.S. Shale Gas Plays ... 71
5.3.1 Barnett Shale Play ... 74
5.3.2 Fayetteville Shale Play ... 75
5.3.3 Haynesville Shale Play ... 77
5.3.4 Marcellus Shale Play ... 79
5.3.5 Woodford Shale Play ... 82
5.3.6 Antrim Shale Play ... 83
5.3.7 New Albany Shale Play ... 85
5.4 Thought-Provoking Discussion Questions ... 86
References and Recommended Reading ... 88

6. Hydraulic Fracturing: The Process ... 91
6.1 Poor Farm Folk to Texas Rich ... 91
6.2 Drilling for Water Versus Drilling for Shale Gas ... 93
6.3 Water Well Systems ... 94
6.3.1 Water Well Site Requirements ... 95
6.3.2 Types of Water Wells ... 95
6.4 Components of a Water Well ... 97
6.4.1 Well Casing ... 97
6.4.2 Grout ... 97
6.4.3 Well Pad ... 97
6.4.4 Sanitary Seal ... 98
6.4.5 Well Screen ... 98
6.4.6 Casing Vent ... 99

6.4.7 Drop Pipe ... 99
6.4.8 Miscellaneous Well Components ... 99
6.5 Shale Gas Drilling Development Technology ... 100
6.6 Drilling ... 100
6.6.1 Well Casing Construction ... 106
6.6.2 Drilling Fluids and Retention Pits ... 109
6.7 Hydraulic Fracturing ... 110
6.7.1 Fracture Design ... 112
6.7.2 Hydraulic Fracturing Process ... 113
6.7.3 Fracturing Fluids ... 117
6.8 Fracking Water Supply ... 121
6.9 Thought-Provoking Discussion Questions ... 126
References and Recommended Reading ... 126

7. Chemicals Used in Hydraulic Fracturing ... 131
7.1 Background ... 131
7.1.1 Types of Fracturing Fluids and Additives ... 131
7.1.2 Naturally Occurring Radioactive Material ... 140
7.2 Fracking Fluids and Their Constituents ... 141
7.2.1 Commonly Used Chemical Components ... 143
7.2.2 Toxic Chemicals ... 144
7.3 Trade Secret and Proprietary Chemicals ... 148
7.4 Thought-Provoking Discussion Questions ... 149
References and Recommended Reading ... 149

8. Environmental Considerations ... 153
8.1 Introduction ... 153
8.2 Access Roads, Well Pads, and Mineral Rights ... 154
8.3 Horizontal Wells ... 156
8.3.1 Reducing Surface Disturbance ... 157
8.3.2 Reducing Wildlife Impacts ... 158
8.3.3 Reducing Community Impacts ... 159
8.3.4 Protecting Groundwater ... 161
8.4 Thought-Provoking Discussion Questions ... 163
References and Recommended Reading ... 163

9. Laws and Regulations Affecting Shale Gas Development and Operations ... 165
9.1 Introduction ... 165
9.1.1 A Network of Confusing and Constraining Rules and Standards ... 166
9.1.2 Costly Modifications of Existing Installations to Meet New Legal Demands ... 166
9.1.3 Inspections, Fines, or Time-Consuming Legal Hearings ... 167

9.1.4 Above All, an Increasingly Burdensome Task of Recordkeeping and Paperwork 168
9.2 Environmental Regulations for Shale Gas Operations 170
9.3 Pollution: Effects Often Easy to See, Feel, Taste, or Smell 172
9.4 Regulatory Structure 174
9.4.1 Federal Environmental Laws and Shale Gas Development 174
9.4.2 State Regulations 175
9.4.3 Local Regulation 177
9.5 Water Quality Regulations 179
9.5.1 Genesis of Clean Water Reform 179
9.6 Clean Water Act 181
9.7 Safe Drinking Water Act 187
9.7.1 SDWA Definitions 189
9.7.2 SDWA Specific Provisions 192
9.7.3 1996 Amendments to SDWA 199
9.7.4 Implementing SDWA 200
9.7.5 Underground Injection Control 201
9.8 Oil Pollution Act of 1990 207
9.9 Air Quality 208
9.9.1 Air Pollution 209
9.9.2 Atmospheric Dispersion, Transformation, and Deposition 210
9.9.3 Dispersion Models 219
9.9.4 Major Air Pollutants 221
9.10 Clean Air Act 229
9.10.1 Clean Air Act Titles 230
9.10.2 Use of Meteorology in Air Quality Regulatory Programs 236
9.11 Clean Air Act and Hydraulic Fracturing Operations 241
9.11.1 Stationary Internal Combustion Engines 242
9.11.2 Reciprocating Internal Combustion Engines 243
9.11.3 Air Quality Permits 244
9.12 Land (Soil) Quality 244
9.12.1 Resource Conservation and Recovery Act 244
9.12.2 Endangered Species Act 247
9.13 Oil and Gas Operations on Public Lands 249
9.14 Superfund (CERCLA) 250
9.15 Emergency Planning and Community Right-to-Know Act 251
9.16 OSHA Oil and Gas Drilling, Servicing, and Storage Standards 252
9.16.1 Occupational Safety and Health Act of 1970 252
9.16.2 SIC Industry Group 138—Oil and Gas Field Services: Process Description 257

9.17 OSHA Standards Applicable to Shale Gas Operations 262
9.17.1 Subpart D, Walking–Working Surfaces 262
9.17.2 Subpart E, Means of Egress 264
9.17.3 Subpart F, Powered Platforms, Manlifts, and Vehicle-Mounted Work Platforms 271
9.17.4 Subpart G, Occupational Health and Environmental Control 278
9.17.5 Subpart H, Hazardous Materials 281
9.17.6 Subpart I, Personal Protective Equipment 285
9.17.7 Subpart J, General Environmental Controls 305
9.17.8 Subpart K, Medical and First Aid 341
9.17.9 Subpart L, Fire Protection 342
9.17.10 Subpart N, Materials Handling and Storage 346
9.17.11 Subpart O, Machinery and Machine Guarding 361
9.17.12 Subpart P, Hand and Portable Powered Tools and Other Hand-Held Equipment 375
9.17.13 Subpart Q, Welding, Cutting, and Brazing 376
9.17.14 Subpart S, Electrical 389
9.17.15 Subpart Z, Toxic and Hazardous Substances 397
9.18 Thought-Provoking Discussion Questions 402
References and Recommended Reading 403

Glossary 411

Appendix A. Chemicals Used in Hydraulic Fracturing 421

Appendix B. Case Study Locations for Hydraulic Fracturing Study 441

Index 445

Preface

With regard to gaining U.S. energy independence and a sustainable, reliable source of energy, the current push to develop renewable energy sources, including hydropower, wave and tidal power, geothermal energy, bioenergy, wind-derived power, solar energy, and fuel cell technology, makes good sense. Although there is considerable argument about which of these potential long-term energy sources is best to develop (i.e., holds the most promise), there can be little doubt as to the need for innovation and the development of viable renewable energy sources.

Just when many of us had decided that the United States was energy poor, that we are running out of hydrocarbon supplies or are lacking relatively easy access to potential hydrocarbon supplies, technology is advancing to the point where natural gas supplies recently thought to be nonexistent or too difficult to mine are now being discovered, mined, and processed and are now available for both industrial and consumer use. Not only is this important for future U.S. natural gas supplies but natural gas also plays a key role in our nation's pursuit of a clean energy future.

Technological development that has made access to newly discovered massive supplies of natural gas is not actually new technology; it has been around for years. Indeed, *hydraulic fracturing*, often called *fracking*, *fracing*, or *hydrofracking* (or even *well stimulation*, as many in the industry like to call it), is a well-known practice that was developed by Haliburton almost 60 years ago. The use of hydraulic fracturing is a double-edged sword, a Dr. Jekyll and Mr. Hyde operation (i.e., both a benefit and a liability). On the one side, the fracturing process, known as a *frack job*, involves the pressurized injection of fluids commonly made up of water and chemical additives into a geologic formation (e.g., gas-bearing shale). The pressure exceeds the rock strength, and the fluid opens or enlarges fractures in the rock. As the formation is fractured, a *propping agent*, such as ceramic or sand beads (even peanut and walnut shells have been used) is pumped into the fractures to keep them from closing as the pumping pressure is released. The fracturing fluids (water and chemical additives) are then returned back to the surface. Natural gas will flow from pores and fractures in the rock into the well for subsequent extraction.

Hydraulic fracturing has proven to be one viable way of accessing vital resources such as natural gas, oil, and geothermal energy. Simply, hydraulic fracturing has helped to expand natural gas production in the United States by unlocking large natural gas supplies in shale and other unconventional formations across the country. The results of hydraulic fracturing are startling. Natural gas production in 2010 reached the highest level in decades. According to the most recent estimates by the Energy Information Administration (EIA),* the United States possesses natural gas resources sufficient to supply the country for more than 110 years. Moreover, the technology has also been used successfully to stimulate water wells, whereby the fluid used is usually pure water (typically water and a chlorine-based disinfectant, such as bleach). Hydraulic fracturing has also been used to remediate waste spills by injecting air, bacteria, or other materials into a subsurface contaminated zone.

On the other edge of the fracking double-edged sword—in line with the old adage that anything that has a good side probably has a few negative aspects lingering in the background or in the bushes somewhere—hydraulic fracturing certainly has its critics, who point out that the process is not exactly a bed of beautiful roses. As the use of hydraulic fracturing has grown, so have concerns about its environmental and public health impacts; for example, a 2011 U.S. House of Representatives Committee Report† observed that hydraulic fracturing raises myriad concerns, including risks to air quality, migration of gases and hydraulic fracturing chemicals to the surface, potential mishandling of wastes (especially wastewaters), land subsidence, and, most importantly, the contamination of ground water. Of these varied concerns the most significant to date is the hydraulic fracturing fluids used to fracture rock formations. These contain numerous chemicals that could harm human health and the environment, especially if they enter and contaminate drinking water supplies. Compounding the concerns of Congress, local politicians, and environmentalists is the resistance of many oil and gas companies to publicly disclosing the chemicals they use.

Environmental Impact of Hydraulic Fracturing provides a fair and balanced discussion and comprehensive guide to every aspect of the process of hydraulic fracturing used to extract natural gas, along with gas exploration and production in various shale fields. The book provides comprehensive coverage of all aspects of the issue, including ongoing controversies about the environmental and operator safety issues arising from possible water pollution, drinking water contamination, on-the-job safety hazards, and harmful chemical exposure to workers and residents near well areas. Several case studies on the fracking process and its ramifications are provided. This book is intended as a reference book for administrators; legal professionals; research engineers; graduate students in chemical, natural gas, petroleum, or mechanical engineering; non-engineering professionals; and the general reader.

---

* EIA, *U.S. Natural Gas Monthly Supply and Disposition Balance*, U.S. Energy Information Administration, Washington, DC, 2012 (online at (http://www.eia.gov/dnav/ng/ng_sum_sndm_s1_m.htm).

† USHR, *Chemicals Used in Hydraulic Fracturing*, Committee on Energy and Commerce, U.S. House of Representatives, Washington, DC, 2011.

Author

Frank R. Spellman, PhD, is a retired U.S. Naval Officer with 26 years of active duty, a retired environmental safety and health manager for a large wastewater sanitation district in Virginia, and a retired assistant professor of environmental health at Old Dominion University, Norfolk, Virginia. He is the author or co-author of 75 books, with more soon to be published. Dr. Spellman consults on environmental matters with the U.S. Department of Justice and various law firms and environmental entities around the globe. He holds a BA in public administration, BS in business management, and MBA, MS, and PhD in environmental engineering. In 2011, he traced and documented the ancient water distribution system at Machu Pichu, Peru, and surveyed several drinking water resources in Amazonia, Ecuador.

Acronyms and Abbreviations

API	American Petroleum Institute
bbl	Barrel, petroleum (42 gallons)
bcf	Billion cubic feet
BLM	Bureau of Land Management
BMP	Best Management Practices
Btu	British thermal unit
CAA	Clean Air Act
CBNG	Coal bed natural gas
CEQ	Council on Environmental Quality
CERCLA	Comprehensive Environmental Response, Compensation, and Liability Act
CFR	Code of Federal Regulations
CH_4	Methane
CO	Carbon monoxide
CO_2	Carbon dioxide
COE	Coefficient of oil extraction
CWA	Clean Water Act
DRBC	Delaware River Basin Commission
EIA	Energy Information Administration
ELG	Effluent Limitation Guidelines
EPA	Environmental Protection Agency
EPCRA	Emergency Planning and Community Right-to-Know Act
FR	*Federal Register*
ft	Foot/feet
FWS	Fish and Wildlife Service
gal	Gallon
GHG	Greenhouse gases
GWPC	Ground Water Protection Council
H_2S	Hydrogen Sulfide
HAP	Hazardous Air Pollutant
HCl	Hydrochloric acid
IOGCC	Interstate Oil and Gas Compact Commission
IR	Infrared
Mcf	1000 cubic feet
MMcf	1,000,000 cubic feet
mrem	Millirem
mrem/yr	Millirem per year
MSDS	Material Safety Data Sheet
NEPA	National Environmental Policy Act
NESHAPS	National Emission Standards for Hazardous Air Pollutants

NETL	National Energy Technology Laboratory
NORM	Naturally occurring radioactive material
NO_x	Nitrogen oxides
NPDES	National Pollution Discharge Elimination System
NYDEC	New York State Department of Environmental Conservation
O_3	Ozone
OPA	Oil Pollution Act
OSHA	Occupational Safety and Health Administration
PM	Particulate matter
ppm	Parts per million
RAPPS	Reasonable and Prudent Practices for Stabilization
RCRA	Resource Conservation and Recovery Act
RP	Recommended Practice
RQ	Reportable quantity
SARA	Superfund Amendments and Reauthorization Act
SCF	Standard cubic feet
SDWA	Safe Drinking Water Act
SO_2	Sulfur dioxide
SPCC	Spill Prevention, Control, and Countermeasures
SRBC	Susquehanna River Basin Commission
STRONGER	State Review of Oil and Natural Gas Environmental Regulations, Inc.
SWDA	Solid Waste Disposal Act
tcf	Trillion cubic feet
TDS	Total dissolved solids
tpy	Tons per year
TRI	Toxic Release Inventory
UIC	Underground Injection Control
USC	United States Code
USDW	Underground source of drinking water
USGS	United States Geological Survey
VOC	Volatile organic compound
WQA	Water Quality Act
yr	Year

1

Introduction

There is nothing new about splitting or cracking rocks to release substances that mankind needs or desires. Consider, for example:

> He split the rocks in the wilderness,
> And gave them drink abundantly as out of the depths.
> He brought streams also out of the rock,
> And caused waters to run down like rivers.
>
> **—Psalm 78:15–18 (World English translation)**

1.1 Shalenanza

Most of us are familiar with the term *bonanza*, which refers to something that is very valuable, profitable, or rewarding. Moreover, many of us have an even more tangible or palatable feeling for the meaning of the glittery term—specifically, those who have witnessed, studied, or experienced a rags-to-riches metamorphosis, whereby they themselves or people they know have been impacted by the discovery of exceptionally large and rich mineral deposits of, for example, ores, precious metals, petroleum, or natural gas.

On the other hand, readers who have noticed the title of this section are scratching their heads and wondering … shale what? Well, I can't say that I am surprised because I just invented the term, so bear with me as I explain what *shalenanza* means. First of all, suppose that you reside in a hard-luck, high-unemployment town such as Youngstown, Ohio, and a new industrial plant is being built that will employ hundreds. This new plant will produce seamless steel pipes for tapping shale formations for oil or natural gas. Thus, a rust belt town with over 11% unemployment will now have at least 450 new jobs. The source of Youngtown's new-found good luck is the vast stores of natural gas in the Marcellus and Utica shale formations that have set off a modern shale gas rush to grab leases and secure permits to drill. The Marcellus boom could offer large numbers of jobs for more than 50 years.

Now, the obvious question: Is the economic windfall enjoyed by Youngstown unique to that area only? The simple and compound answer to this question is no. Similar hopes are alive for other hard-luck towns in Ohio, Pennsylvania, West Virginia, New York, Kentucky, Tennessee, and Alabama. Consider, for example, Lorain, Ohio, where U.S. Steel will add 100 jobs with a $100 million upgrade of a plant that makes seamless pipe for construction or gas exploration and production industries (Sheeran, 2011). Also, in Pittsburgh, U.S. Steel will eventually add a significant number of jobs with construction of a multibillion dollar petrochemical refinery; Ohio had been in the running for the plant. The refinery will convert natural gas liquids to other chemicals used in heating fuel, power generation, transportation fuel, plastics, tires, fabrics, glass, paint, and antifreeze (Anon., 2012). Beneficiaries of shale development and the manufacture of ancillary equipment also include railcar industries; the shale industry has caused an increased need for freight cars, and at present producers of such rail cars have a backlog of orders.

What we have here is a modern-day gold rush or, more correctly, a shale gas rush, which I have termed a *shalenanza*, which has created many *shaleionaires* (a term invented by someone else). Simply, in a tough economy, when families are struggling to make ends meet and business owners are facing declining revenues and tough choices, the windfall generated by and garnered from this *shalenanza* can only be characterized as an economic blessing. One group of folks who know exactly what I am referring to here and who know the exact meaning of the term *shalenanza* are those poor dirt farmers in the hill country of western Pennsylvania. These are folks who have been working hard to eke out a living from the land while at the same time drawing unemployment compensation and food stamps because of lost manufacturing jobs in a depressed economic region. Investors and speculators knocking down their doors to have them sign lucrative leases to the mineral rights on their property is a morale booster that might leave most landowners speechless, scratching their brains cells for the right words to say. May I suggest one word: *shalenanza*?

1.2 Other Side of the Coin

(*Note:* Many of the issues discussed in this section are presented only briefly here but are developed to a greater extent later in the text.) Like many bonanzas or gold rushes in the past, the shale gas shalenanza or rush has made a few people rich and others desperate or downright miserable. The problem? Take your pick. A few landowners in shale-rich areas have received thousands of dollars an acre in upfront payments (with the promise of thousands more to come in royalties) for the right to drill under their property. These folks are the so-called shaleionaires.

At the same time, some of their neighbors—many of whom are also shaleionaires—have drawn water from their taps that can only be characterized as brown, smelly, and, on occasion, explosive. Because shale gas is imbedded in dense, low-permeability rock, drillers use a mixture of water, sand, ceramic beads, peanut shells, and chemicals to open up fissures in the stone through which the gas can escape up. This, of course, is the process that this text is concerned with—*hydraulic fracking* or, more colloquially, *fracking* or *fracing*. Critics of fracking have pointed the finger of blame for contaminated tapwater (and other environmental issues) at former Vice President Dick Cheney, who largely authored the 2005 Energy Bill (the so-called Halliburton Loophole), which explicitly exempted fracking from federal review under the Safe Drinking Water Act (SDWA). Under this provision, drilling companies are under no obligation to make public which chemicals they use, although many of them are recognized or suspected carcinogens. The U.S. Environmental Protection Agency (USEPA) is currently investigating cases of suspected contamination in towns located near fracking activities.

Beyond contaminated tapwater, other issues of concern are related to environmental contamination of water, air, and the land. With regard to tapwater and other water contamination events related to shale natural gas extraction, the primary concerns are threefold: (1) contamination of surface water due to erosion and groundcover removal during drilling site extraction; (2) proper management of separate but multiple users in a single watershed to protect water quality and ensure adequate water resources to meet the needs of the watershed stakeholders; and (3) treatment and safe disposal of the water produced.

With regard to air quality, natural gas is often lauded as the cleanest of all fossil fuels, and its air quality benefits are rarely disputed; however, it is the *production* of natural gas that is the problem. Its extraction from shale impacts air quality and releases greenhouse gases into the atmosphere. The impact of these air emissions released during drilling and production varies, depending on the phase of the drilling operation. And then there is a potential air emission problem when drilling and fracking are completed, production begins, and permanent emission sources are established (e.g., compressor engines, venting or leaking condensate tanks, collection ponds). Fluids brought to the surface can include a mixture of natural gas, water, and hydrocarbon liquids—the greater the amount of water and hydrocarbon liquids, the "wetter" the gas. Wet gas must be dehydrated to separate the gases from the water and hydrocarbons, resulting in the production of a *condensate*. Hydrocarbons can be released from the condensate whenever it is stored and transported from the site to refineries for incorporation into liquid fuels. In addition, fugitive and intermittent emissions from equipment and transmissions also occur.

The surrounding drilling and hydraulic fracturing land area can also be subject to contamination. Specifically, surrounding farmland and forests near the wellhead can be severely impacted not only by drilling and fracking activities but also by the movement of heavy drilling and fracking equipment over the land areas, causing soil compaction (League of Women Voters, 2009). Another recent issue with hydraulic fracking for oil and natural gas is earthquakes. Some are beginning to question whether the magnitude

5.6 earthquake that rocked Oklahoma on November 5, 2011, was caused by fracking. The following section provides a basic understanding of earthquakes and their causes.

1.3 Earthquakes

> *It's been raining a lot, or very hot—it must be earthquake weather!*
> *FICTION:* Many people believe that earthquakes are more common in certain kinds of weather. In fact, no correlation with weather has been found. Earthquakes begin many kilometers (miles) below the region affected by surface weather. People tend to notice earthquakes that fit the pattern and forget the ones that don't. Also, every region of the world has a story about earthquake weather, but the type of weather is whatever they had for their most memorable earthquake.
>
> **—USGS (2009)**

1.3.1 What Causes Earthquakes?

Anyone who has witnessed or studied one of the over a million or so earthquakes that occur each year on Earth is unlikely to forget such occurrences. Even though most earthquakes are insignificant, a few thousand of them produce noticeable effects such as tremors and ground shaking. The passage of time has shown that about 20 earthquakes each year cause major damage and destruction. It is estimated that about 10,000 people die each year because of earthquakes. Over the millennia, the effects of damaging earthquakes have been obvious to those who witness the results; however, the cause of earthquakes has not been as obvious. The cause of earthquakes has shifted from being the wrath of mythical beasts to the wrath of gods, from being unexplainable magical occurrences to normal, natural phenomena occasionally required to retain Earth's structural integrity. We can say, overall, that an earthquake on Earth provides our planet with a sort of a geological homeostasis required to maintain life as we know it. Through the ages earthquakes have come under the scrutiny of some of the world's greatest writers. Consider, for example, Voltaire's classic satirical novel, *Candide*, published in 1759, in which he mercilessly satirizes science and, in particular, earthquakes. Voltaire based his observations on the great Lisbon, Portugal, earthquake that occurred in 1759 and caused the deaths of more than 60,000 people. Upon viewing the total devastation of Lisbon, Dr. Pangloss says to Candide:

> ... the heirs of the dead will benefit financially; the building trade will enjoy a boom. Private misfortune must not be overrated. These poor people in their death agonies, and the worms about to devour them, are playing their proper and appointed part in God's master plan.

Although we still do not know what we do not know about earthquakes and their causes, we have evolved from using witchcraft or magic to explain their origins to the scientific methods employed today. In the first place, we do know that earthquakes are caused by the sudden release of energy along a fault. Earthquakes are usually followed by a series of smaller earthquakes that we refer to as *aftershocks*. Aftershocks represent further adjustments of rock along the fault. There are currently no reliable methods for predicting when earthquakes will occur.

We have developed a couple of theories with regard to the origin of earthquakes. One of these theories suggests that earthquakes occur via *elastic rebound*. According to the elastic rebound theory, subsurface rock masses subjected to prolonged pressures from different directions will slowly bend and change shape. Continued pressure sets up strains so great that the rocks will eventually reach their elastic limit and rupture (break), suddenly snapping back into their original unstrained state. It is the snapping back (elastic rebound) that generates the seismic waves radiating outward from the break. The greater the stored energy (strain), the greater the release of energy. The coincidence of many active volcanic belts with major belts of earthquake activity indicates that volcanoes and earthquakes may have a common cause. Plate interactions commonly cause both earthquakes (tectonic earthquakes) and volcanoes.

1.3.2 Seismology

Seismology is generally considered to be the study of earthquakes, but it is actually the study of how seismic waves behave in the Earth. The source of an earthquake is the *hypocenter* or *focus* (i.e., the exact location within the Earth where seismic waves are generated). The *epicenter* is the point on the Earth's surface directly above the focus. Seismologists want to know where the focus and epicenter are located so a comparative study of the behavior of the earthquake event can be made with previous events to further our understanding of earthquakes. Seismologists use instruments to detect, measure, and record seismic waves. Generally, the instrument used is the *seismograph*, which has been around for a long time. Modern improvements have upgraded these instruments from paper or magnetic tape strips to electronically recorded digital data. The relative arrival times of the various types of waves at a single location can be used to determine the distance to the epicenter. To determine the exact epicenter location, records from at least three widely separated seismograph stations are required.

1.3.2.1 Seismic Waves

Some of the energy released by an earthquake travels through the Earth. The speed of these seismic waves depends on the density and elasticity of the materials through which they travel. Seismic waves come in several types:

- *P-waves*—Primary, pressure, or push–pull waves arrive first and are the first waves to be detected by a seismograph. They are compressional waves that expand and contract to travel through solids, liquids, or gases at speeds varying from 3.4 to 8.6 miles per second. P-waves move faster at depth, depending on the elastic properties of the rock through which they travel. P-waves are the same thing as sound waves.
- *S-waves*—Secondary or shear waves travel at a velocity between 2.2 and 4.5 miles per second, depending on the rigidity and density of the material through which they travel. They are the second set of waves to arrive at the seismograph and will not travel through gases or liquids; thus, the velocity of S-waves through gases or liquids is zero.
- *Surface waves*—Several types of surface waves travel along the Earth's outer layer or on layer boundaries within the Earth. These rolling, shaking waves are the slowest waves but the ones that cause damage in large earthquakes.

1.3.3 Earthquake Magnitude and Intensity

The size of an earthquake is measured using two parameters—energy released (*magnitude*) and damage caused (*intensity*).

1.3.3.1 Earthquake Magnitude

The size of an earthquake is usually given in terms of its *Richter magnitude*, which was devised by Charles Richter. The Richter magnitude measures the amplitude (height) of the largest recorded wave at a specific distance from the earthquake. A better measure is the *Richter scale*, which measures the total amount of energy released by an earthquake as recorded by seismographs. The amount of energy released is related to the Richter scale by the equation:

$$\text{Log}\, E = 11.8 + 1.5M$$

where

Log = logarithm to base 10
E = energy released (in ergs)
M = Richter magnitude

When using the equation to calculate the Richter magnitude, it quickly becomes apparent that each increase of 1 in the Richter magnitude yields a 31-fold increase in the amount of energy released. Thus, a magnitude 6 earthquake releases 31 times more energy than a magnitude 5 earthquake. A magnitude 9 earthquake releases 31 × 31 or 961 times more energy than a magnitude 7 earthquake.

TABLE 1.1
Modified Mercalli Intensity Scale

Intensity	Description
I	Not felt except under unusual conditions
II	Felt by only a few on upper floors
III	Felt by people lying down or seated
IV	Felt by many indoors, by few outside
V	Felt by everyone, people awakened
VI	Trees sway, bells ring, some objects fall
VII	Causes alarm, walls and plaster crack
VIII	Chimneys collapse, poorly constructed buildings are seriously damaged
IX	Some houses collapse, pipes break
X	Ground cracks, most buildings collapse
XI	Few buildings survive, bridges collapse
XII	Total destruction

1.3.3.2 Earthquake Intensity

Earthquake intensity is a rough measure of an earthquake's destructive power—that is, its size and strength, or how much the Earth shook at a given place near the source of an earthquake. To measure earthquake intensity, Mercalli in 1902 devised an intensity scale of earthquakes based on the impressions of people involved, movement of furniture and other objects, and damage to buildings. The shock is most intense at the epicenter, which, as noted earlier, is located on the surface directly above the focus. Mercalli's intensity scale uses a series of numbers (on a scale of 1 to 12) to indicate different degrees of intensity (see Table 1.1). Keep in mind that this scale is somewhat subjective, but it provides both a qualitative and systematic evaluation of earthquake damage.

DID YOU KNOW?

Although it is correct to say that for each increase of 1 in Richter magnitude, there is a tenfold increase in amplitude of the wave, it is incorrect to say that each increase of 1 in Richter magnitude represents a tenfold increase in the size of the earthquake.

1.4 Internal Structure of Earth

Information obtained from seismographs and other instruments indicate that the lithosphere may be divided into three zones: crust, mantle, and core (see Figure 1.1):

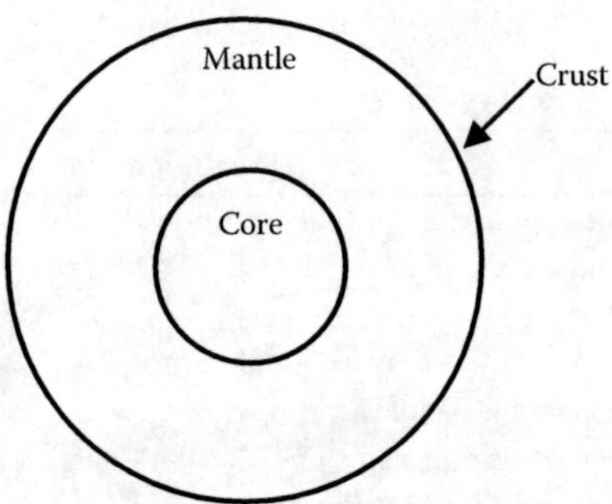

FIGURE 1.1
Internal structure of Earth.

- *Earth's crust*—The outmost and thinnest layer of the lithosphere is the crust. There are two different types of crust: (1) thin (as little as 4 miles in places) oceanic crust composed primarily of basalt that underlies the ocean basins, and (2) thicker continental crust (primarily granite 20 to 30 miles thick) that underlies the continents.
- *Earth's mantle*—Beneath the crust is an 1800-mile-thick intermediate, dense, hot zone of semisolid rock known as the mantle, which is thought to be composed mainly of olivine-rich rock.
- *Earth's core*—Earth's core is about 4300 miles in diameter. It is thought to be composed of a very hot, dense iron and nickel alloy. The core is divided into two different zones. The outer core is a liquid because the temperatures there are adequate to melt the iron–nickel alloy. The highly pressurized inner is core solid because the atoms are tightly crowded together.

As shown in the box on the next page, the controversy over whether earthquakes are generated because of wastewater injection and other fracking practices is bound to continue for some time. At present, most experts, including this author, are of the mind that the jury is still out. Scientists do not like to live by conjecture alone. There may be some truth to the earthquake fracking wastewater injection process, but the correlation has yet to be proven.

1.5 Key Definitions

Every branch of science, every profession, and every engineering process has its own language for communication. Hydraulic fracturing and shale-oil mining are no different. To work even at the edge of oil-shale fracking, you must acquire a fundamental vocabulary for the processes involved.

DID YOU KNOW?

In December 2011, the Associated Press reported that several officials believed that the 4.0 earthquake that struck northeast Ohio, outside Youngstown, was the 11th in a series of minor earthquakes in the area, many of which struck near the Youngstown injection well. Environmentalists and property owners living near the gas drilling well have questioned the safety of fracking to the environmental and public health. Federal regulators, however, have declared the technology safe. On January 1, 2012, an official in Ohio said that the underground disposal of wastewater from natural gas drilling operations would be halted in the Youngstown area until scientists could analyze data from the most recent string of earthquakes there (Fountain, 2012).

As Voltaire said, "If you wish to converse with me, define your terms." In this section, we define the terms and concepts used by fracking practitioners in applying their skills to make their technological endeavors bear fruit. These terms and concepts are presented early in the text so readers can become familiar with them now, before the text addresses the issues these terms describe. The practicing fracking engineer or student of fracking should understand these terms and concepts; otherwise, it will be difficult (if not impossible) to practice or understand fracking.

Hydraulic fracturing and shale gas drilling have an extensive and unique terminology that is generally well defined, but a few terms are not only poorly defined but also defined from different and conflicting points of view. Anytime we look to a definition for meaning, we are wise to remember the words of Voltaire, as well as those of another great philosopher, Yogi Berra, who defined things in his own unique way—for example, "95% of baseball is pitching, the other 50% is hitting."

1.5.1 Fracking Terminology

- *Air quality*—A measure of the amount of pollutants emitted into the atmosphere and the dispersion potential of an area to dilute those pollutants.
- *Aquifer*—A body of rock that is sufficiently permeable to conduct groundwater and to yield economically significant quantities of water to wells and springs.
- *Basin*—A closed geologic structure in which the beds dip toward a center location; the youngest rocks are at the center of a basin and are partly or completely ringed by progressively older rocks.
- *Bcf*—Billion cubic feet, a gas measurement equal to 1,000,000,000 cubic feet.

- *Biogenic gas*—Natural gas produced by living organisms or biological processes.
- *Btu*—British thermal unit, the amount of energy required to heat 1 pound of water by 1°F.
- *Casing*—Steel piping positioned in a wellbore and cemented in place to prevent the soil or rock from caving in. It also serves to isolate fluids, such as water, gas, and oil, from the surrounding geologic formations.
- *Coal bed methane (CBM)/coal bed methane gas (CBMG)*—A clean-burning natural gas found deep inside and around coal seams. The gas has an affinity to coal and is held in place by pressure from groundwater. CBMG is produced by drilling a wellbore into the coal seams, pumping out large volumes of groundwater to reduce the hydrostatic pressure, and allowing the gas to dissociate from the coal and flow to the surface.
- *Completion*—The activities and methods required to prepare a well for production following drilling, including installation of equipment for production from a gas well.
- *Corridor*—A strip of land through which one or more existing or potential utilities may be co-located.
- *Directional drilling*—The technique of drilling at an angle from a surface location to reach a target formation not located directly underneath the well pad.
- *Disposal well*—A well that injects produced water into an underground formation for disposal.
- *Drill rig*—The mast, draw works, and attendant surface equipment of a drilling or workover unit.
- *Emission*—Air pollution discharge into the atmosphere, usually specified by mass per unit time.
- *Endangered species*—Those species of plants or animals classified by the Secretary of the Interior or the Secretary of Commerce as endangered pursuant to Section 4 of the Endangered Species Act of 1973, as amended.
- *Exploration*—The process of identifying a potential subsurface geologic target formation and the active drilling of a borehole designed to assess the natural gas or oil.
- *Flow line*—A small-diameter pipeline that generally connects a well to the initial processing facility.
- *Formation (geologic)*—A rock body distinguishable from other rock bodies and useful for mapping or description. Formations may be combined into groups or subdivided into members.

- *Fracturing fluids*—A mixture of water and additives used to hydraulically induce cracks in the target formation.

DID YOU KNOW?

One well can be fracked 10 or more times, and there can be up to 30 wells on one pad. An estimated 50 to 60% of the fracking fluid is returned to the surface during well completion and subsequent production, bringing with it toxics gases, liquids, and solid material that are naturally present in underground gas deposits. Under some circumstances, none of the injection fluid is recovered (B.C. Oil & Gas Commission, 2001).

- *Flowback*—The fracture fluids that return to the surface after a hydraulic fracture is completed.
- *Frac*—Hydraulic fracturing, as adapted by the petroleum industry.
- *Groundwater*—Subsurface water that is in the zone of saturation and is the source of water for wells, seepage, and springs. The top surface of the groundwater is the *water table.*
- *Habitat*—The area in which a particular species lives. In wildlife management, the major elements of a habitat are considered to the food, water, cover, breeding space, and living space.
- *Horizontal drilling*—A drilling procedure in which the wellbore is drilled vertically to a kick-off depth above the target formation and then angled through a wide 90-degree arc such that the producing portion of the well extends horizontally through the target formation.
- *Hydraulic fracturing*—Injecting fracturing fluids into the target formation at a force exceeding the parting pressure of the rock, thus inducing a network of fractures through which oil or natural gas can flow to the wellbore.
- *Hydrostatic pressure*—The pressure exerted by a fluid at rest due to its inherent physical properties and the amount of pressure being exerted on it from outside forces.
- *Injection well*—A well used to inject fluids into an underground formation for either enhanced recovery or disposal.
- *Lease*—A legal document that conveys to an operator the right to drill for oil and gas. Also, the tract of land on which a lease has been obtained and where producing wells and production equipment are located.
- *Mcf*—A natural gas measurement unit for 1000 cubic feet.
- *MMcf*—A natural gas measurement unit for 1,000,000 cubic feet.

DID YOU KNOW?

Natural gas is generally priced and sold in units of 1000 cubic feet (Mcf, using the Roman numeral for one thousand). Units of a trillion cubic feet (Tcf) are often used to measure large quantities, as in resources or reserves in the ground or annual nation energy consumption. A Tcf is equal to 1 billion Mcf and is enough natural gas to

- Heat 15 million homes for one year
- Generate 100 billion kilowatt-hours of electricity
- Fuel 12 million natural gas-fired vehicles for one year

- *NORM*—Naturally occurring radioactive materials; includes naturally occurring uranium-235 and daughter products such as radium and radon.
- *Oil-equivalent gas (OEG)*—The volume of natural gas needed to generate the equivalent amount of heat as a barrel of crude oil. Approximately 6000 cubic feet of natural gas are equivalent to one barrel of crude oil.
- *Particulate matter (PM)*—A small particle of solid or liquid matter (e.g., soot, dust, mist). PM_{10} refers to particulate matter having a size diameter of less than 10 millionths of a meter (micrometer, or µm); $PM_{2.5}$ is less than 2.5 µm in diameter.
- *Permeability/porosity*—The capacity of a rock to transmit a fluid, which depends on the size and shape of pores and interconnecting pore throats. A rock may have significant porosity (many microscopic pores) but have low permeability if the pores are not interconnected (see Figures 1.2 to 1.5). Permeability may also exist or be enhanced through fractures that connect the pores. Though shales may be as porous as other sedimentary rocks, their extremely small pore sizes make them relatively impermeable to gas flow, unless natural or artificial fractures occur. Porosity is the percent volume of the rock that is not occupied by solids. Again, permeability is a measure of the ease with which a fluid can flow through a rock; the greater the permeability of a rock, the easier it is for the fluid to flow through the rock. Permeability is measured in units of *darcies* (D) or *millidarcies* (mD). A darcy is the permeability that will allow a flow of 1 cubic centimeter per second of a fluid with 1 centipoise viscosity (resistance to flow) through a distance of 1 centimeter through an area of 1 square centimeter under a differential pressure of 1 atmosphere (atm). In naturally occurring materials, permeability values range over many orders of magnitude.

DID YOU KNOW?

In 1850, Henri Darcy, the City Water Engineer for Dijon, France, received endless complaints about the filthy water coming from the city mains. Darcy installed sand filters to purify the system. While purifying water in the mains, Darcy experimented with fluid flow through porous materials and developed equations to describe it. For his efforts, Darcy earned immortality via the universally used *darcy*, a unit representing how easily fluid flows through porous media.

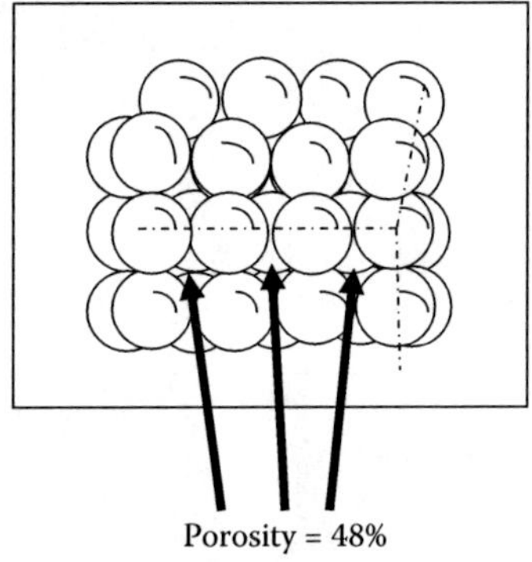

FIGURE 1.2
When spheres are stacked within a box, the empty space between the spheres equals 48% of the total volume. (Adapted from Raymond, M.D. and Leffler, W.L., *Oil and Gas Production in Nontechnical Language*, PennWell Corporation, Tulsa, OK, 2006.)

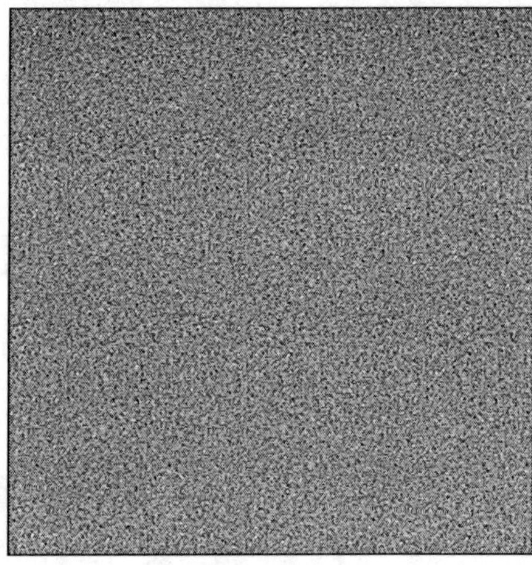

FIGURE 1.3
Well-sorted sand grains. (Adapted from Raymond, M.D. and Leffler, W.L., *Oil and Gas Production in Nontechnical Language*, PennWell Corporation, Tulsa, OK, 2006.)

FIGURE 1.4
Poorly sorted sand grains. (Adapted from Raymond, M.D. and Leffler, W.L., *Oil and Gas Production in Nontechnical Language*, PennWell Corporation, Tulsa, OK, 2006.)

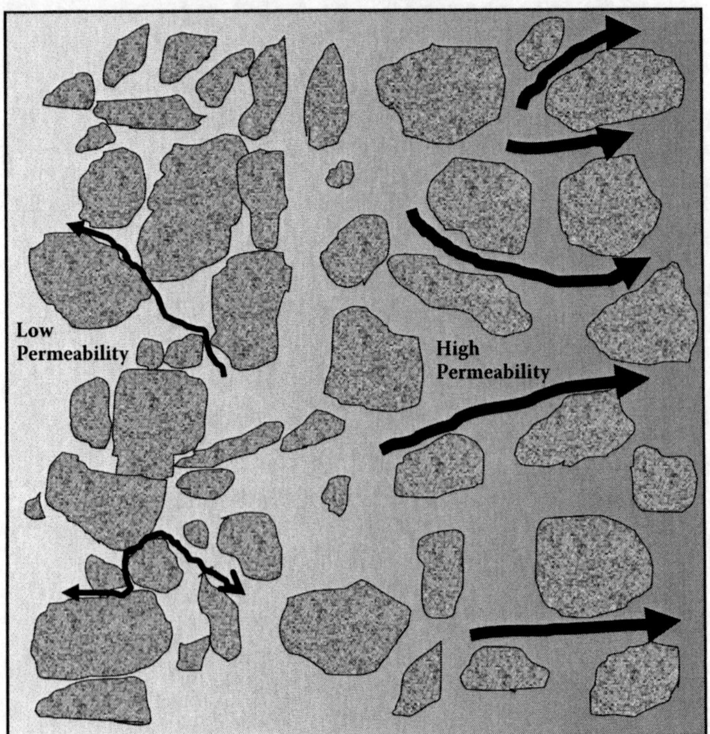

FIGURE 1.5
High- and low-permeability rock. (Adapted from Raymond, M.D. and Leffler, W.L., *Oil and Gas Production in Nontechnical Language*, PennWell Corporation, Tulsa, OK, 2006.)

- *Primacy*—A right that can be granted to states by the federal government that allows state agencies to implement programs with federal oversight. Usually, the states develop their own set of regulations. By statute, states may adopt their own standards; however, these must be at least as protective as the federal standards they replace and may be even more protective in order to address local conditions. Once these state programs are approved by the relevant federal agency (usually the USEPA), the state then has primary jurisdiction.
- *Produced water*—Water produced from oil and gas wells.
- *Propping agents/proppant*—Silica sand or other particles pumped into a formation during a hydraulic fracturing operation to keep fractures open and maintain permeability.
- *Proved reserves*—That portion of recoverable resources that is demonstrated by actual production or conclusive formation tests to be technically, economically, and legally producible under existing economic and operating conditions.
- *Reclamation*—Rehabilitation of a disturbed area to make it acceptable for designated uses. This normally involves regrading, replacement of topsoil, revegetation, and other work necessary to restore the area.
- *Setback*—The distance that must be maintained between a well or other specified equipment and any protected structure or feature.
- *Shale gas*—Natural gas produced from low-permeability shale formations.
- *Slickwater*—Water-based fracturing fluid mixed with a friction-reducing agent, commonly potassium chloride.
- *Split estate*—Condition that exists when the surface rights and mineral rights of a given area are owned by different persons or entities; also referred to as *severed estate.*
- *Stimulation*—Any of several processes used to enhance near-wellbore permeability and reservoir permeability.
- *Stipulation*—A condition or requirement attached to a lease or contract, usually dealing with protection of the environment or recovery of a mineral.
- *Sulfur dioxide* (SO_2)—A colorless gas formed when sulfur oxidizes, often as a result of burning trace amounts of sulfur in fossil fuels.
- *Tcf*—A natural gas measurement unit for one trillion cubic feet.
- *Technically recoverable resources*—The total amount of resources, discovered and undiscovered, thought to be recoverable with available technology, regardless of economics.

- *Thermogenic gas*—Natural gas that is formed by the combined forces of high pressure and temperature (both from deep burial within the Earth's crust), resulting in the natural cracking of the organic matter in the source rock matrix.
- *Thixotrophy*—The property of a gel to become fluid when disturbed (as by shaking).
- *Threatened and endangered species*—Plant or animal species that have been designated as being in danger of extinction.
- *Tight gas*—Natural gas trapped in a hardrock, sandstone, or limestone formation that is relatively impermeable.
- *Tight sand*—A very low or no permeability sandstone or carbonate.
- *Total dissolved solids (TDS)*—The dry weight of dissolved material, organic and inorganic, contained in water and usually expressed in parts per million.
- *Underground Injection Control (UIC) program*—A program administered by the USEPA, primacy state, or Indian tribe under the Safe Drinking Water Act to ensure that subsurface emplacement of fluids does not endanger underground sources of drinking water.
- *Underground source of drinking water (USDW)*—As defined by 40 CFR §144.3, a USDW is an aquifer or its portion:

  (a) (1) Which supplies any public water system; or

  (2) Which contains a sufficient quantity of groundwater to supply a public water system; and

  (i) Currently supplies drinking water for human consumption; or

  (ii) Contains fewer than 10,000 mg/L total dissolved solids; and

  (b) Which is not an exempted aquifer.
- *Water quality*—The chemical, physical, and biological characteristics of water with respect to its suitability for a particular use.
- *Watershed*—All lands that are enclosed by a continuous hydrologic drainage divide and lay upslope from a specified point on a stream.
- *Whipstock*—A wedge-shaped piece of metal placed downhole to deflect the drill bit.
- *Workover*—To perform one or more remedial operations on a producing or injection well to increase production. Deepening, plugging back, pulling, and resetting the line are examples of worker operations.

1.6 Purpose of Text

Although the title of this book, *Environmental Impacts of Hydraulic Fracturing,* makes clear the intention of the author to discuss hydraulic fracturing along with its impact on the environment, it is important to point out that it is not an attack on hydraulic fracturing or on the mining of natural gas, oil, or natural resources. Moreover, it is not an attack on our ongoing attempt to find and produce our own energy resources necessary to make the United States energy independent, maintain (at the minimum) our present living standards, and ensure their sustainability for the generations to follow. I am an advocate for mining and using our own natural resources.

Producing and processing the sustainable energy supplies needed to satisfy future needs are not always clean, pristine, or non-intrusive activities. The truth is that the processes used for mining coal or drilling for oil and natural gas can cause or release toxic byproducts. These chemicals and waste products can contaminate the air, soil, and water.

A classic example of the significant environmental impact of mining an energy source can be seen in the practice of mountaintop mining. Mountaintop mining is a form of surface coal mining in which explosives are used to access coal seams, generating large volumes of waste that bury adjacent streams. The resulting mountaintop waste that then fills valleys and streams can significantly compromise water quality, often causing permanent damage to ecosystems and rendering streams unfit for drinking, fishing, and swimming. It is estimated that almost 2000 miles of Appalachian headwater systems have been buried by mountaintop coal mining (USEPA, 2011).

The Bottom Line: As a practicing environmental professional, I recognize that we must achieve a balance between protecting the air, water, soil, and ecosystems that life on Earth depends on and utilizing the natural resources that Earth possesses. It is from this perspective—maintaining a balance between resource mining and environmental protection—that I present the material in this text. When we split rocks in the wilderness we must be cognizant of the surroundings, whether they be visible on the surface or invisible far below the ground. Obtaining what we need from Mother Earth should be done without injuring the source.

1.7 Thought-Provoking Discussion Questions

1. Do you believe that fracking operations cause earthquakes?
2. If it is proven that earthquakes are caused by fracking operations, do you think we should still frack for natural gas?

3. Is the environment just one of those things that will have to take a few bruises while we explore for and extract more hydrocarbons that we need to sustain our way of life?
4. Are environmental concerns real or just talking points for radical groups?
5. Can we always protect the environment from human-caused damage?
6. Should we focus our attention on renewable sources of energy instead of drilling for more oil and natural gas?
7. Has the Solyndra scandal stymied our efforts to switch from hydrocarbon fuel sources to renewable sources?
8. Is natural gas really as clean as the experts say it is?
9. Does fracking contribute to global warming? How?
10. Who do you think is more responsible for pollution: individuals, companies, or the government?

References and Recommended Reading

Anon. (2012). Shell will spend billions on a chemical refinery in Pennsylvania. *The Plain Dealer*, March 15 (http://www.cleveland.com/business/index.ssf/2012/03/shell_will_spend_billions_on_a.html).

AP. (2011). 4.0 earthquake strikes in northeast Ohio. *USA Today*, December 31 (www.usatoday.com/news/nation/story/2011-12-31/northeast-ohio-earthquake/52307134/1).

B.C. Oil & Gas Commission. (2001). *Fracturing (Fracking) and Disposal of Fluids*, Information Sheet 15. B.C. Oil & Gas Commission, British Columbia (www.bctwa.org/Frk-BCOil&GasCom-Fracking.pdf).

Fountain, H. (2012). Ohio quakes halt well dumping of fracking fluids. *The New York Times*, January 5.

League of Women Voters. (2009). *Study Guide II: Marcellus Shale Natural Gas: Environmental Impact*. League of Women Voters of Indiana County, Pennsylvania (www.palwv.org/indiana).

Raymond, M.D. and Leffler, W.L. (2006). *Oil and Gas Production in Nontechnical Language*. PennWell Corporation, Tulsa, OK.

Sheeran, T.J. (2001). Ohio shale drilling spurs jobs hopes in Rust Belt. Associated Press, November 27 (www.daytondailynews.com/news/ohio-news/ohio-shale-drilling-spurs-job-hopes-in-rust-belt-1290369.html).

USEPA. (2011). *Final Guidance to Protect Water Quality in Appalachian Communities from Impacts of Mountaintop Mining*. U.S. Environmental Protection Agency, Washington, DC (yosemite.epa.gov/opa/admpress.nsf/bd4379a92ceceeac8525735900400c27/1dabfc17944974d4852578d400561a13!OpenDocument).

USGS. (2009). *Earthquake Facts & Earthquake Fantasy*. U.S. Geological Survey, Reston, VA (http://earthquake.usgs.gov/learn/topics/megaqk_facts_fantasy.php).

2

Shale Gas

Drill here. Drill now. Pay less.

—Newt Gingrich

2.1 Importance of Shale Gas

Natural gas, coal, and oil supply about 85% of the nation's energy, with natural gas supplying about 22% of the total (USEIA, 2008a) (Figure 2.1). Natural gas plays a key role in meeting U.S. energy demands, and its percent contribution to the U.S. energy supply is expected to remain fairly constant for the next 20 years. The United States has abundant natural gas resources—more than 1744 trillion cubic feet (Tcf) of technically recoverable natural gas, including over 300 Tcf of proved reserves (the discovered, economically recoverable fraction of the original gas in place) (USEIA, 2008b, 2010). Technically recoverable unconventional gas (shale gas, tight sands, and coa bed natural gas) accounts for 60% of the onshore recoverable resource. At the 2010 U.S. production rate, approximately 26.86 Tcf, it is estimated that the

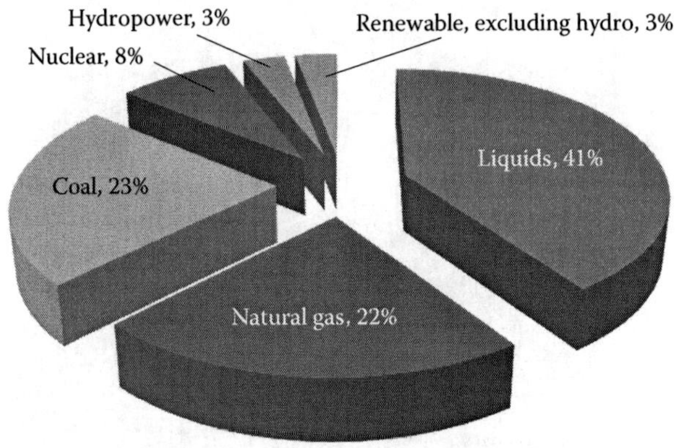

FIGURE 2.1
U.S. energy consumption by fuel. (From USEIA, *Annual Energy Outlook 2008 with Projections to 2030*, U.S. Energy Information Administration, Washington, DC, 2008; www.eia.gov/oiaf/aeo/pdf/0383(2008).pdf.)

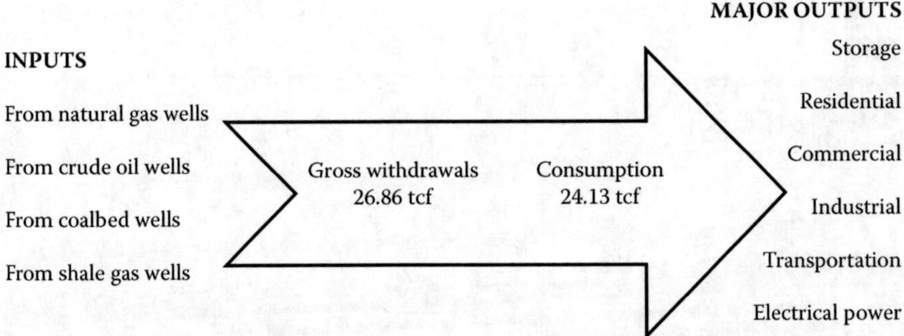

FIGURE 2.2
Natural gas flows in 2010 (trillion cubic feet). (From USEIA, *Total Energy: Natural Gas Flow, 2010 (Trillion Cubic Feet)*, U.S. Energy Information Administration, Washington, DC, 2012; http://www.eia.gov/totalenergy/data/annual/diagram3.cfm.)

current recoverable resource could provide enough natural gas to supply the United States for the next 110 years (USEIA, 2012) (Figure 2.2). Note that, historically, estimates of the size of the total recoverable resource have grown over time as knowledge of the resource has improved and recovery technology has advanced. Unconventional gas resources are a prime example of this trend; for example, U.S. proved reserves of wet natural gas increased by 11% in 2009 to 284 Tcf (USEIA, 2010).

As shown in Figure 2.3, natural gas use is distributed across several sectors of the economy. In addition to serving a vital role in residential heating, natural gas is an important energy source for the industrial, commercial, and electrical generation sectors. Although forecasts vary in the outlook for

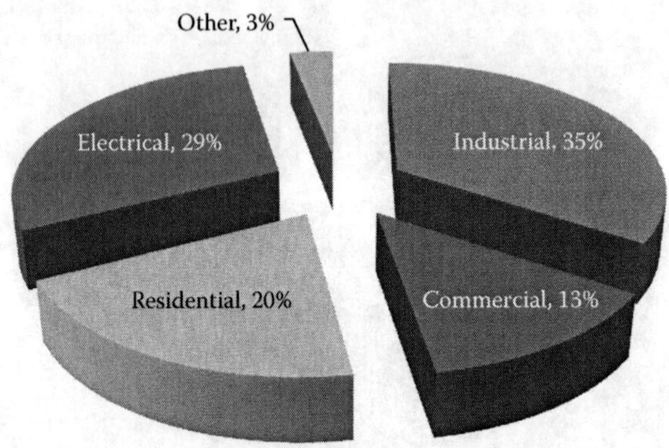

FIGURE 2.3
Natural gas use by sector. (From USEIA, *March 2009 Monthly Energy Review*, U.S. Energy Information Administration, Washington, DC, 2009; www.eia.gov/totalenergy/data/monthly/previous.cfm#2009.)

DID YOU KNOW?

The use of *unconventional* to describe a gas resource is open to interpretation; as technology advances and discrete reservoirs become limited, the reserves considered unconventional a few decades ago are more commonly viewed as conventional by modern standards (Santoro et al., 2011). In this book, *conventional* refers specifically to discrete reservoirs of associated or unassociated natural gas, and *unconventional* refers to tight-gas formations.

future demand for natural gas, they all have one thing in common: Natural gas will continue to play a significant role in the U.S. energy picture for some time to come (USDOE, 2009; USEIA, 2010).

Natural gas, due to its clean-burning nature and economical availability, has become a very popular fuel for the generation of electricity (NaturalGas.org, 2008a). In the 1970s and 1980s, the choice for the majority of electrical utility generators was primarily coal or nuclear power, but, due to economic, environmental, technological, and regulatory changes, natural gas has become the fuel of choice for many new power plants. In 2007, natural gas accounted for 39.1% of electric industry productive capacity (USEIA, 2008b).

Natural gas is the fuel of choice for a wide range of industries (USEIA, 2002). It is a major fuel source for pulp and paper, metals, chemicals, petroleum refining, and food processing. These five industries alone account for almost three quarters of industrial natural gas use and together employ 4 million people in the United States (Bureau of Labor Statistics, 2007). Natural gas is also a feedstock for a variety of products, including plastics, chemicals, and fertilizers. Industrial use of natural gas accounted for 6.63 Tcf of demand in 2007 and was expected to grow to 6.82 Tcf by 2010. It is interesting to note that for many products, there is no economically viable substitute for natural gas.

A look at basic natural gas statistics, however, reveals that natural gas is being consumed by the U.S. economy at a rate that exceeds domestic production, and the gap is increasing (USEIA, 2011). Despite possessing a large resource endowment, the United States consumes natural gas at a rate requiring rapid replacement of reserves. Ambrose et al. (2008) estimated that the gap between demand and domestic supply will grow to nearly 9 Tcf by the year 2025. The good news is that many believe that unconventional natural gas resources such as shale gas can significantly alter that balance.

DID YOU KNOW?

Half of the natural gas consumed today is produced from wells drilled within the last 3.5 years (IPAMS, 2008).

Without domestic shale gas and other unconventional gas production, the gap between demand and domestic production will widen even more, leaving imports to fill the need. Worldwide consumption of natural gas is also increasing; therefore, the United States can anticipate facing an increasingly competitive market for these imports. This increased reliance on foreign sources of energy could pose at least two problems for the United States: (1) it would serve to decrease our energy security, and (2) it could create a multibillion dollar outflow to foreign interests, thus making such funds unavailable for domestic investment.

2.2 Versatility of Natural Gas

From lighting streetlamps and houses in the 1800s and early 1900s, natural gas usage has advanced by leaps and bounds because of its versatility. Because of its high Btu content, an extensive and improved distribution network, and advancements in technology, natural gas has become easy to use in various applications and is now the energy source of choice of many. Natural gas is also reliable—84% of the natural gas consumed in the United States is produced in the United States, and 97% of the gas used in this country is produced in North America (USEIA, 2008b).

Although our supply of natural gas will eventually be depleted, during this depletion process (about 100 years or more) American ingenuity and innovation should be applied to developing a viable hydrogen fuel cell or other energy replacement system that can power our trucks, cars, ships, and airplanes of the future. For the time being, though, it must be all about innovation and natural gas. Moreover, energy innovation must be the first step taken toward replacing lost American jobs. Consider our current economic conditions: bankruptcies, foreclosures, and high unemployment rates. Although it is true that unemployment wounds, it is also true that these wounds can be healed. Again, we must begin this process by converting from gasoline to natural gas. Natural gas conversion is not the panacea for all our energy needs; however, it can be a lifeline to get us off foreign oil and other energy imports and on our way to innovation via renewable energy. Keep in mind that the United States has natural gas reserves (both dry and liquid natural gas) that exceed 200 Tcf, which should be enough natural gas to get us through this century.

In hindsight, ideally, if the Obama administration had taken that TARP or stimulus money and put it to work on natural gas conversion, we would be in better shape today. In January 2009, the new administration could have sat down with the troubled automobile manufacturers and told them, first, that the government was going to bail them out by giving them money to stabilize their situation and, second, that all new cars must be built to run on natural gas only. At the same time, the government should have paid for the approximately 122,000 gas stations in the United States (Census Bureau, 2002) to convert to natural gas. This conversion to natural gas would have made us less dependent on unstable foreign countries and the delivery system less subject to interruption.

Natural gas conversion offers several advantages: Natural gas is efficient and clean burning. Of all the fossil fuels, natural gas is by far the cleanest burning. It emits approximately half the carbon dioxide (CO_2) compared to coal, and the levels of other air pollutants are low. Along with emitting low levels of carbon dioxide, natural gas also emits water vapor; thus, combustion of natural gas produces the same compounds that people exhale when breathing. Oil and coal are composed of much more complex organic molecules with greater nitrogen and sulfur content. Their combustion byproducts include larger quantities of CO_2 (see Table 2.1), nitrogen oxides (NO_x), sulfur dioxide (SO_2), and particulate ash (see Table 2.2). By comparison, the combustion of natural gas liberates very small amounts of SO_2 and NO_x, virtually no ash, and lower levels of CO_2, carbon monoxide (CO), and other hydrocarbons (NaturalGas.org, 2008b).

DID YOU KNOW?

Of all the fossil fuels, natural gas is by far the cleanest burning.

Many environmental activists, with regard to so-called benign fossil fuel sources, view natural gas as the White Knight who will rescue the planet from the pollution of the Red Knight industrialists. Because natural gas emits only half as much CO_2 as coal and approximately 30% less than fuel oil, it is generally considered to be central to energy plans focused on the reduction of greenhouse gas (GHG) emissions (Navigant Consulting, 2008). Because CO_2 makes up a large fraction of U.S. GHG emissions, increasing the role of natural gas in U.S. energy supply relative to other fossil fuels would result in lower GHG emissions.

The need for the United States to reduce its dependence on foreign sources of fossil fuels and fossil fuels in general is increasing; however, the transition to sustainable renewable energy sources will no doubt require considerable time, effort, and investment in order for these sources to become economical enough to supply a significant portion of the nation's energy consumption. It has been estimated that fossil fuels (oil, gas, and coal) will supply 82.1% of the nation's energy needs in 2030 (USEIA, 2008a). Because natural gas is the cleanest burning of the fossil fuels, an environmental benefit could be realized by shifting toward proportionately greater reliance on natural gas until such time as sources of alternative energy are more efficient, economical, and widely available. Moreover, the move toward sustainable renewable energy sources, such as those discussed below, requires that a supplemental

TABLE 2.1

U.S. Carbon Dioxide Emissions from Energy and Industry, 1990–2008

	Carbon Dioxide Emissions (Million Metric Tons Carbon Dioxide)									
Fuel Type/Process	**1990**	**1995**	**2000**	**2002**	**2003**	**2004**	**2005**	**2006**	**2007**	**2008**
Petroleum	2185.9	2208.4	2461.3	2469.9	2516.7	2605.4	2625.7	2594.9	2588.6	2436.0
Coal	1803.4	1899.9	2138.1	2077.2	2115.6	2140.3	2161.0	2129.9	2154.5	2125.2
Natural gas	1024.7	1183.7	1240.6	1229.5	1194.6	1195.4	1176.1	1157.1	1231.2	1241.0
Renewables[a]	6.1	10.2	10.4	13.0	11.7	11.4	11.5	11.8	11.6	11.6
Energy total	5020.1	5302.3	5850.4	5789.6	5838.6	5952.5	5974.3	5893.7	5986.4	5814.4

Source: USEIA, *Emissions of Greenhouse Gases Report*, U.S. Energy Information Administration, Washington, DC, 2009 (www.eia.gov/oiaf/1605/ggrpt/carbon.html).

[a] Includes emissions from electricity generation using nonbiogenic municipal solid waste and geothermal energy.

TABLE 2.2

Combustion Emissions (Pounds/Billion Btu of Energy Input)

	Combusted Source		
Air Pollutant	**Natural Gas**	**Oil**	**Coal**
Carbon dioxide (CO_2)	117,000	164,000	208,000
Carbon monoxide (CO)	40	33	206
Nitrogen oxides (NO_x)	92	448	457
Sulfur dioxide (SO_2)	0.6	1122	2591
Particulates (PM)	7.0	84	2744
Formaldehyde	0.750	0.220	0.221
Mercury (Hg)	0.000	0.007	0.016

Source: USEIA, *Natural Gas 1998: Issues and Trends*, U.S. Energy Information Administration, Washington, DC, 1999 (www.eia.gov/oil_gas/natural_gas/analysis_publications/natural_gas_1998_issues_and_trends/it98.html).

energy source be available when weather conditions and electrical storage capacity prove challenging. Such a backstop energy source and a temporary bridge until renewable energy sources are refined must be widely available on near instantaneous and continuous demand. The availability of extensive natural gas transmission and distribution pipeline systems makes natural gas uniquely suitable for this role; therefore, natural gas is an integral part of the effort to move forward with alternative energy options. With the current emphasis on the potential effects of air emissions on global climate change, air quality, and visibility, cleaner fuels such as natural gas are important to our nation's energy future (IPAMS, 2008; Navigant Consulting, 2008).

2.3 Alternative or Renewable Energy*

Before continuing our discussion of shale-derived natural gas, it is important to touch upon alternatives to all fossil fuels, or renewable energy. As mentioned, the worldwide use of liquid fossil fuels and their decreasing availability along with the politics involved and other economic forces are pushing for substitute, alternative, or renewable fuel sources. This is the case, of course, because of the current and future economic problems that $4+/gal gasoline have generated (especially in the United States) and because of the perceived crisis developing with high carbon dioxide emissions, the major contributing factor of global climate change. It is important to make a clear distinction between alternative and renewable energy. *Alternative energy* is an umbrella term that refers to any source of usable energy intended to replace fuel sources without the undesired consequences of the replaced fuels. The term "alternative" presupposes an undesirable connotation for fossil fuels (for many people, the term "fossil fuel" has joined that endless list of four-letter words). Alternative energy is fuel energy that does not use up natural resources or harm the environment. Examples of alternative fuels include petroleum as an alternative to whale oil, coal as an alternative to wood, alcohol as an alternative to fossil fuels, and coal gasification as an alternative to petroleum. The key point in understanding alternative energy is that these fuels need not be renewable.

Renewable energy is energy generated from natural resources—such as sunlight, wind, water (hydro), ocean thermal, wave and tide action, biomass and geothermal heat—that are naturally replenished and thus renewable. Renewable energy resources are virtually inexhaustible—they are

* Much of the information and data in this section are from USEIA, *Renewable Energy Trends 2004*, U.S. Energy Information Administration, Washington, DC, 2005 (ftp://ftp.eia.doe.gov/renewables/062804.pdf); USEIA, *How Much Renewable Energy Do We Use?*, U.S. Energy Information Administration, Washington, DC, 2010 (www.eia.gov/energy in_brief/renewable_energy.cfm).

replenished at the same rate as they are used—but they are limited in the amount of energy that is available per unit time. If we have not come full circle in our cycling from renewable to nonrenewable back to renewable, we are getting close. Consider, for example, that in 1850, about 90% of the energy consumed in the United States came from renewable energy resources (e.g., hydropower, wind, burning wood). Today, though, the United States is heavily reliant on nonrenewable fossil fuels (natural gas, oil, and coal). In 2009, about 7% of all energy consumed and about 8.5% of total electricity production was from renewable energy resources.

Currently, most of the renewable energy is used to generate electricity, provide heat for industrial processes, and heat and cool buildings, as well as for transportation fuels. Electricity producers (utilities, independent produces, and combined heat and power plants) accounted for 51% of the total U.S. renewable energy consumed in 2007 for producing electricity. Most of the rest of the remaining renewable energy consumed was biomass, which was used for industrial applications (principally papermaking) by plants producing only heat and steam. Biomass is also used for transportation fuels (ethanol) and to provide residential and commercial space heating. The largest share of renewable-generated electricity comes from hydroelectric energy (71%), followed by biomass (16%), wind (9%), geothermal (4%), and solar (0.2%). Wind-generated electricity increased by almost 21% in 2007 over 2006, more than any other energy source. Its growth rate was followed closely by solar, which increased by over 19% in 2007 compared to 2006.

From Table 2.3 it is obvious that currently there are five primary forms of renewable energy: solar, wind, biomass, geothermal, and hydroelectric. Each of these holds promise and poses challenges regarding future development. The United States imports more than 50% of its oil. Replacing some of our petroleum with fuels provided from solar power or made from organic plant matter, for example, could save money and strengthen our energy security.

Renewable energy is plentiful, and the technologies are improving all the time. There are many ways to use renewable energy. Our main focus should be on finding and developing a renewable source of liquid fuels (e.g., developed from biomass), because our economy runs on liquid fuels. Most of the non-liquid renewable energies will not (at the present time) provide power for airplanes and heavy trucks; that is, neither fuel cells nor solar, wind, hydroelectric, geothermal, or wave and tidal energy can power the main transportation vehicles we use today. Note that trains have not been mentioned. Many trains today are powered by diesel or diesel–electric systems. If necessary, trains could be retrofitted to steam power developed by burning coal and wood products (a step back into the past); however, this cannot be done to power heavy trucks and airplanes.

Most Americans still do not understand that we are running short of the fuels that we use every day—the fuels that made us what we are today. We built the world's greatest economy on oil that sold for $3 to $4 per barrel

TABLE 2.3

U.S. Energy Consumption by Energy Source, 2009

Energy Source	Energy Consumption (quadrillion Btu)
Total	94.820
Fossil fuels (coal, coke, natural gas, petroleum)	78.631
Electricity net imports	0.116
Nuclear electric power	8.328
Renewable energy	7.745
Biomass (biofuels, waste, wood, and wood-derived)	3.884
Biofuels	1.546
Waste	0.447
Wood-derived fuels	1.891
Geothermal energy	0.373
Hydroelectric, conventional	2.682
Solar thermal/photovoltaic energy	0.109
Wind energy	0.697

Source: USEIA, *U.S. Energy Consumption by Energy Source*, U.S. Energy Information Administration, Washington, DC, 2010 (www.eia.gov/cneaf/alternate/page/renew_energy_consump/table1.html).

of oil. That is no longer the case. In order to maintain our current level of living, the so-called good life, we must find and develop renewable energy sources to power and secure our future. In the meantime, we should mine our reserves of natural gas to the fullest extent possible and hope that by the time we exhaust our natural gas supplies we will have discovered practical sources of sustainable renewable energy (Spellman and Bieber, 2011).

DID YOU KNOW?

Natural gas used by consumers is composed almost entirely of methane (USDOT, 2011).

2.4 The 411 on Natural Gas

Natural gas is a combination of hydrocarbon gases consisting primarily of molecules that range from one to four carbon atoms in length. The gas with one carbon atom in the molecule is methane (CH_4), with two is ethane (C_2H_6), with three is propane (C_3H_8), and with four is butane (C_4H_{10}). All are paraffin-type hydrocarbon molecules. A typical natural gas composition, including

TABLE 2.4
Typical Composition of Natural Gas

Methane	70 to 90%
Ethane	1 to 10%
Propane	Trace to 5%
Butane	Trace to 2%
Carbon dioxide	Trace
Oxygen	Trace
Nitrogen	Trace
Hydrogen sulfide	Trace
Rare gases (Ar, He, Ne, Xe)	Trace

five lesser components, is shown in Table 2.4. The percentages shown in the table vary among gas fields, but methane gas is by far the most common hydrocarbon. Many natural gas fields contain almost pure methane, and the gas from pipelines that is burned in homes and industry is methane gas (Hyne, 2001).

Natural gas is odorless and colorless. When ignited, it releases a significant amount of energy. It is found in rock formations (reservoirs) beneath the surface of the Earth; in some cases, it may be associated with oil deposits (NaturalGas.org, 2008c). Natural gas was formed millions of years ago when the remains of plants and animals (diatoms) decayed and built up in thick layers. This decayed matter from plants and animals is called *organic material*—it was once alive. Over time, the sand and silt changed to rock, covered the organic material, and trapped it beneath the rock. Pressure and heat changed some of this organic material into coal, some into oil (petroleum), and some into natural gas—tiny bubbles of odorless gas. This process is shown in Figure 2.4.

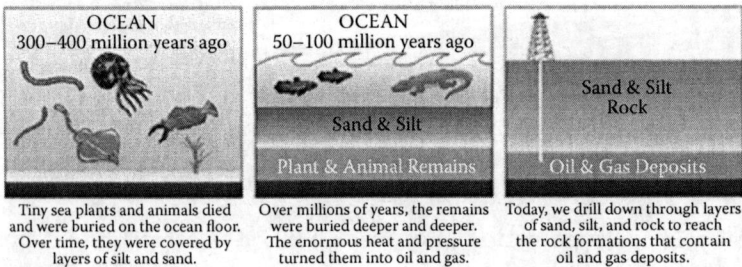

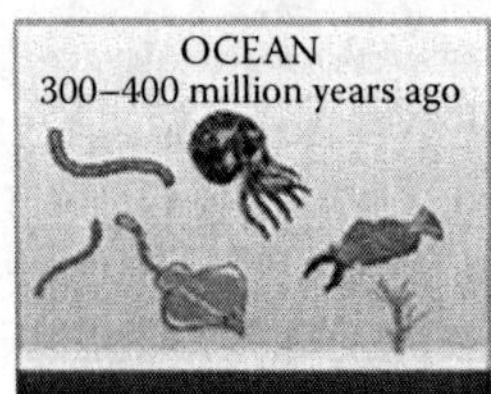

Tiny sea plants and animals died and were buried on the ocean floor. Over time, they were covered by layers of silt and sand.

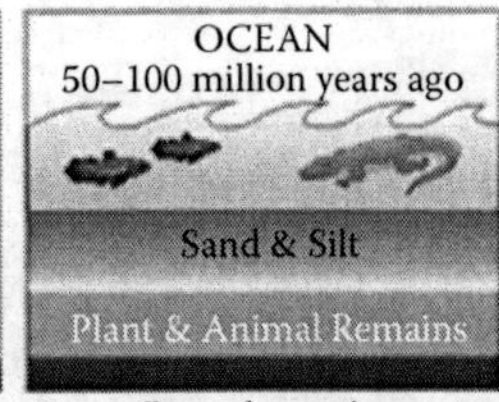

Over millions of years, the remains were buried deeper and deeper. The enormous heat and pressure turned them into oil and gas.

Today, we drill down through layers of sand, silt, and rock to reach the rock formations that contain oil and gas deposits.

FIGURE 2.4
Petroleum and natural gas formation. (From USEIA, *Natural Gas Explained*, U.S. Energy Information Administration, Washington, DC, 2011; http://www.eia.gov/energyexplained/index.cfm?page=natural_gas_home.)

DID YOU KNOW?

The hydrogen sulfide (H_2S) listed in Table 2.4 is sometimes referred to as *well gas*. It is found mixed with natural gas or by itself. It is a very poisonous gas that is lethal in low concentrations. The gas has the foul odor of rotten eggs and can be detected in extremely small amounts. Although it is can be smelled initially, it quickly paralyzes the olfactory senses and can no longer be detected; thus, it can be extremely dangerous.

2.5 The 411 on Processing and Refining Natural Gas

When natural gas is fracked from shale formations and brought to the surface for storage and shipping from the drilling site, it must be processed or refined before being used by the final consumer. Natural gas comes out of the ground as a complex hydrocarbon mixture that must be treated before it is turned into the natural gas that is used in our homes and many industries. In addition, thousands of products are made from petrochemicals; for example, natural gas is the feedstock for the production of ammonia, via the Haber process, that is used in fertilizer production. Natural gas processing plants are complex, but they generally utilize four main processes for removing impurities before the gas is sent to the consumer as pipeline-quality dry natural gas:

- Remove oil and condensates.
- Remove water.
- Separate the nature gas liquids from the natural gas.
- Remove sulfur and carbon dioxide.

DID YOU KNOW?

Natural gas as it is used by consumers is much different from the natural gas that is brought up from underground to the well head (USDOT, 2011).

2.5.1 Natural Gas Refining: The Process*

A typical natural gas processing plant operation is shown in Figure 2.5. The figure illustrates the various unit processes that convert raw natural gas into the gas that is ultimately pipelined to end users. Notice that the resulting

* Based on information from USEPA, in *AP 42, Fifth Edition*, Vol. I, U.S. Environmental Protection Agency, Washington, DC, 2012 (www.epa.gov/ttnchie1/ap42/ch05/index.html).

natural gas liquids (NGLs) include ethane, propane, butane, iso-butane, and natural gasoline (denoted as pentanes+). Natural gas from high-pressure wells is usually passed through field separators at the well to remove hydrocarbon condensate and water (see Figure 2.5). Natural gasoline, butane, and propane are usually present in the gas and must be removed by gas processing plants. Natural gas is considered "sour" if hydrogen sulfide (H_2S), the gas with the characteristic rotten egg odor, is present in amounts greater than 5.7 milligrams per normal cubic meter (mg/Nm^3), or 0.25 grains per 100 standard cubic feet (gr/100 scf). The H_2S must be removed (called "sweetening" the gas or acid gas removal) before the gas can be utilized. If H_2S is present, the gas is usually sweetened by absorption of the H_2S in an amine solution. Amine processes are common unit processes that are used for over 95% of all gas sweetening in the United States. Other methods, such as carbonate processes, solid bed absorbents, and physical absorption, are employed in some sweetening plants.

DID YOU KNOW?

Some field processing can be accomplished at or near the well head; however, the complete processing of natural gas takes place at a processing plant, usually located in a natural gas producing region (USDOT, 2011).

The major emission sources in the natural gas processing industry are compressor engines, acid gas wastes, fugitive emissions from leaking process equipment, flares, and, if present, glycol dehydrator vent streams. Moreover, regeneration of the glycol solutions used for dehydrating natural gas can release significant quantities of benzene, toluene, ethylbenzene, and xylene, as well as a wide range of less toxic organics.

As noted earlier, the amine process is the most widely used method for H_2S removal. The process is summarized below:

$$2RNH_2 + H_2S \rightarrow (RNH_3)_2S \tag{2.1}$$

where

R = mono-, di-, or tri-ethanol

N = nitrogen

H = hydrogen

S = sulfur

The recovered hydrogen sulfide gas stream may be (1) vented, (2) flared in waste gas flares or modern smokeless flares, (3) incinerated, or (4) utilized for the production of elemental sulfur or sulfuric acid. If the recovered H_2S gas stream is not to be utilized as a feedstock for commercial applications, the

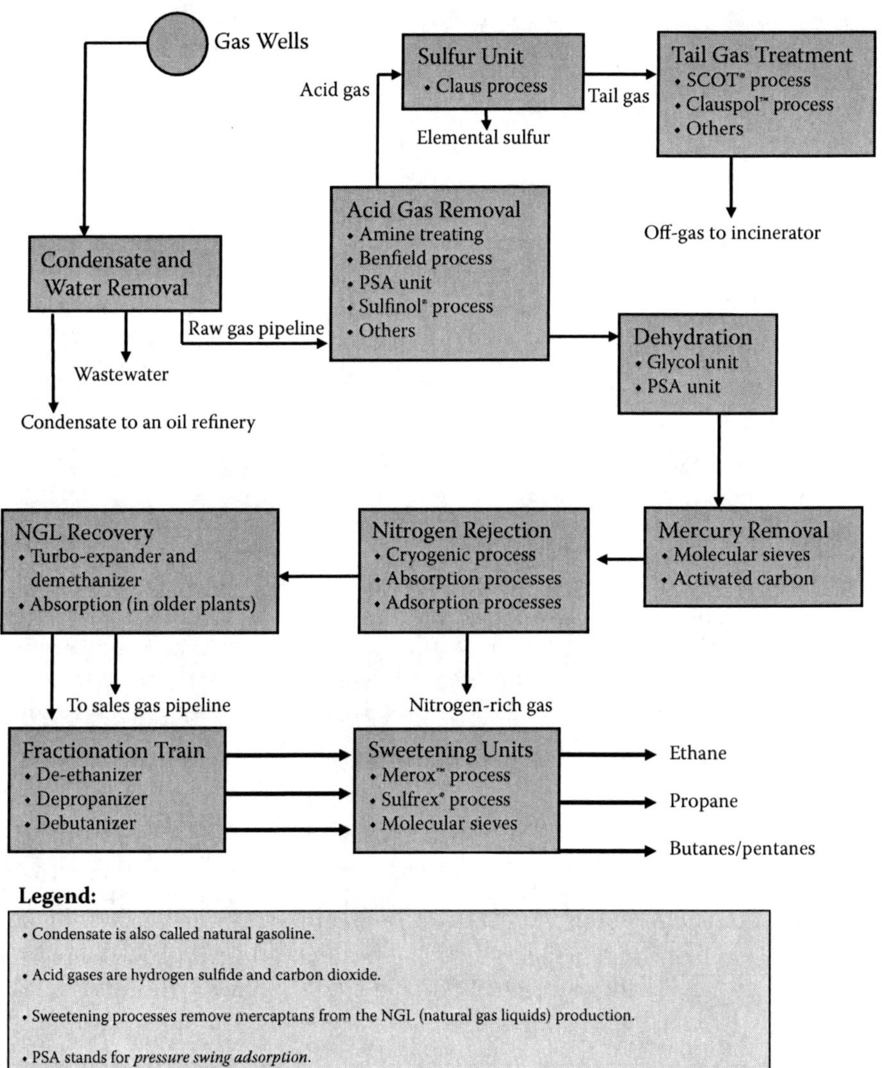

FIGURE 2.5
Schematic flow diagram of a typical natural gas processing plant. (Adapted from http://en.wikipedia.org/wiki/File:NatGasProcessing.svg.)

DID YOU KNOW?

Natural gas must be processed to produce pipeline-quality dry natural gas (USDOT, 2011).

gas is usually passed to a tail gas incinerator in which the H_2S is oxidized to SO_2 and then passed to the atmosphere through a stack. Emissions will result from gas sweetening plants only if the acid waste gas from the amine process is flared or incinerated. Most often, the acid waste gas is used as feedstock in nearby sulfur recovery or sulfuric acid plants (USEPA, 1974).

When flaring or incineration is practiced, the major pollutant of concern is SO_2. Most plants employ elevated smokeless flares or tail gas incinerators for complete combustion of all waste gas constituents, including virtually 100% conversion to H_2S to SO_2. Little smoke and few particulates or hydrocarbons result from these devices; because gas temperatures do not usually exceed 650°C (1200°F), significant quantities of nitrogen oxides are not formed.

DID YOU KNOW?

From the well head, natural gas is transported to processing plants through a network of small-diameter, low-pressure gathering pipelines (USDOT, 2011).

Note that some plants still utilize older, less efficient waste gas flares. Because these flares usually burn at temperatures lower than necessary for complete combustion, larger emissions of hydrocarbons and particulates, as well as H_2S, can occur. No data are available to estimate the magnitude of these emissions from waste gas flares (USEPA, 1974).

DID YOU KNOW?

A complex gathering system can consist of thousands of miles of pipes, interconnecting the processing plant to over 100 wells in the area (USDOT, 2011).

2.6 Thought-Provoking Discussion Questions

1. The emissions from waste gas flares—are they really a problem?
2. How would you eliminate or reduce fugitive emissions from shale gas processing equipment?
3. Is replacing gasoline with natural gas a practical solution to our energy problems?
4. The author's position on the practicality of using renewable energy to produce electricity is limited by our on-demand requirements—we want it when we want it. Do you agree? Explain.

5. Do the advantages of transformation to renewable energy sources outweigh the disadvantages? Explain.
6. Do you think natural gas is cleaner than other sources of energy?
7. Do you believe in the accuracy of the estimates of natural gas supplies that will be available to us in the future?
8. Does natural gas production contribute to global climate change? Explain.
9. If natural gas production is a contributor to global warming, is this a bad thing? Explain.
10. Would a shift to global cooling be better than a shift to global warming? If we undergo global cooling, where would we get the energy necessary to keep us warm?

References and Recommended Reading

Ambrose, W.A., Potter, E.C., and Briceno, R. (2008). An unconventional future for natural gas in the United States. *Biotimes*, 53(2):37–41 (www.geotimes.org/feb08/article.html?id=feature_gas.html).

Bureau of Labor Statistics. (2007). *Monthly Labor Review Online*, 130(11) (www.bls.gov/opub/mlr/2007/11/contents.htm).

Census Bureau. (2002). *SIC 5541 Gasoline Service Stations*. U.S. Census Bureau, Washington, DC.

Ecology Audits. (1975). *Atmospheric Emissions Survey of the Sour Gas Processing Industry*, EPA-450/3-75-076. U.S. Environmental Protection Agency, Research Triangle Park, NC.

Hyne, N.J. (2001). *Nontechnical Guide to Petroleum Geology, Exploration, Drilling, and Production*, 2nd ed. PennWell, Tulsa, OK.

IPAMS. (2008). *America's Independent Natural Gas Producers: Producing Today's Clean Energy, Ensuring Tomorrow's Innovation*. Independent Petroleum Association of Mountain States, Denver, CO (www.ipams.org/mediadocs/Callupdraft10.pdf).

Katz, D.K. (1959). *Handbook of Natural Gas Engineering*. McGraw-Hill, New York.

Kirk, R.E and Othmer, D.F., Eds. (1951). *Encyclopedia of Chemical Technology*, Vol. 7. Interscience Encyclopedia, New York.

Maddox, R.N. (1974). *Gas and Liquid Sweetening*, 2nd ed. Campbell Petroleum Series, Norman, OK.

NaturalGas.org. (2008a). *Electrical Generation Using Natural Gas*, www.naturalgas.org/overview/uses_eletrical.asp.

NaturalGas.org. (2008b). *Natural Gas and the Environment*, www.naturalgas.org/environment/naturalgas.asp.

NaturalGas.org. (2008c). *Overview of Natural Gas*, www.naturalgas.org/overview/overview.asp.

Navigant Consulting. (2008). *North American Natural Gas Supply Assessment*. American Clean Skies Foundation, Washington, DC.

Santoro, R.L., Howarth, R.W., and Ingraffea, A.R. (2011). *Indirect Emissions of Carbon Dioxide from Marcellus Shale Gas Development*. Cornell University, Ithaca, NY.

Spellman, F.R. and Bieber, R. (2011). *The Science of Renewable Energy*. CRC Press, Boca Raton, FL.

USDOE. (2009). *Modern Shale Gas Development in the United States: A Primer*. U.S. Department of Energy, Washington, DC.

USDOT. (2011). *Fact Sheet: Natural Gas Processing Plants*. U.S. Department of Transportation, Washington, DC (http://primis.phmsa.dot.gov/comm/FactSheets/FSNaturalGasProcessingPlants.htm)

USEIA. (2002). *Manufacturing Energy Consumption Survey*. U.S. Energy Information Administration, Washington, DC (www.eia.gov/emeu/mecs/contents.html).

USEIA. (2004). *Greenhouse Gases, Climate Change, and Energy*. U.S. Energy Information Administration, Washington, DC (www.eia.gov/oiaf/1605/ggccebro/chapter1.html).

USEIA. (2008a). *Annual Energy Outlook 2008 with Projections to 2030*. U.S. Energy Information Administration, Washington, DC (www.eia.gov/oiaf/aeo/pdf/0383(2008).pdf).

USEIA. (2008b). *Annual Energy Review 2007*. U.S. Energy Information Administration, Washington, DC (ftp://ftp.eia.doe.gov/multifuel/038407.pdf).

USEIA. (2009a). *March 2009 Monthly Energy Review*. U.S. Energy Information Administration, Washington, DC (www.eia.gov/totalenergy/data/monthly/previous.cfm#2009).

USEIA. (2009b). *Natural Gas Year-in-Review 2008*. U.S. Energy Information Administration, Washington, DC (http://www.eia.gov/pub/oil_gas/natural_gas/feature_articles/2009/ngyir2008/ngyir2008.html).

USEIA. (2010). *U.S. Crude Oil, Natural Gas, and Natural Gas Liquids Reserves 2009*. U.S. Energy Information Administration, Washington, DC (www.eia.gov/oil_gas/natural_gas/data_publications/crude_oil_natural_gas_reserves/cr.html).

USEIA. (2011). *Natural Gas Explained*. U.S. Energy Information Administration, Washington, DC (www.eia.gov/energyexplained/index.cfm?page=natural_gas_ home).

USEIA. (2012). *Total Energy: Natural Gas Flow, 2010 (Trillion Cubic Feet)*. U.S. Energy Information Administration, Washington, DC (http://www.eia.gov/totalenergy/data/annual/diagram3.cfm).

USEPA. (1972). *Federal Air Quality Control Regions*. U.S. Environmental Protection Agency. Rockville, MD.

USEPA. (1974). *Sulfur Compound Emissions of the Petroleum Production Industry*, EPA-650/2-75-030. U.S. Environmental Protection Agency, Cincinnati, OH.

USEPA. (2012). Petroleum industry. In *AP 42, Fifth Edition*, Vol. I. U.S. Environmental Protection Agency, Washington, DC (www.epa.gov/ttnchie1/ap42/ch05/index.html).

3

Shale Gas Geology

No! There's the land. (Have you seen it?)
It's the cussedest land that I know,
From the big, dizzy mountains that screen it
To the deep, deathlike valleys below.
Some say God was tired when He made it;
Some say it's a fine land to shun;
Maybe; but there's some as would trade it
For no land on earth—and I'm one.

—Robert W. Service (*The Spell of the Yukon*)

Geologists can read these layers of sedimentary rock like the pages of a history book.

—U.S. Geological Service (2006)

A sedimentary rock is a clue to the past; it is filled with memories of splendor and horror.

—Frank R. Spellman

3.1 Introduction

To get even closer to the margin of understanding shale gas and hydraulic fracturing one must have a fundamental, foundational, basic knowledge of sedimentary geology. For this reason, this chapter presents very basic coverage of sedimentary geology.

3.2 Sedimentation

Exposed rocks on the surface of the Earth are especially vulnerable to the surface agents of erosion (weathering, erosion, rain, streamflow, wind, wave action, ocean circulation). When eroded, these rock fragments (called *detrilus*) are commonly picked up and transported by wind, water, or ice.

These rock fragments are generally referred to as *sediments* when they have been dropped by the transporting agents. The components of sediments include small fragments (gravel, sand, or silt size), new minerals (mainly clays), and dissolved portions of the source rock (e.g., dissolved salts in river and ocean water).

Sediments on the surface of the Earth may form in several ways: (1) by mere mechanical accumulation (via wind or water), such as gravel and sand deposits in a river or sand dunes in a desert; (2) by chemical precipitation, such as salt and calcite precipitation in shallow seas and lakes; and (3) by activity of organisms, such as carbonate accumulation in coral reefs or the accumulation of organic matter in swamps (coal precursor). Sediments are typically deposited in layers or beds called *strata*. When sediments become compacted and cemented together (a process known as *lithification*), they form sedimentary rocks. This compaction or lithification of sedimentary materials into stratified layers is probably the most significant feature of sedimentary rocks.

These stratified layers are like pages in the ultimate history book—Earth's history—where each page is dedicated to a particular time frame, earliest to present. Sediments of any particular time period form a distinct layer that is underlain and overlain by equally distinct layers of older and younger times, respectively. These layers, composed of such common rock types as sandstone, shale, and limestone, make up about 75% of the rocks exposed on the surface of the Earth. Geologists can study sedimentary rocks in the making; therefore, they probably know more about the origin of this type of rock than igneous and metamorphic rocks combined.

3.3 Types of Sedimentary Rocks

Several different types of sedimentary rocks can be distinguished according to the source of rock materials that form them:

- Clastic (or detrital) sedimentary rocks
  - Conglomerates
  - Sandstones
  - Mudstones/shales
- Chemical and biochemical sedimentary rocks
  - Limestone/dolostone
  - Evaporites
  - Carbonaceous rocks

TABLE 3.1

Classification of Clastic Sedimentary

Name of Particle	Size Range (mm)	Loose Sediment	Consolidated Rock
Boulder	>256	Gravel	Conglomerate
Cobble	64–256	Gravel	Conglomerate
Pebble	2–64	Gravel	Conglomerate
Sand	1/16–2	Sand	Sandstone
Silt	1/256–1/16	Silt	Siltstone
Clay	<1/256	Clay	Claystone, mudstone, and shale

Source: Adapted from Fichter, L.S., *Sedimentary Rocks*, James Madison University, Harrisonburg, VA, 2000 (http://csmres.jmu.edu/geollab/fichter/SedRx/).

3.3.1 Clastic Sedimentary Rocks

Clastic sedimentary rocks are the group of rocks most people think of when they think of sedimentary rocks. Clastic sedimentary rocks are made up of pieces (fragmented material from other rocks, or *clasts*) of preexisting rocks. Pieces of rock are loosened by weathering and then transported by water, wind, gravity, or glacial action to some basin or depression, where the sediment is trapped. If the sediment is buried deeply, it becomes compacted and cemented, forming sedimentary rock. Depending on grain size, sedimentary rocks are subdivided into conglomerate, sandstone, siltstone, and shale (see Table 3.1).

The formation of a clastic sedimentary rock involves the following processes:

- *Transportation*—Sediments move to their final destination by sliding down slopes, being picked up by the wind, or by being carried by running water in streams, rivers, or ocean currents. During transport, the sediment particles will be sorted according to size and density and will be rounded by abrasion. The distance the sediment is transported and the energy of the transporting medium all leave clues in the final sediment that tell us something about the mode of transportation.
- *Deposition*—Sediment is deposited when the energy of the transporting medium becomes too low to continue the transport process. In the deposition process, when the velocity of the transporting medium becomes too low to transport sediment, the sediment falls out and becomes deposited.
- *Diagenesis*—This is the process of chemical and physical change that turns sediment into rock; that is, the term describes all of the changes that sediment undergoes after deposition and before the transition to metamorphism. The first step in the process is compaction, which

occurs when the weight of the overlying material increases. As the grains of the material are compacted together, pore space is reduced and water is eliminated from the substance. The free water usually carries mineral components in solution, and these constituents may later precipitate as new minerals in the pore spaces. This causes cementation, which binds the individual particles together and can be seen in quartz, calcite, iron oxide, clay, glauconite, and feldspar. The next stage of diagenesis involves alteration. Limestone and feldspar, for example, are converted to dolomite and albite, respectively. This alteration occurs when carbonate rock begins to dissolve under pressure, due to either deep burial or tectonic squeezing. In addition, an absence of oxygen during the compaction process may cause other alterations to the original sediment.

Some of the more common types of clastic sedimentary rocks are described below:

- *Shale*—Shale consists of consolidated clay and other fine particles (mud) that have hardened into rock. It is the most abundant of all sedimentary rocks, comprising about 60 to 70% of the sedimentary rocks on Earth. Characteristically fine-grained and thinly bedded, shale is split easily along dividing (bedding) planes. Shale is classified or typed by composition; for example, shale containing large amounts of clay is referred to as *argillaceous shale*, and shale containing appreciable amounts of sand is known as *arenaceous shale*. Shale high in organic matter is typically black in color and referred to as *carbonaceous shale*. Shale that contains large amounts of lime is known as *calcareous shale* and is used in the manufacture of Portland cement. Another type of shale, *oil shale*, is currently of great interest worldwide because of the supply and demand and increasing cost of crude oil. Oil shale contains kerogen, a fossilized insoluble organic material that is converted into petroleum products. Oil shale may be a short-term solution to crude oil shortage problems (see Case Study 3.1).
- *Sandstone*—Sandstones, composed essentially of cemented sand, comprise about 30% of all sedimentary rocks. The most abundant mineral in sandstone is quartz, along with lesser amounts of calcite, gypsum, and various iron compounds. Sandstone is used as an abrasive (for sandpaper) and as a building stone.
- *Conglomerate*—The least abundant sediment type, conglomerates are consolidated gravel deposits with variable amounts of sand and mud between the pebbles. Conglomerates accumulate in stream channels, along the margins of mountain ranges, and on beaches. Conglomerates composed largely of angular pebbles are called *breccias*; those formed in glacial deposits are called *tillites*.

Case Study 3.1. U.S. Oil-Shale Deposits*

Oil shale is commonly defined as a fine-grained sedimentary rock containing organic matter that yields substantial amounts of oil and combustible gas upon destructive distillation. Most of the organic matter is insoluble in ordinary organic solvents; therefore, it must be decomposed by heating to release such materials. Underlying most definitions of oil shale is its potential for the economic recovery of energy, including shale oil and combustible gas, as well as a number of byproducts. A deposit of soil shale having economic potential is generally one that is at or near enough to the surface to be developed by open-pit or conventional underground mining or by *in situ* methods.

Oil shales range widely in organic content and oil yield. Commercial grades of oil shale, as determined by their yield of shale oil, range from about 100 to 200 liters per metric ton (L/t) of rock. The U.S. Geological Survey has used a lower limit of about 40 L/t for classification of federal oil shale lands. Others have suggested a limit as low as 25 L/t.

Deposits of oil shale can be found in many parts of the world. These deposits, which range in age from Cambrian to Tertiary, may occur as minor accumulations of little or no economic value or as giant deposits that occupy thousands of square kilometers and reach thicknesses of 700 m or more. Oil shales were deposited in a variety of depositional environments, including lakes ranging from freshwater to highly saline, epicontinental marine basins and subtidal shelves, and limnic and coastal swamps, commonly in association with deposits of coal.

In terms of mineral and elemental content, oil shale differs from coal in several distinct ways. Oil shale typically contains much larger amounts of inert mineral matter (60 to 90%) than coals, which have been defined as containing less than 40% mineral matter. The organic matter of oil shale is the source of liquid and gaseous hydrocarbons; it typically has a higher hydrogen and lower oxygen content than that of lignite and bituminous coal.

In general, the precursors of the organic matter in oil shale and coal also differ. Much of the organic matter in oil shale is of algal origin but may also include remains of vascular land plants that more commonly compose much of the organic matter in coal. The origin of some of the organic matter in oil shale is obscure because of the lack of recognizable biologic structures that would help identify the precursor organism. Such materials may be of bacterial origin or the product of bacterial degradation of algae or other organic matter. The mineral component of some oil shale is composed of carbonates, including calcite, dolomite, and siderite, with lesser amounts of aluminosilicates. For other oil shale, the

* The material presented in this section is adapted from Dyni, J.R., *Geology and Resources of Some World Oil-Shale Deposits*, U.S. Geological Survey Central Region, Denver, CO, 2005; Spellman, F.R., *Geology for Non-Geologists*, Government Institutes Press, Lanham, MD, 2009.

reverse is true—silicates including quartz, feldspar, and clay minerals are dominant and carbonates are a minor component. Many oil shale deposits contain small, but ubiquitous, amounts of sulfides, including pyrite and marcasite, indicating that the sediments probably accumulated in dysaerobic to anoxic waters that prevented the destruction of the organic matter by burrowing organisms and oxidation.

In the past, shale oil in the world market was not competitive with petroleum, natural gas, or coal; however, this may be changing with prices of oil pushing toward $150 per barrel. Even before shale oil began to become somewhat competitive with current world market prices for crude oil, it was being used in several countries that possess easily exploitable deposits of oil shale but lack other fossil fuel resources. Some oil shale deposits contain minerals and metals that add byproduct value, such as alum, nahcolite (sodium bicarbonate), dawsonite (sodium aluminum carbonate hydroxide), sulfur, ammonium sulfate, vanadium, zinc, copper, and uranium. Tectonic events and volcanism have altered some deposits. Structural deformation may impair the mining of an oil shale deposit, whereas igneous intrusions may have thermally degraded the organic matter. Thermal alteration of this type may be restricted to a small part of the deposit or it may be widespread, making most of the deposit unfit for recovery of shale oil.

The gross heating value of oil shale on a dry-weight basis ranges from about 500 to 4000 kilocalories per kilogram (kcal/kg) of rock. The higher grade kukersite oil shale of Estonia, which fuels several electric power plants, has a heating value of about 2000 to 2200 kcal/kg. By comparison, the heating value of lignitic coal ranges from 3500 to 4600 kcal/kg on a dry, mineral-free basis (ASTM, 1966).

3.3.2 Recoverable Resources

The commercial development of an oil shale deposit depends upon many factors. The geologic setting and the physical and chemical characteristics of the resource are of primary importance. Roads, railroads, power lines, water, and available labor are among the factors to be considered in determining the viability of an oil shale operation. Oil shale lands that could be mined may be preempted by current land usage, such as population centers, parks, and wildlife refuges. Development of new *in situ* mining and processing technologies may allow an oil shale operation in previously restricted areas without causing damage to the surface or posing problems of air and water pollution. The availability and price of petroleum could ultimately bring about the viability of a large-scale oil shale industry. Today, few if any deposits can be economically mined and processed for shale oil in competition with petroleum. Nevertheless, some countries with oil shale resources but lacking petroleum reserves have found it expedient to operate an oil shale industry. As suppliers of petroleum diminish in future years and petroleum costs increase, greater use of oil shale for the production of electric power, transportation fuels, petrochemicals, and other industrial products seems likely.

3.3.3 Determining Grade of Oil Shale

The grade of oil shale has been determined by many different methods with the results expressed in a variety of units. The heating value of the oil shale may be determined using a calorimeter. Values obtained by this method are reported in English or metric units, such as British thermal units (Btu) per pound of oil shale, calories per gram (cal/g) of rock, kilocalories per kilogram (kcal/kg) of rock, megajoules per kilogram (MJ/kg) of rock, and other units. The heating value is useful for determining the quality of an oil shale that is burned directly in a power plant to produce electricity. Although the heating value of a given oil shale is a useful and fundamental property of the rock, it does not provide information on the amounts of shale oil or combustible gas that would be yielded by retorting (destructive distillation).

The grade of oil shale can be determined by measuring the yield of oil of a shale sample in a laboratory retort. This is perhaps the most common type of analysis that is currently used to evaluate an oil shale resource. The method commonly used in the United States is the *modified Fischer assay*, first developed in Germany and then adapted by the U.S. Bureau of Mines for analyzing oil shale of the Green River formation in the western United States (Stanfield and Frost, 1949). The technique was subsequently standardized as the American Society for Testing and Materials (ASTM) Method D-3904-80 (ASTM, 1984). Some laboratories have further modified the Fischer assay method to better evaluate different types of oil shale and different methods of oil-shale processing.

The standardized Fischer assay method consists of heating a 100-g sample crushed to a –8 mesh (2.38-mm mesh) screen in a small aluminum retort to 500°C at a rate of 12°C per minute and holding it at that temperature for 40 minutes. The distilled vapors of oil, gas, and water are passed through a condenser cooled with ice water into a graduated centrifuge tube. The oil and water are then separated by centrifuging. The quantities reported are the weight percentages of shale oil (and its specific gravity), water, shale residue, and "gas plus loss" by difference.

The Fischer assay method does not determine the total available energy in an oil shale. When oil shale is retorted, the organic matter decomposes into oil, gas, and a residuum of carbon char that remains in the retorted shale. The amounts of individual gases—chiefly hydrocarbons, hydrogen, and carbon dioxide—are not normally determined but are reported collectively as "gas plus loss," which is the difference of 100 weight percent (wt%) minus the sum of the weights of oil, water, and spent shale. Some oil shale may have a greater energy potential than that reported by the Fischer assay method, depending on the components of the "gas plus loss."

The Fischer assay method also does not necessarily indicate the maximum amount of oil that can be produced by a given oil shale. Other retorting methods, such as the Tosco II process, are known to yield in excess of 100% of the yield reported by the Fischer assay. In fact, special methods of retorting, such as the Hytort process, can increase oil yields of some oil shales by as much as three to four times the yield obtained by the Fischer assay method (Dyni et al., 1990; Schora et al., 1983). At best, the Fischer assay method only approximates the energy potential of an oil shale deposit. Newer techniques for evaluating oil shale resources include the Rock-Eval and material balance Fisher assay methods. Both give more complete information about the grade of oil shale, but they are not widely used. The modified Fischer assay, or close variations thereof, is still the major source of information for most deposits. It would be useful to develop a simple and reliable assay method for determining the energy potential of an oil shale that would include the total heat energy and the amounts of oil, water, combustible gases including hydrogen, and char in sample residue.

3.3.4 Origin of Organic Matter

Organic matter in oil shale includes the remains of algae, spores, pollen, plant cuticle, and corky fragments of herbaceous and woody plants, as well as other cellular remains of lacustrine (i.e., sedimentary environment of a lake), marine, and land plants. These materials are composed chiefly of carbon, hydrogen, oxygen, nitrogen, and sulfur. Some organic matter retains enough biological structures so that specific types can be identified as to genus and even species. In some oil shales, the organic matter is unstructured and is best described as amorphous (bituminite). The origin of this amorphous material is not well known, but it is likely a mixture of degraded algal or bacterial remains. Small amounts of plant resins and waxes also contribute to the organic matter. Fossil shell and bone fragments composed of phosphatic and carbonate minerals, although of organic origin, are excluded from the definition of organic matter used herein and are considered to be part of the mineral matrix of the oil shale. Most of the organic matter in oil shale is derived from various types of marine and lacustrine algae. It may also include varied admixtures of biologically higher forms of plant debris that depend on the depositional environment and geographic position. Bacterial remains can be volumetrically important in many oil shales, but they are difficult to identify As mentioned, most of the organic matter in oil shale is insoluble in ordinary organic solvents, whereas some is bitumen, which is soluble in certain organic solvents. Solid hydrocarbons, including gilsonite, wurtzlite, grahamite, ozokerite, and albertite, are present as veins or pods in some oil shale. These hydrocarbons have somewhat varied chemical and physical characteristics, and several have been mined commercially.

3.3.5 Thermal Maturity of Organic Matter

The thermal maturity of an oil shale refers to the degree to which the organic matter has been altered by geothermal heating. If the oil shale is heated to a high enough temperature, as may be the case if the oil shale were deeply buried, the organic matter may thermally decompose to form oil and gas. Under such circumstances, oil shales can be source rocks for petroleum and natural gas. The Green River formation oil shale, for example, is presumed to be the source of the oil in the Red Wash field in northeastern Utah. On the other hand, oil shale deposits that have economic potential for their shale oil and gas yields are geothermally immature and have not been subjected to excessive heating. Such deposits are generally close enough to the surface to be mined by open-pit mining, underground mining, or *in situ* methods.

The degree of thermal maturity of an oil shale can be determined in the laboratory by several methods. One technique is to observe the changes in color of the organic matter in samples collected from varied depths in a borehole. Assuming that the organic matter is subjected to geothermal heating as a function of depth, the colors of certain types of organic matter change from lighter to darker colors. These color differences can be noted by a petrographer and measured using photometric techniques.

The geothermal maturity of organic matter in oil shale is also determined by the reflectance of vitrinite (a common constituent of coal derived from vascular land plants), if present in the rock. Vitrinite reflectance is commonly used by petroleum explorationists to determine the degree of geothermal alteration of petroleum source rocks in a sedimentary basin. A scale of vitrinite reflectance has been developed that indicates when the organic matter in a sedimentary rock has reached temperatures high enough to generate oil and gas. However, this method can pose a problem with respect to oil shale, because the reflectance of vitrinite may be depressed by the presence of lipid-rich organic matter. Vitrinite may be difficult to recognize in oil shale because it resembles other organic material of algal origin which may not have the same reflectance response as vitrinite, thus leading to erroneous conclusions. For this reason, it may be necessary to measure vitrinite reflectance from laterally equivalent vitrinite-bearing rocks that lack the algal material.

In areas where the rocks have been subjected to complex folding and faulting or have been intruded by igneous rocks, the geothermal maturity of the oil shale should be evaluated for proper determination of the economic potential of the deposit.

3.3.6 Classification of Oil Shale

Oil shale has received many different names over the years, such as cannel coal, boghead coal, alum shale, stellarite, albertite, kerosene shale, bituminite, gas coal, algal coal, wollongite, schistes bitumineux, torbanite, and kukersite. Some of these names are still used for certain types of oil shale.

Recently, however, attempts have been made to systematically classify the many different types of oil shale on the basis of the depositional environment of the deposit, the petrographic character of the organic matter, and the precursor organisms from which the organic matter was derived. A useful classification of oil shale was developed by Hutton (1987, 1988, 1991), who pioneered the use of blue/ultraviolet fluorescent microscopy in the study of oil shale deposits in Australia. Adapting petrographic terms from coal terminology, Hutton developed a classification of oil shale based primarily on the origin of the organic matter. His classification has proved to be useful for correlating different kinds of organic matter in oil shale with the chemistry of the hydrocarbons derived from oil shale. Hutton (1991) visualized oil shale as one of three broad groups of organic-rich sedimentary rocks: (1) humic coal and carbonaceous shale, (2) bitumen-impregnated rock, and (3) oil shale.

Hutton divided oil shale into three groups based on their environments of deposition—terrestrial, lacustrine, and marine. Terrestrial oil shale includes those composed of lipid-rich organic matter such as resin spores, waxy cuticles, and corky tissue of roots and stems of vascular plants commonly found in coal-forming swamps and bogs. Lacustrine oil shales include lipid-rich organic matter derived from algae that lived in freshwater, brackish, or saline lakes. Marine oil shales are composed of lipid-rich organic matter derived from marine algae, acritarchs (unicellular organisms of questionable origin), and marine dinoflagellates (commonly regarded as algae).

Several quantitatively important petrographic components of the organic matter in oil shale—telalginite, lamalginite, and bituminite—are adapted from coal petrography. Telalginite is organic matter derived from large colonial or thick-walled unicellular algae, typified by genera such as *Botrycoccus*. Lamalginite includes thin-walled colonial or unicellular algae that occur as laminae with little or no recognizable biologic structures. Telalginite and lamalginite fluoresce brightly in shades of yellow under blue/ultraviolet light. Bituminite, on the other hand, is largely amorphous, lacks recognizable biologic structures, and weakly fluoresces under blue light. It commonly occurs as an organic groundmass with fine-grained mineral matter. The material has not been fully characterized with respect to composition or origin, but it is commonly an important component of marine oil shales. Coaly materials including vitrinite and inertinite are rare to abundant components of oil shale; both are derived from humic matter of land plants and have moderate and high reflectance, respectively, under the microscope.

Within his threefold grouping of oil shale (terrestrial, lacustrine, and marine), Hutton (1991) recognized six specific oil-shale types: cannel coal, lamosite, marinite, torbanite, tasmanite, and kukersite. The most abundant and largest deposits are marinites and lamosites. Cannel coal is brown to black oil shale composed of resins, spores, waxes, and cutinaceous and corky materials derived from terrestrial vascular plants together with varied amounts of vitrinite and inertinite. Cannel coals originate in oxygen-deficient ponds or shallow lakes in peat-forming swamps and bogs (Stach et al., 1975).

Lamosite is pale, grayish brown and dark gray to black; the chief organic constituent of lamosite is lamalginite derived from lacustrine planktonic algae. Other minor components in lamosite include vitrinite, inertinite, telalginite, and bitumen. The Green River formation oil shale deposits in the western United States and a number of Tertiary lacustrine deposits in eastern Queensland, Australia, are lamosites.

Marinite is gray to dark gray to black; it is of marine origin and the chief organic components are lamalginite and bituminite, derived chiefly from marine phytoplankton. Marinite may also contain small amounts of bitumen, telalginite, and vitrinite. Marinites are deposited typically in epeiric seas (large but shallow bodies of saltwater) such as on broad shallow marine shelves or inland seas where wave action is restricted and currents are minimal. The Devonian–Mississippian oil shales of the eastern United States are typical marinites. Such deposits are generally widespread, covering hundreds to thousands of square kilometers, but they are relatively thin, often less than about 100 m.

Torbanite, tasmanite, and kukersite are related to specific kinds of algae from which the organic matter was derived; the names are based on local geographic features. Torbanite, named after Torbane Hill in Scotland, is a black oil shale whose organic matter is composed mainly of telalginite derived largely from lipid-rich *Botryococcus* and related algal forms found in lakes ranging from freshwater to brackish water. It also contains small amounts of vitrinite and inertinite. The deposits are commonly small but can be extremely high grade. Tasmanite, named from oil-shale deposits in Tasmania, is a brown to black oil shale. The organic matter consists of telalginite derived chiefly from unicellular tasmanitid algae of marine origin and lesser amounts of vitrinite, lamalginite, and inertinite. Kukersite, which takes its name from Kukruse Manor near the town of Kohtla-Jarve, Estonia, is a light brown marine oil shale. Its principal organic component is telalginite derived from the green alga, *Gloeocapsomorpha prisca*. The Estonian oil shale deposits in northern Estonia along the southern coast of the Gulf of Finland and its eastern extension into Russia, the Leningrad deposit, are kukersites.

3.3.7 Evaluation of Oil Shale Resources

Relatively little is known about many of the world's deposits of oil shale, and much exploratory drilling and analytical work need to be done. Early attempts to determine the total size of world oil shale resources were based on few facts, and estimates of the grade and quantity of many of these resources were speculative, at best. The situation today has not greatly improved, although much information has been published in the past decade or so, notably for deposits in Australia, Canada, Estonia, Israel, and the United States. Evaluation of world oil shale resources is especially difficult because of the wide variety of analytical units that are reported. The grade of a deposit is variously expressed in U.S. or Imperial gallons of shale oil per short ton (GPT) or rock, liters of shale oil per metric ton (L/t) of rock, barrels, short or metric tons of shale oil, kilocalories per kilogram (kcal/kg) of oil shale, or gigajoules (GJ) per unit weight of oil shale.

To bring some uniformity into this assessment, oil shale resources in this discussion are given in both metric tons of shale oil and in equivalent U.S. barrels of shale oil, and the grade of oil shale, where known, is expressed in liters of shale oil per metric ton (L/t) of rock. If the size of the resource is expressed only in volumetric units (barrels, liters, cubic meters, and so on), the density of the shale oil must be known or estimated to convert these values to metric tons. Most oil shales produce shale oil that ranges in density from about 0.85 to 0.97 by the modified Fischer assay method. In cases where the density of the shale oil is unknown, a value of 0.910 is assumed for estimating resources. Byproducts may add considerable value to some oil shale deposits. Uranium, vanadium, zinc, alumina, phosphate, sodium carbonate minerals, ammonium sulfate, and sulfur are some of the potential byproducts. The spent shale after retorting is used to manufacture cement, notably in Germany and China. The heat energy obtained by the combustion of the organic matter in oil shale can be used in the cement-making process. Other products that can be made from oil shale include especially carbon fibers, adsorbent carbons, carbon black, bricks, construction and decorative blocks, soil additives, fertilizers, rock wool insulating material, and glass. Most of these uses are still small or in experimental stages, but the economic potential is large.

3.3.8 U.S. Oil Shale Deposits

Numerous deposits of oil shale, ranging from Precambrian to Tertiary age, are present in the United States. The two most important deposits are in the Eocene Green River formation in Colorado, Wyoming, and Utah and in the Devonian–Mississippian black shales in the eastern United States. Oil shale associated with coal deposits of Pennsylvanian age is also found in the eastern United States. Other deposits are known to be in Nevada, Montana, Alaska, Kansas, and elsewhere, but these are either too small or too low grade, or they have not yet been well enough explored (Russell, 1990). Because of their size and grade, most investigations have focused on the Green River and the Devonian Mississippian deposits.

3.3.8.1 Green River Formation

3.3.8.1.1 Geology

Lacustrine sediments of the Green River formation were deposited in two large lakes that occupied 65,000 km^2 in several sedimentary/structural basins in Colorado, Wyoming, and Utah during early through middle Eocene time. The Uinta Mountain uplift and its eastward extension, the Axial Basin anticline, separate these basins. The Green River lake system was in existence for more than 10 million years during a time of a warm temperate to subtropic climate. During parts of their history, the lake basins were closed, and the waters became highly saline. Fluctuations in the amount of inflowing stream waters caused large expansions and contractions of the lakes as evidenced by widespread intertonguing of marly lacustrine strata with beds of land-derived sandstone and siltstone. During arid times, the lakes contracted, and the waters became increasingly saline and alkaline. The lake-water content of soluble sodium carbonates and chloride increased, whereas the less-soluble divalent Ca+Mg+Fe carbonates were precipitated with organic-rich sediments. During the driest periods, the lake waters reached salinities sufficient to precipitate beds of nahcolite, halite, and trona. The sediment pore waters were sufficiently saline to precipitate disseminated crystals of nahcolite, shorite, and dawsonite along with a host of other authigenic (generated where it was found or observed) carbonate and silicate minerals (Milton, 1977).

A noteworthy aspect of the mineralogy is the complete lack of authigenic sulfate minerals. Although sulfate was probably a major anion in the stream waters entering the lakes, the sulfate ion was presumably totally consumed by sulfate-reducing bacteria in the lake and sediment waters according to the following generalized oxidation–reduction reaction:

$$2CH_2O + SO_4^{-2} \rightarrow 2HCO_3^{-1} + H_2S \text{ (hydrogen sulfide)}$$

Note that two moles of bicarbonate are formed for each mole of sulfate that is reduced. The resulting hydrogen sulfide could either react with available Fe^{++} to precipitate as ion sulfide minerals or escape from the sediments as a gas (Dyni, 1998). Other major sources of carbonate include calcium-carbonate-secreting algae, hydrolysis of silicate minerals, and direct input from inflowing streams. The warm alkaline waters of the Eocene Green River lakes provided excellent conditions for the abundant growth of blue–green algae (cyanobacteria) that are thought to be the major precursor of the organic matter in the oil shale. During times of freshening waters, the lakes hosted a variety of fishes, rays, bivalves, gastropods, ostracodes, and other aquatic fauna. Areas peripheral to the lakes supported a large and varied assemblage of land plants, insects, amphibians, turtles, lizards, snakes, crocodiles, birds, and numerous mammalian animals (Grande, 1984; MacGinitie, 1969; McKenna, 1960).

3.3.8.1.2 Historical Developments

The occurrence of oil shale in the Green River formation in Colorado, Utah, and Wyoming has been known for many years. During the early 1900s, it was clearly established that the Green River deposits were a major resource of shale oil (Gavin, 1924; Winchester, 1916; Woodruff and Day, 1914). During this early period, the Green River and other deposits were investigated, including oil shale of the marine Phosphoria formation of Permian age in Montana (Bowen, 1917; Condit, 1919) and oil shale in Tertiary lake beds near Elko, Nevada (Winchester, 1923). In 1967, the U.S. Department of Interior began an extensive program to investigate the commercialization of the Green River oil shale deposits. The dramatic increases in petroleum prices resulting from the OPEC oil embargo of 1973–1974 triggered another resurgence of oil shale activities during the 1970s and into the early 1980s. In 1974, several parcels of public oil shale lands in Colorado, Utah, and Wyoming were put up for competitive bid under the federal Prototype Oil Shale Leasing Program. Two tracts were leased in Colorado and two in Utah to oil companies.

Large underground mining facilities, including vertical shafts, room-and-pillar entries, and modified *in situ* retorts, were constructed on the Colorado tracts, but little or no shale oil was produced. During this time, Unocal Oil Company was developing its oil shale facilities on privately owned land on the south side of the Piceance Creek Basin. The facilities included a room-and-pillar mine with surface entry, a 10,000-barrel/day (1460 ton/day) retort, and an upgrading plant. A few miles north of the Unocal property, Exxon Corporation opened a room-and-pillar mine with surface entry, haulage roads, a waste-rock dumpsite, and a water-storage reservoir and dam. In the late 1970s, the U.S. Bureau of Mines opened an experimental mine in the northern part of the Piceance Creek Basin that included a shaft 723 m deep and several room-and-pillar entries. The goal was to conduct research on the deeper deposits of oil shale, which are commingled with nahcolite and dawsonite. The site was closed in the late 1980s.

On the Utah tracts, three energy companies spent about $80 million to sink a 313-m-deep vertical shaft and inclined haulage way to a high-grade zone of oil shale and to open several small entries. Other facilities included a mine services building, water and sewage-treatment plants, and a water-retention dam. The Seep Ridge project sited south of the Utah tracts, funded by Geokinetics, Inc., and the U.S. Department of Energy, produced shale oil by a shallow *in situ* retorting method. Several thousand barrels of shale oil were produced.

The Unocal oil shale plant was the last major project to produce shale oil from the Green River formation. Plant construction began in 1980, and capital investment for constructing the mine, retort, upgrading plant, and other facilities was $650 million. Unocal produced 657,000 tons (about 4.4 million barrels) of shale oil, which were shipped to Chicago for refining into transportation fuels and other products under a program partly subsidized by the U.S. government. The average rate of production in the last months of operation was about 875 tons (about 5900 barrels) of shale oil per day; the facility was closed in 1991. In the past few years, Shell Oil Company began an experimental field project to recover shale oil by a proprietary *in situ* technique, and the results to date appear to favor continued research.

DID YOU KNOW?

We stated that there are 42 gallons in each barrel of oil (42 gal/bbl). Have you ever wondered why we abbreviate barrels as "bbl"? Should it not be "bl" instead? In the 1860s, the oil industry scrambled to find containers of that particular size. Standard Oil Company began manufacturing their own barrels to that specification and painted them blue to identify them; consequently, transactions referred to the oil as being delivered in blue barrels, or bbls.

3.3.8.1.3 Shale Oil Resources

As the Green River formation oil shale deposits in Colorado became better known, estimates of the resource increased from about 20 billion barrels in 1916, to 900 billion barrels in 1961, and to 1.0 trillion barrels (~147 billion tons) in 1989 (Donnell, 1961; Pitman et al., 1989; Winchester, 1916). The Green River oil shale resources in Utah and Wyoming are not as well known as those in Colorado. Trudell et al. (1983) calculated the measured and estimated resources of shale oil in an area of about 5200 km^2 in eastern Uinta Basin, Utah, to be 214 billion barrels (31 billion tons) of which about one third is in the rich Mahogany oil shale zone. Culbertson et al. (1980) estimated the oil shale resources in the Green River formation in the Green River Basin in southwest Wyoming to be 244 billion barrels (35 billon tons) of shale oil. Additional resources are also in the Washakie Basin east of the Green River Basin in southwest Wyoming. Trudell et al. (1973) reported that several members of the Green River formation on Kinney Rim on the west side of the Washakie Bain contain sequences of low to moderate grades of oil shale in three core holes. Two sequences of oil shale in the Laney Member, 11 and 42 m thick, average 63 L/t and represent as much as 8.7 million tons of *in situ* shale oil per square kilometer.

3.3.8.1.4 Other Mineral Resources

In addition to fossil energy, the Green River oil shale deposits in Colorado contain valuable resources of sodium carbonate minerals, including nahcolite ($NaHCO_3$) and dawsonite ($NaAl(OH)_2CO_3$). Both minerals are commingled with high-grade oil shale in the deep northern part of the basin. Dyni (1974) estimated the total nahcolite resource at 29 billion tons. Beard et al. (1974) estimated nearly the same amount of nahcolite and 17 billion tons of dawsonite. Both minerals have value for soda as (Na_2CO_3), and dawsonite also has potential value for its alumina (Al_2O_3) content. The latter mineral is most likely to be recovered as a byproduct of an oil shale operation. One company is solution mining nahcolite for the manufacture of sodium bicarbonate in the northern part of the Piceance Creek Basin at depths of about 600 m (Day, 1998). Another company stopped solution mining nahcolite in 2004 but now processes soda ash obtained from the Wyoming trona (hydrated sodium bicarbonate carbonate) deposits to manufacture sodium bicarbonate.

The Wilkins Peak Member of the Green River formation in the Green River Basin in southwestern Wyoming contains not only oils but also the world's largest known source of natural sodium carbonate as trona ($Na_2CO_3 \cdot NaHCO_3 \cdot 2H_2O$). The trona resource is estimated at more than 115 billion tons in 22 beds ranging from 1.2 to 9.8 m in thickness (Wiig et al., 1995). In 1997, trona production from five mines was 16.5 million tons (Harris, 1997). Trona is refined into soda ash (Na_2CO_3) used in the manufacture of bottle and flat glass, baking soda, soap and detergents, waste treatment chemicals, and many other industrial chemicals. One ton of soda ash is obtained from about 2 tons of trona ore. Wyoming trona supplies about 90% of U.S. soda ash needs; in addition, about one third of the total Wyoming soda ash produced is exported. In the deeper part of the Piceance Creek Basin, the Green River oil shale contains a potential resource of natural gas, but its economic recovery is questionable (Cole and Daub, 1991). Natural gas is also present in the Green River oil shale deposits in southwest Wyoming and probably in the oil shale in Utah, but in unknown quantities.

3.3.8.2 Eastern Devonian–Mississippian Oil Shale

3.3.8.2.1 Depositional Environment

Black organic-rich marine shale and associated sediments of Late Devonian and Early Mississippian age underlie about 725,000 km^2 in the eastern United States. These shales have been exploited for many years as a resource of natural gas but have also been considered as a potential low-grade resource of shale oil and uranium (Conant and Swanson, 1961; Roen and Kepferle, 1993). Over the years, geologists have applied many local names to these shales and associated rocks, including Chattanooga, New Albany, Ohio, Sunbury, and Antrim, among others. Papers detailing the stratigraphy, structure, and gas potential of these rocks in the eastern United States have been published by the U.S. Geological Survey (Roen and Kepferle, 1993).

The black shales were deposited during Late Devonian and Early Mississippian time in a large epeiric (inland) sea that covered much of middle and eastern United States east of the Mississippi River. The area includes the broad, shallow Interior Platform on the west that grades eastward into the Appalachian Basin. The depth to the base of the Devonian–Mississippian black shales ranges from surface exposures on the Interior Platform to more than 2700 m along the depositional axis of the Appalachian Basin (De Witt et al., 1993). The Late Devonian sea was relatively shallow with minimal current and wave action, much like the environment in which the Alum Shale of Sweden was deposited in Europe. A large part of the organic matter in the black shale is amorphous bituminite, although a few structured fossil organisms such as *Tasmanites, Botryococcus, Foerstia,* and others have been recognized. Conodonts (extinct primitive fishlike chordates with cone like teeth) and linguloid (small, oval, inarticulate) brachiopods are sparingly distributed through some beds. Although much of the organic matter is amorphous and of uncertain origin, it is generally believed that much of it was derived from planktonic algae.

In the distal (furthest away) parts of the Devonian sea, the organic matter accumulated very slowly along with very fine-grained clayey sediments in poorly oxygenated waters free of burrowing organisms. Conant and Swanson (1961) estimated that 30 cm of the upper part of the Chattanooga Shale deposited on the Interior Platform in Tennessee could represent as much as 150,000 years of sedimentation The black shales thicken eastward into the Appalachian Basin due to increasing amounts of clastic sediments that were shed into the Devonian sea from the Appalachian highland lying to the east of the basin. Pyrite and marcasite are abundant authigenic materials, but carbonate minerals are only a minor fraction of the mineral matter.

3.3.8.2.2 Resources

The oil shale resource is in that part of the Interior Platform where the black shales are the richest and closest to the surface. Although long known to produce oil upon retorting, the organic matter in the Devonian–Mississippian black shale yields only about half as much as the organic matter of the Green River oil shale, which is thought to be attributable to differences in the type of organic matter (or type of organic carbon) in each of the oil shales. The Devonian–Mississippian oil shale has a higher ratio of aromatic to aliphatic organic carbon than Green River oil shale and is shown by material balance Fischer assays to yield much less shale oil and a higher percentage of carbon residue (Miknis, 1990).

Hydroretorting Devonian–Mississippian oil shale can increase the oil yield by more than 200% of the value determined by Fischer assay. In contrast, the conversion of organic matter to oil by hydroretorting is much less for Green River oil shale, about 130 to 140% of the Fischer assay value. Other marine oil shales also respond favorably to hydroretorting, with yields as much as or more than 300% of the Fischer assay (Dyni et al., 1990). Mathews et al. (1980) evaluated the Devonian–Mississippian oil shales in areas of the Interior Platform where the shales are rich enough in organic matter and close enough to the surface to be mined by open pit. Results of investigations in Alabama, Illinois, Indiana, Kentucky, Ohio, Michigan, eastern Missouri, Tennessee, and West Virginia indicate that 98% of the near-surface mineable resources are in Kentucky, Ohio, Indiana, and Tennessee (Matthews, 1983). Following are the criteria for the evaluation of the Devonian–Mississippian oil shale resource used by Matthews et al. (1980):

1. Organic carbon content, ≥10 wt%
2. Overburden, ≤200 m
3. Stripping ratio, ≤2.5:1
4. Thickness of shale bed, ≥3 m
5. Open-pit mining and hydroretorting

On the basis of these criteria, the total Devonian–Mississippian shale oil resources were estimated to be 423 billion barrels (61 billion tons).

3.4 Chemical and Organic Sedimentary Rocks

Chemical and organic sedimentary rocks are the other main group of sediments besides clastic sediments. They are formed by weathered material in solution precipitating from water or as biochemical rocks made of dead marine organisms. Usually special conditions are required for these rocks to form, such as high temperature, high evaporation, and high organic activity. Some chemical sediment is deposited directly from the water in which the material is dissolved—for example, solution upon evaporation of seawater. Such deposits are generally referred to as *inorganic chemical sediments.* Chemical sediments that have been deposited by or with the assistance of plants or animals are said to be *organic* or *biochemical sediments.* Accumulated carbon-rich plant material may form coal. Deposits made mostly of animal shells may form limestone, chert, or coquina.

3.4.1 Chemical Sedimentary Rocks

Sedimentary rocks formed from sediments created by inorganic processes are discussed below:

- *Limestone*—Calcite ($CaCO_3$) is precipitated by organisms usually to form a shell or other skeletal structure. Accumulation of these skeletal remains results in the most common type of chemical sediment, limestone. Limestone may form by inorganic precipitation as well as by organic activity.
- *Dolomite*—Dolomite consists of carbonate mineral known as magnesium limestone ($CaMg(CO_3)_2$) and occurs in more or less the same settings as limestone. Dolomite is formed when some of the calcium in limestone is replaced by magnesium.
- *Evaporites*—These are sedimentary rocks (true chemical sediments) that are derived from minerals precipitated from seawater. Rock salt, which is composed of halite (NaCl), and rock gypsum

($CaSO_4 \cdot 2H_2O$), are the most common types of evaporites. High evaporation rates cause concentration of solids to increase due to water loss by evaporation.

3.4.2 Biochemical Sedimentary Rocks

Biochemical sedimentary rocks consist of sediments formed from the remains or secretions of organisms. They include *fossiliferous limestone*, *coquina* (limestone composed of shells and coarse shell fragments), *chalk* (porous, fine-textured variety of limestone composed of calcareous shells), *lignite* (brown coal), and *bituminous* (soft) *coal*.

Case Study 3.2. Coal*

America has more coal than any other fossil fuel resource. The United States also has more coal reserves than any other single country in the world. In fact, one quarter of all the known coal in the world is in the United States. The United States has more coal that can be mined than the rest of the world has oil that can be pumped from the ground. Currently, coal is mined in 26 of the 50 states. Coal is used primarily in the United States to generate electricity. In fact, it is burned in power plants to produce more than half of the electricity we use. A stove uses about half a ton of coal a year. A water heater uses about 2 tons of coal a year. And a refrigerator accounts for another half ton a year. Even though you many never see the coal, you use several tons of it every year. The material that formed fossil fuels varied greatly over time as each layer was buried. As a result of these variations and length of time the coal was forming, several types of coal were created. Depending on its composition, each type of coal burns differently and releases different types of emissions. The four types (or ranks) of coal mined today are anthracite, bituminous, sub-bituminous, and lignite:

- *Lignite*—The largest portion of the world's coal reserves is made up of lignite, a soft, brownish-black coal that forms the lowest level of the coal family. You can even see the texture of the original wood in some pieces of lignite found primarily west of the Mississippi River in the United States.
- *Sub-bituminous*—Next up the scale is sub-bituminous coal, a dull black coal. It gives off a little more energy (heat) than lignite when it burns. It is mined mostly in Montana, Wyoming, and a few other western states.

* The material presented in this section is adapted from USDOE, *Coal, Our Most Abundant Fuel*, U.S. Department of Energy, Washington, DC, 2012 (http://fossil.energy.gov/education/energylessons/coal/gen_coal.html).

- *Bituminous*—Still more energy is packed into bituminous coal, sometimes referred to as *soft coal*. In the United States, it is found primarily east of the Mississippi River in midwestern states such as Ohio and Illinois and in the Appalachian mountain range from Kentucky to Pennsylvania.
- *Anthracite*—Anthracite is the hardest coal and gives off a great amount of heat when it burns. Unfortunately, in the United States, as elsewhere in the world, there is little anthracite coal to be mined. The U.S. reserves of anthracite are located primarily in Pennsylvania.

DID YOU KNOW?

A basin is a large area with thick sedimentary rocks. It is where most natural gas is found.

3.5 Physical Characteristics of Sedimentary Rocks

Sedimentary rocks possess definite physical characteristics and display certain features that make them readily distinguishable from igneous or metamorphic rocks. Some of the most important sedimentary characteristics include the following:

- *Stratification*—Probably the most characteristic feature of sedimentary rocks is their tendency to occur in strata or beds. These strata are formed when geological agents such as wind, water, or ice gradually deposit sediment.
- *Cross-bedding*—Refers to sets of beds that are inclined relative to one another. The beds are inclined in the direction that the wind or water was moving at the time of deposition. Boundaries between sets of cross beds usually represent an erosional surface. They are very common in beach deposits, sand dunes, and river-deposited sediment.
- *Texture*—The size, shape, and arrangements of materials derived by processes of weathering, transportation, deposition, and diagenesis determine the texture of sedimentary rocks. Again, the textures we find in sediment and sedimentary rocks are dependent on processes that occur during each stage of formation. These include source materials, the nature of wind and water currents present,

the distance that materials were transported or time spent in the transportation process, biological activity, and exposure to various chemical environments.

- *Graded bedding*—In a stream, as current velocity wanes, first the larger or denser particles are deposited followed by smaller particles. This results in bedding showing a decrease in grain size from the bottom of the bed to the top of the bed (fine sediment on top and coarse at the bottom).
- *Ripple marks*—These are characteristic of shallow water deposition and are caused by small waves or winds that leave ripples of sand on the surface of a beach or on the bottom of a stream. Ripples of this type have also been preserved in certain sedimentary rocks and may provide geologists with information about the conditions of deposition when the sediment was originally deposited.
- *Mud cracks*—It is not uncommon to find mud cracks resulting from the drying out of wet sediment on the bottom of dried-up lakes, ponds, or stream beds. These many-sided (polygonal) shapes give a honeycomb appearance on the surface. If preserved in sedimentary rocks, such shapes suggest that the rock was subjected to alternating periods of flooding and drying.
- *Concretions*—These spherical or flattened masses of rock enclosed in some shales or limestones are generally harder than the rock enclosing them. Because concretions are usually harder than the enclosing rock, they are often left behind after the surrounding rock has been eroded away.
- *Fossils*—Fossils are the remains or evidence of once living organisms that have been preserved in the Earth's crust. Because life has evolved, fossils give clues to the relative age of the sediment and can be important indicators of past climates.
- *Color*—Hematites (iron oxides) produce a pink or red color in such areas as the Grand Canyon and Painted Desert.

3.6 Sedimentary Rock Facies

A sedimentary facies is a group of characteristics that describe an accumulation of deposits that have distinctive characteristics and grade laterally into other sedimentary deposits (they are not deposited uniformly) as a result of changing environments and original deposits (see Figure 3.1). Each facies is a distinctive portion of the rock layer. The change between rock types is called a *facies change*.

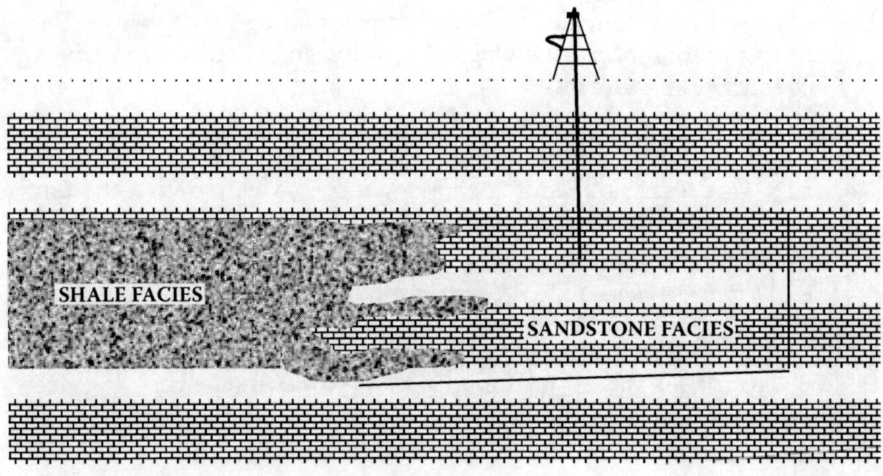

FIGURE 3.1
Sedimentary rock facies.

3.7 Thought-Provoking Discussion Questions

1. Does a student or practitioner of hydraulic fracturing need to have a background in geology? Explain.
2. Should the United States mine the energy resources contained in the Green River formation?

References and Recommended Reading

ASTM. (1966). *ASTM D388-12: Standard Classification of Coals by Rank*. American Society for Testing and Materials, West Conshohocken, PA.

ASTM. (1984). *ASTM D3904-80: Standard Test Method for Oil from Oil Shale*. American Society for Testing and Materials, West Conshohocken, PA.

Beard, T.M., Tait, D.B., and Smith, J.W. (1974). Nahcolite and dawsonite resources in the Green River formation, Piceance Creek Basin, Colorado, in Murray, D.K., Ed., *Rocky Mountain Association of Geologists Guidebook for Energy Resources of the Piceance Basin*, Rocky Mountain Association of Geologists, Denver, CO, pp. 101–109.

Bowen, F.F. (1917). Phosphatic oil shales near Dell and Dillon, Beaverhead Country, Montana. *U.S. Geological Survey Bulletin*, 661:315–320.

Cole, R.D. and Daub, G.T. (1991). Methane occurrences and potential resources in the lower Parachute Creek Meander of Green River formation, Piceance Creek Basin, Colorado. *24th Oil Shale Symposium Proceedings: Colorado School of Mines Quarterly*, 83(4):1–7.

Conant, L.C. and Swanson, V.E. (1961). *Chattanooga Shale and Related Rocks of Central Tennessee and Nearby Areas*, USGS Professional Paper 357. U.S. Geological Survey, Washington, DC.

Condit, D.D. (1919). Oil shale in western Montana, southeastern Idaho, and adjacent parts of Wyoming and Utah. *U.S. Geological Survey Bulletin*, 711:15–40.

Cross, T.A. and Homewood, P.W. (1997). Amanz Gressly's role in founding modern stratigraphy. *Geological Society of America Bulletin*, 109(12):1617–1630.

Culbertson, W.C., Smith, J.W., and Trudell, L.G. (1980). *Oil Shale Resources and Geology of the Green River Formation in the Green River Basin, Wyoming*, NTIS Report LETC/RI-80-6. U.S Department of Energy, Laramie Energy Technology Center, Laramie, WY.

Day, R.L. (1998). Solution mining of Colorado nahcolite, in *Proceedings of the First International Soda Ash Conference*, Rocks Springs, Wyoming, June 10–12, 1997, Wyoming State Geological Survey Public Information Circular 40. Wyoming State Geological Survey, Laramie.

de Witt, Jr., W., Roen J.B., and Wallace, L.G. (1993). Stratigraphy of Devonian black shales and associated rocks in the Appalachian Basin. *U.S. Geological Survey Bulletin*, 1909-B:B1–B57.

Donnell, J.R. (1961). Tertiary geology and oil-shale resources of the Piceance Creek Basin between the Colorado and White Rivers, northwestern Colorado. *U.S. Geological Survey Bulletin*, 1082-L:835–891.

Dyni, J.R. (1974). Stratigraphy and nahcolite resources of the saline facies of the Green River formation in northwest Colorado, in Murray, D.K., Ed., *Rocky Mountain Association of Geologists Guidebook for Energy Resources of the Piceance Basin*, Rocky Mountain Association of Geologists, Denver, CO, pp. 111–122.

Dyni, J.R. (1998). Prospecting for Green River-type sodium carbonate deposits, in *Proceedings of the First International Soda Ash Conference*, Rocks Springs, Wyoming, June 10–12, 1997, Wyoming State Geological Survey Public Information Circular 40. Wyoming State Geological Survey, Laramie.

Dyni, J.R., Anders, D.E., and Rex, Jr., R.C. (1990). Comparison of hydro-retorting, Fischer assay, and Rock-Eval analyses of some world oil shales, in *Proceedings 1989 Eastern Oil Shale Symposium*. University of Kentucky Institute of Mining and Minerals Research, Lexington, pp. 270–286.

Fichter, L.S. (2000). *Sedimentary Rocks*. James Madison University, Harrisonburg, VA (http://csmres.jmu.edu/geollab/fichter/SedRx/).

Gavin, M.J. (1924). Oil shale, an historical, technical, and economic study. *U.S. Bureau of Mines Bulletin*, 210:1–215.

Grande, L. (1984). Paleontology of the Green River formation with a review of the fish fauna. *Geological Survey of Wyoming Bulletin*, 63:1–333.

Harris, R.E. (1997). Fifty years of Wyoming trona mining, in McCutcheon, T.J. and McCutcheon, J.A., Eds., *Prospect to Pipeline: 48th Annual Field Conference Guidebook*. Wyoming Geological Association, Laramie, pp. 177–182.

Hutton, A.C. (1987). Petrographic classification of oil shales. *International Journal of Coal Geology*, 8:203–231.

Hutton, A.C. (1988). *Organic Petrography of Oil Shales*, USGS short course, January 25–29, Denver CO.

Hutton, A.C. (1991). Classification, organic petrography and geochemistry of oil shale, in *Proceedings 1990 Eastern Oil Shale Symposium*. University of Kentucky Institute for Mining and Minerals Research, Lexington, pp. 16–172.

MacGinitie, H.D. (1969). *The Eocene Green River Flora of Northwestern Colorado and Northeastern Utah*. University of California Press, Berkeley, 203 pp.

Matthews, R.D. (1983). The Devonian–Mississippian oil shale resource of the United States, in Gary, H.H., Ed., *Sixteenth Oil Shale Symposium Proceedings*. Colorado School of Mines Press, Golden, pp. 14–25.

Matthews, R.D., Janka, J.C., and Dennison, J.M. (1980). Devonian Oil Shale of the Eastern United States: A Major American Energy Resource, paper presented at the American Association of Petroleum Geologists Meeting, Evansville, IN, October 1–3, 43 pp.

McKenna, M.C. (1960). *Fossil Mammalia from the Early Wasatchian Four Mile Fauna, Eocene of Northwest Colorado*. University of California Press, Berkeley, 130 pp.

Miknis, F.P. (1990). Conversion characteristics of selected foreign and domestic oil shales, in *Twenty-Third Oil Shale Symposium Proceedings*. Colorado School of Mines Press, Golden, pp. 100–109.

Milton, C. (1977). Mineralogy of the Green River formation. *The Mineralogy Record*, 8:368–379.

Pitman, J.K., Pierce, F.W., and Grundy, W.D. (1989). *Thickness, Oil Yield, and Kriged Resource Estimates for the Eocene Green River Formation, Piceance Creek Basin, Colorado*, Oil and Gas Investigations Chart OC-132. U.S. Geological Survey, Washington, DC.

Reading, H.G., Ed. (1996). *Sedimentary Environments and Facies*. Blackwell Scientific, New York.

Roen, J.B. and Kepferle, R.C., Eds. (1993). Petroleum geology of the Devonian and Mississippian black shale of eastern North America. *U.S. Geological Survey Bulletin*, 1909-B:1–434.

Russell, P.L. (1990). *Oil Shales of the World, Their Origin, Occurrence and Exploitation*. Pergamon Press, New York.

Schora, F.C., Janka, J.C., Lynch, P.A., and Feldkirchner, H. (1983). Progress in the commercialization of the Hytort Process, in *Proceedings 1982 Eastern Oil Shale Symposium*. University of Kentucky Institute for Mining and Minerals Research, Lexington, pp. 183–190.

Stach, E., Taylor, G.H., Machowsky, M.-Th., Chandra, D., Teichmuller, M., and Teichmuller, R. (1975). *Stach's Textbook of Coal Petrology*. Gebruder Borntraeger, Berlin.

Stanfield, K.E. and Frost, I.C. (1949). *Method of Assaying Oil Shale by a Modified Fischer Retort*, Report of Investigations 4477. U.S. Bureau of Mines, Washington, DC.

Trudell, L.G., Roehler, H.W., and Smith, J.W. (1973). Geology of Eocene rocks and oil yields of Green River oil shales on part of Kinney Rim, Washakie Basin, Wyoming, Report of Investigations 7775. U.S. Bureau of Mines, Washington, DC.

Trudell, L.G., Smith, J.W., Beard, T.N., and Mason, G.M. (1983). *Primary Oil-Shale Resources of the Green River Formation in the Eastern Uinta Basin, Utah*, Report No. DOE/LC/RI-82-4. U.S. Department of Energy, Washington, DC.

USGS. (2004). *Sedimentary Rocks*. U.S. Geological Survey, Washington, DC (http://education.usgs.gov/lessons/schoolyard/RockSedimentary.html).

USGS. (2006). *The Making of Sedimentary Rocks*. U.S. Geological Survey, Washington, DC (http://education.usgs.gov/index.html).

Wiig, S.V., Grundy, W.D., and Dyni, J.R. (1995). *Trona Resources in the Green River Formation, Southwest Wyoming*, USGS Open-File Report 95-476. U.S. Geological Survey, Washington, DC, 88 pp.

Winchester, D.E. (1916). Oil shale in northwestern Colorado and adjacent areas. *U.S. Geological Survey Bulletin*, 641-F:139–198.

Winchester, D.E. (1923). Oil shale of the Rocky Mountain region. *U.S. Geological Survey Bulletin*, 729:1–204.

Woodruff, E.G. and Day, D.T. (1914). Oil shales of northwestern Colorado and northeastern Utah. *U.S. Geological Survey Bulletin*, 581:1–21.

4

Shale Gas Plays

Until recently, many oil and gas companies considered natural gas locked in tight, impermeable shale uneconomical to produce. Advanced drilling and reservoir stimulation methods have dramatically increased gas production from unconventional shales. The Barnett Shale formation in Texas has experienced the most rapid development. The Marcellus Shale formation of the Appalachian basin, in the northeastern United States, potentially represents the largest unconventional gas resource in the United States. Other shale formations, such as the Haynesville Shale, straddling Texas and Louisiana, have also attracted interest, as have some formations in Canada. The resource potential of these shales has significantly increased the natural gas reserve estimates in the United States (Andrews, 2009).

4.1 Unconventional Natural Gas Resources

Unconventional gas shales are fine-grained, organic-rich sedimentary rocks. The shales are both the source of and the reservoir for natural gas, unlike conventional petroleum reservoirs. In the shales, gas occupies pore spaces, and organic matter adsorbs gas on its surface. The Society of Petroleum Engineers has described *unconventional resources* as petroleum accumulations that are pervasive throughout a large area and that are not significantly affected by hydrodynamic influences (they are also referred to as *continuous-type deposits*). In contrast, conventional petroleum and natural gas occur in porous sandstone and carbonate reservoirs. Under hydrodynamic pressure exerted by water, petroleum migrated upward from its organic source until impermeable cap rock—such as shale, shaly rocks, micrite (lime mud), chalk, permafrost, and salt layers—trapped it in the reservoir rock. The "gas cap" that accumulated over the petroleum has been the source of most produced natural gas. Once the gas and oil migrate into the gas cap (trap), they separate according to density. Being lightest, the gas goes to the top of the trap to form the gas cap, where the pores of the reservoir rock are occupied by gas. The oil goes to the middle of the trap, the oil reservoir. Saltwater, being the heaviest, goes to the bottom (Hyne, 2001) (Figure 4.1).

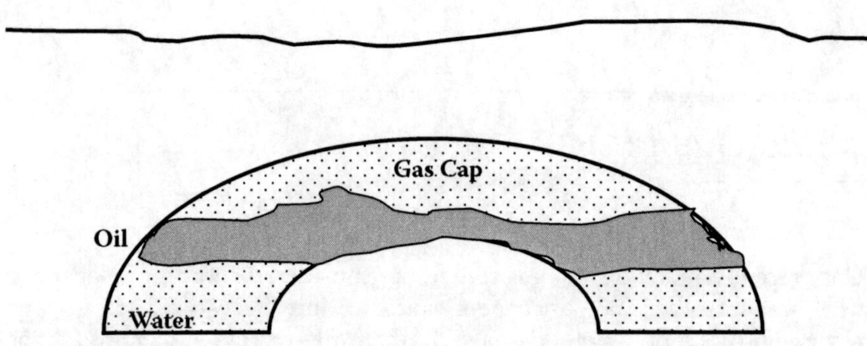

FIGURE 4.1
Hydrocarbon trap.

Exploration and production companies explore for these deposits by using complex technologies to identify prospective drilling locations. Once extracted, the natural gas is processed to eliminate other gases, water, sand, and other impurities. Some hydrocarbon gases, such as butane and propane, are captured and marketed separately. Once it has been processed, the cleaned natural gas is distributed through a system of pipelines across thousands of miles (USEIA, 2007). It is through these pipelines that natural gas is transported to its endpoint for residential, commercial, and industrial use.

DID YOU KNOW?

Of all the fossil fuels, natural gas is by far the cleanest burning.

Although the focus of this text is on unconventional natural shale gas, it is important to understand that there are other sources of unconventional natural gas currently being mined. What is considered unconventional natural gas changes over time with economics and the development of new technologies. Today, the six main categories of unconventional natural gas are deep natural gas, tight natural gas, shale gas, coal bed methane, geopressurized zones, and methane hydrates (NaturalGas.org, 2011):

- *Deep natural gas*—Natural gas that exists in deposits very far underground, beyond conventional drilling depths (typically, 15,000 feet or deeper).
- *Tight natural gas*—Gas that is stuck in a very tight formation underground, trapped in unusually impermeable hard rock or in a sandstone or limestone formation that is unusually nonporous (tight sand).

- *Shale gas*—Certain shale basins contain natural gas, usually when two thick, black shale deposits sandwich a thinner area of shale.
- *Coal bed methane*—Many coal seams contain natural gas, either within the seam itself or the surrounding rock.
- *Geopressurized zones*—Areas formed by layers of clay that are deposited and compacted very quickly on the top of more porous, absorbent material such as sand or silt. Water and natural gas trapped within this clay are squeezed out by the rapid compression of the clay and enter the more porous sand or silt deposits.
- *Methane hydrates*—Formations made up of a lattice of frozen water, which forms a sort of cage around molecules of methane.

4.1.1 Natural Gas Measurement

Natural gas is measured in either volumetric or energy units. As a gas, it is measured by the volume it displaces at standard temperatures and pressures, usually expressed in cubic feet. Gas companies generally measure natural gas in thousands of cubic feet (Mcf), millions of cubic feet (MMcf), or billions of cubic feet (Bcf), and they estimate resources such as original gas-in-place in trillions of cubic feet (Tcf).

4.1.2 Natural Gas Calculation

Calculating and tracking natural gas by volume can be useful, but natural gas can also be measured as a source of energy. Similar to other forms of energy, natural gas can be computed and presented in British thermal units (Btu). One Btu is the quantity of heat required to raise the temperature of 1 pound of water by 1°F at normal pressure (Spellman, 2009). There are about 1000 Btu in 1 cubic foot of natural gas delivered to the consumer (NaturalGas.org, 2008). Natural gas distribution companies typically measure the gas delivered to a residence in *therms* for billing purposes. A therm is equal to 100,000 Btu (approximately 100 cubic feet) of natural gas. The therm is equal to about 105.5 megajoules, 25,200 kilocalories, or 29.3 kilowatt-hours.

DID YOU KNOW?

It has been estimated that burning natural gas to produce 1 therm of heat releases 13,446 lb (6.099 kg) of carbon dioxide into the atmosphere (www.pge.com/about/environment/calculator/assumptions.shtml#averagecalifornian).

DID YOU KNOW?

Unconventional production now accounts for 46% of the total U.S. production (Navigant Consulting, 2008).

4.1.3 U.S. Natural Gas Reserves

The U.S. increased its natural gas reserves by 6% from 1970 to 2006, producing approximately 725 Tcf of gas during that period. This increase is primarily a result of advancements in technology, resulting in an increase in economically recoverable reserves (reserves becoming proven) that were previously thought to be uneconomic (BP, 2008; Navigant Consulting, 2008). Overall, unconventional natural gas is anticipated to represent an ever-increasing portion of U.S. proved reserves, while conventional gas reserves are declining. Over the last decade, production of unconventional sources has increased almost 65%, from 5.4 trillion cubic feet per year (Tcf/yr) in 1998 to 8.9 Tcf/year in 2007 (USEIA, 2008).

4.2 Distribution of Unconventional Natural Shale Gas

As shown in Figure 4.2, the lower 48 states have a wide distribution of highly organic shales containing vast resource of natural gas. Already, the relatively new Barnett Shale play in Texas accounts for 6% of all natural gas produced in the lower 48 states (USEIA, 2011). Improved drilling and fracking technologies have contributed considerably to the economic potential of shale gas. The domestic energy outlook in the United States is brightening because of the potential for production in the known onshore shale basins, coupled with other (tight gas and coal bed natural gas) unconventional gas plays. Figure 4.3 shows the projected contribution of shale gas to overall unconventional gas production in the United States.

Shale gas production has become economically feasible because of three factors: (1) advances in horizontal drilling, (2) advances in hydraulic fracturing, and, perhaps most importantly, (3) rapid increases in natural gas prices in the last several years as a result of significant supply and demand pressures. One thing is certain—without the advances in the preexisting technologies of directional drilling and hydraulic fracturing, many of the unconventional natural gas plays and the current shalenanza would not be economical, practical, viable, or in progress. The accuracy of this statement can be seen when one reviews the record: As recently as the late 1990s, only 40 drilling rigs in the United States (6% of total active rigs) were capable of onshore horizontal drilling; that number grew to 519 rigs (28% of total active rigs) by May 2008 (USEIA, 2009).

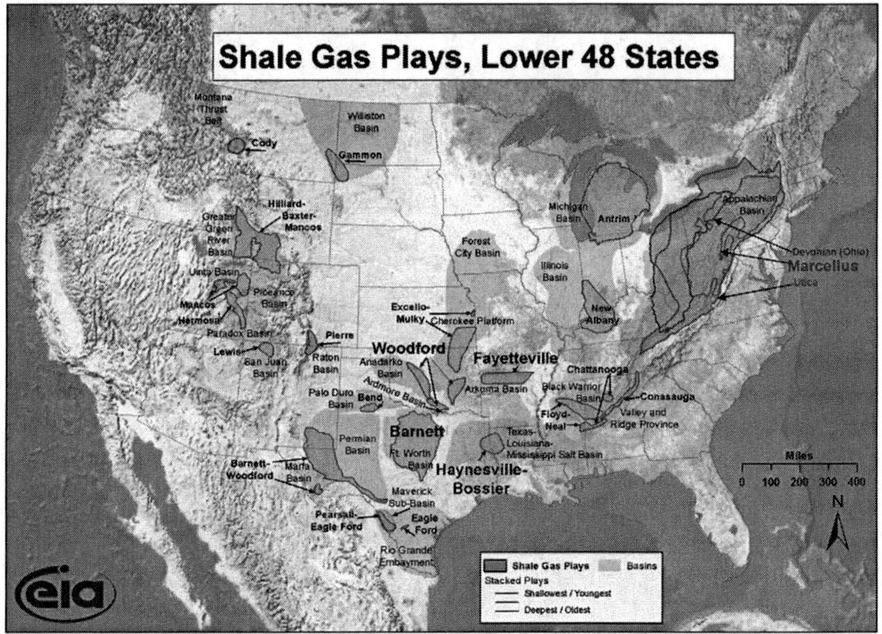

FIGURE 4.2
U.S. shale gas plays. (Adapted from U.S. Energy Information Administration, www.eia.gov/oil_gas/rpd/shaleusa2.pdf.)

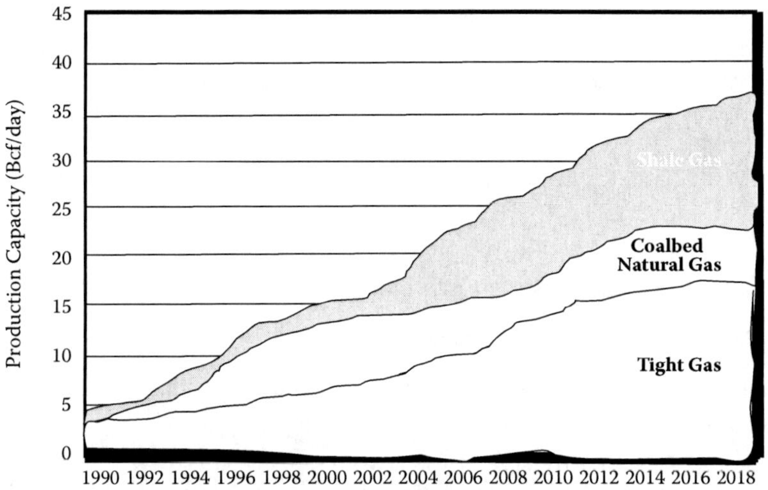

FIGURE 4.3
U.S. unconventional gas outlook (Bcf/day). (Adapted from Arthur, J.D. et al., *An Overview of Modern Shale Gas Development in the United States*, ALL Consulting, Tulsa, OK, 2008.)

DID YOU KNOW?

Shalenanza—Shale gas was supporting more than 600,000 American jobs in 2010; by 2015, shale gas is predicted to support nearly 870,000 jobs and contribute $118.2 billion to the GDP (IHS, 2011).

It has been suggested that the rapid growth of unconventional natural gas plays has not been captured by recent resource estimates compiled by the U.S. Energy Information Administration (USEIA) and that, therefore, their resource estimates do not accurately reflect the contribution of shale gas (Navigant Consulting, 2008). Since 1998, annual production has consistently exceeded the USEIA's forecasts of unconventional gas production. A great deal of this increase is attributable to the 27% growth of U.S. shale gas production by 2010. In 2011, the growth in production was 34%. It should reach 43% by 2015 and increase still further by 2035 to 60% (IHS, 2011).

DID YOU KNOW?

Shale gas resource estimates are likely to change as new information, additional experience, and advances in technology become available.

A significant benefit of shale gas plays is that many exist in areas previously developed for natural gas production; therefore, much of the necessary pipeline infrastructure is already in place. Many of these areas are also proximal to the nation's population centers, thus potentially facilitating transportation to consumers; however, additional pipelines will have to be built to access development in areas that have not seen gas production before (Muhlfelder, 2009).

4.3 Fast Forward to the Future

Because natural gas burns cleanly, the nation's domestic natural gas resources, and the presence of supporting infrastructure, the development of domestic shale gas reserves will be an important component of the United States' energy coffer for many years. Recent successes in a variety of geologic basins have created the opportunity for shale gas to be a strategic part of the nation's energy and economic growth (IPAMS, 2008).

4.4 Thought-Provoking Discussion Questions

1. Do we really know what the actual distribution and quantity of natural gas reserves are in the United States?
2. Why can't best management practices be universally applied and required?

References and Recommended Reading

Andrews, A., Ed. (2009). *Unconventional Gas Shales: Development, Technology, and Policy Issues*. Congressional Research Service, Washington, DC (www.fas.org/sgp/crs/misc/R40894.pdf).

BP. (2008). *BP Statistical Review of World Energy, June 2008*. BP, London (www.bp.com/statisticalreview).

IHS. (2011). *The Economic and Employment Contributions of Shale Gas in the United States*. IHS Global Insight, Englewood, CO (www.ihs.com/EconomicContributionofShaleGasintheUS).

IPAMS. (2008). *America's Independent Natural Gas Producers: Producing Today's Clean Energy, Ensuring Tomorrow's Innovation*. Independent Petroleum Association of Mountain States, Denver, CO (www.ipams.org/mediadocs/Callupdraft10.pdf).

Muhlfelder, T. (2009). *The Shale Gale: The Implications for North American Natural Gas Pipeline Development*. HIS Global Insight, Englewood, CO (www.ihs.com/products/cera/energy-report.aspx?id=106592110).

NaturalGas.org. (2008). *Overview of Natural Gas: Background*, www.naturalgas.org/overview/unconvent_ng_resource.asp.

NaturalGas.org. (2011). *Overview of Natural Gas: Unconventional Natural Gas Resources*, http://naturalgas.org/overview/unconventinal_ng_resource.asp.

Navigant Consulting. (2008). *North American Natural Gas Supply Assessment*. American Clean Skies Foundation, Washington, DC.

Spellman, F.R. (2009). *Physics for Non-Physicists*. Government Institutes Press, Lanham, MD.

USEIA. (2007). *About U.S. Natural Gas Pipelines—Transporting Natural Gas*. U.S. Energy Information Administration, Washington, DC (ftp://ftp.eia.doe.gov/pub/test-dmr/Pipeline_a.pdf).

USEIA. (2008). *Annual Energy Outlook 2008 with Projections to 2030*. U.S. Energy Information Administration, Washington, DC (www.eia.gov/oiaf/aeo/pdf/0383(2008).pdf).

USEIA. (2009). *Modern Shale Gas Development in the United States: A Primer*. U.S. Department of Energy, Washington, DC.

USEIA. (2011). *Natural Gas Explained: Where Our Natural Gas Comes From*. U.S. Energy Information Administration, Washington, DC (http://www.eia.gov/energyexplained/index.cfm?page=natural gas_where).

5

Shale Gas Sources

> We need to know whether the natural gas located underneath the surface is a real source of fuel for the next generation, or a speculative bubble hyped by the oil and gas industry, and echoed by the federal government's energy experts.
>
> **—Congressman Ed Markey (Massachusetts, 2010)**

> Natural gas is hemispheric. I like to call it hemispheric in nature because it is a product that we can find in our neighborhoods.
>
> **—Former President George W. Bush**

5.1 Shale Gas in the United States

If shale rock is a capable source rock, there is potential for it to store economic quantities of gas. The potential of a shale formation to contain economic quantities of gas can be evaluated by identifying specific source rock characteristics such as total organic carbon (TOC), thermal maturity, and kerogen analysis. Together, these factors can be used to predict the likelihood of the prospective shale to produce economically viable volumes of natural gas. A number of wells may have to be analyzed in order to sufficiently characterize the potential of a shale formation, particularly if the geologic basin is large and there are variations in the target shale zone. Total organic carbon, thermal maturity, and kerogen analysis are important gas shale characteristics, and it is essential to have an understanding of these parameters:

- *Total organic carbon (TOC)* is the remnant of ancient life preserved in sedimentary rocks, after degradation by bacterial and chemical processes and further modified by temperature, pressure, and time or thermal maturation. The latter step, thermal maturation, is a function of depth (burial history) and proximity to thermal sources. Maturation provides the chemical reactions in the pathway (see Figure 5.1) required to produce gas.
- The *thermal maturity* of a rock is a measure of the degree to which organic metamorphous has progressed. It gives a crude indication of the maximum temperature the rock has experienced.

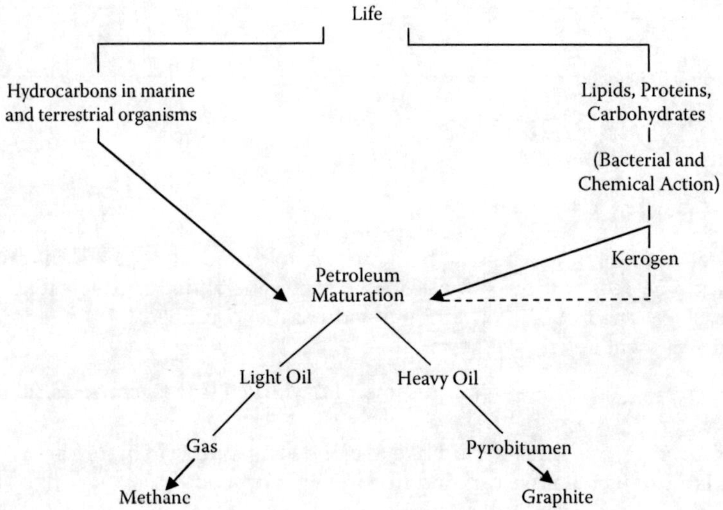

FIGURE 5.1
Pathways that convert living organisms to organic carbon. (Adapted from Chilingar, G.V. and Yen, T.F., Eds., *Bitumens, Asphalts, and Tar Sands*, Elsevier, Amsterdam, 1978.)

- *Kerogen* is a mixture of organic chemical compounds that make up a portion of the organic matter in sedimentary rocks. Some types of kerogen release crude oil or natural gas, depending on how each is heated within the Earth's crust.

5.2 Geologic Time Scale

In order to grasp the information pertinent to the nature of geologic shale formations discussed in this text it is important to have an understanding of the geologic time scale. Table 5.1 is provided on the following pages to help the reader correlate the relationship between each shale gas formation discussed and where each resides within the spectrum of the geologic time scale.

DID YOU KNOW?

Proved reserves—That portion of recoverable resources that is demonstrated by actual production or conclusive formation tests to be technically, economically, and legally producible under existing economic and operating conditions.

5.3 Active U.S. Shale Gas Plays

As shown earlier in Figure 4.2 in Chapter 4, shale gas is present across much of the lower 48 states. Figure 4.2 shows the approximate locations of current producing gas shales and prospective shales. To date, the most active of these shales are the Barnett Shale, Haynesville/Bossier Shale, Antrim Shale, Fayetteville Shale, Marcellus Shale, and New Albany Shale. This chapter does not discuss all of the unconventional gas shales; rather, discussion here is limited to these most active shale gas plays. Each of these shale gas plays or basins is different, and each has a unique set of exploration criteria and operational challenges. Because of these differences, the development of shale gas resources in each of these areas faces potentially unique challenges. The Antrim and New Albany Shales, for example, are shallower shales that produce significant volumes of formation water, unlike most of the other gas shales. Development of the Fayetteville Shale is occurring in rural areas of north-central Arkansas, while development of the Barnett Shale is focused in the area of Forth Worth, Texas, in an urban and suburban environment.

DID YOU KNOW?

Technically recoverable resources—The total amount of resource, discovered and undiscovered, that is thought to be recoverable with available technology, regardless of economics.

With the development and refinement of new technologies, shale gas plays once believed to have limited economic viability are now being reevaluated. In the following sections, the key characteristics of each of the major gas plays are summarized using 2008 data. This information provides a means to compare some of the key characteristics and evaluate the different shale gas plays. Also, note that estimates of the shale gas resources, especially the portion that is technically recoverable, are likely to increase over time as new data become available from additional drilling, as experience is gained in producing shale gas, as our understanding of the resource characteristics increases, and as recovery technologies improve.

DID YOU KNOW?

Original gas in-place—The entire volume of gas contained in the reservoir, regardless of the ability to produce it.

TABLE 5.1
Geologic Time Scale

Erathem or Era	System, Subsystem or Period, Subperiod	Series or Epoch
Cenozoic ("Age of Recent Life") 65 million years ago to Present	**Quaternary** 1.8 million years ago to the Present	**Holocene** 11,477 years ago (±85 years) to the Present (Greek *holos* for "present"; *ceno* for "new")
		Pleistocene (Great Ice Age) 1.8 million to approximately 11,477 (±85 years) years ago (Greek *pleistos* for "most"; *ceno* for "new")
	Tertiary 65.5 to 1.8 million years ago	**Pliocene** 5.3 to 1.8 million years ago (Greek *pleion* for "more"; *ceno* for "new")
		Milocene 23.0 to 5.3 million years ago (Greek *meion* for "less"; *ceno* for "new")
		Oligocene 33.9 to 23.0 million years ago (Greek *oligos* for "little" or "few"; *ceno* for "new")
		Eocene 55.8 to 33.9 million years ago (Greek *eos* for "dawn"; *ceno* for "new")
		Paleocene 65.5 to 58.8 million years ago (Greek *palaois* for "old"; *ceno* for "new")
Mesozoic ("Age of Medieval Life") 251.0 to 65.5 million years ago	**Cretaceous** ("The Age of Dinosaurs") 145.5 to 65.5 million years ago	Late or Upper Early or Lower
	Jurassic 199.6 to 145.5 million years ago	Late or Upper
	Triassic 251.0 to 199.6 million years ago	Late or Upper Middle Early or Lower

Paleozoic ("Age of Ancient Life") 542.0 to 251.0 million years ago	**Permian** 299.0 to 251.0 million years ago	Lopingian Guadalupian Cisuralian
	Pennsylvanian ("The Coal Age") 318.1 to 299.0 million years ago	Late or Upper Middle Early or Lower
	Mississippian 359.2 to 318.1 million years ago	Late or Upper Middle Early or Lower
	Devonian 416.0 to 359.2 million years ago	Late or Upper Middle Early or Lower
	Silurian 443.7 to 416.0 million years ago	Pridoli Ludlow Wenlock Llandovery
	Ordovician 488.3 to 443.7 million years ago	Late or Upper Middle Early or Lower
	Cambrian 542.0 to 488.3 million years ago	Late or Upper Middle Early or Lower
Precambrian Approximately 4 billion years ago to 542.0 million years ago		

Source: Adapted from USGS, *The Geologic Time Scale*, U.S. Geologic Service, Washington, DC, 2008 (http://vulcan.wr.usgs.gov/Glossary/geo_time_scale.html).

5.3.1 Barnett Shale Play

The Barnett Shale formation is located in the Fort Worth Basin of north-central Texas. It is a Mississippian-age black shale occurring at a depth of 6500 to 8500 feet (Table 5.2) and is bounded by limestone formations above (Marble Falls Limestone) and below (Chappel Limestone) (Figure 5.2). With 15,306 wells drilled or pending, the Barnett Shale was the most prominent shale gas play in the United States, but the Haynesville Shale gas play has surpassed Barnett in shale gas production (Railroad Commission of Texas, 2009;

DID YOU KNOW?

Infill drilling is used to increase the production rate by drilling between producing wells in an established field.

TABLE 5.2

Barnett Shale Gas Play Data (2005–2008)

Parameter	Data	Ref.
Estimated basin area (square miles)	5000	—
Depth (feet)	6500–8500	Hayden and Pursell (2005)
Net thickness (feet)	100–600	Hayden and Pursell (2005)
Depth to base of treatable water (feet)[a]	~1200	—
Rock column thickness between top of play and bottom of treatable water (feet)	5300–7300	—
Total organic carbon (%)	4.5	Hayden and Pursell (2005)
Total porosity (%)	4–5	Hayden and Pursell (2005)
Gas content (scf/ton)	300–350	Hayden and Pursell (2005)
Water production (barrels/day)	NA	—
Well spacing (acres)	60–160	Hayden and Pursell (2005)
Original gas-in-place (Tcf)	327	Navigant Consulting (2008)
Technically recoverable resources (Tcf)	44	Navigant Consulting (2008)

Note: Information presented in this table, such as original gas-in-place and technically recoverable resources, is presented for general comparative purposes only. The numbers provided are based on the sources shown and this research did not include a resource evaluation. Rather, publicly available data was obtained from a variety of sources and is presented for general characterization and comparison with other shale gas plays. Resource estimates for any basin may vary greatly depending on individual company experience, data available at the time the estimate was performed, and other factors. Furthermore, these estimates are likely to change as production methods and technologies improve.

Abbreviations: Mcf = thousands of cubic feet; NA, data not available; scf = standard cubic feet; Tcf = trillion cubic feet.

[a] Depth to base of treatable water data are based on depth data from state oil and gas agencies and state geological survey data.

<table>
<tr><th>Period</th><th></th><th>Group/Unit</th></tr>
<tr><td rowspan="3">Permian</td><td rowspan="2">Leonardian</td><td>Clear Fork Group</td></tr>
<tr><td>Wichita Group</td></tr>
<tr><td>Wolfcampian</td><td rowspan="2">Cisco Group</td></tr>
<tr><td rowspan="5">Pennsylvanian</td><td>Virgilian</td></tr>
<tr><td>Missourian</td><td>Canyon Group</td></tr>
<tr><td>Desmoinesian</td><td>Strawn Group</td></tr>
<tr><td>Atokan</td><td>Bend Group</td></tr>
<tr><td>Morrowan</td><td>Marble Falls Limestone</td></tr>
<tr><td rowspan="3">Mississippian</td><td>Chesterian–Meramecian</td><td rowspan="2">Barnett Shale</td></tr>
<tr><td rowspan="3">Osagean</td></tr>
<tr><td>Chappel Limestone</td></tr>
<tr><td rowspan="3">Ordovician</td><td>Viola Limestone</td></tr>
<tr><td rowspan="2">Canadian</td><td>Simpson Group</td></tr>
<tr><td>Ellenburger Group</td></tr>
</table>

FIGURE 5.2
Stratigraphy of the Barnett Shale. (Adapted from Hayden, J. and Pursell, D., *The Barnett Shale: Visitor's Guide to the Hottest Gas Play in the US*, Pickering Energy Partners, Inc., Houston, TX, 2005; AAPG, *Correlation of Stratigraphic Units of North America (CASUNA) Project*, American Association of Petroleum Geologists, Tulsa, OK, 1987.)

USEIA, 2011); however, Barnett still remains the showcase for modern tight-reservoir development typical of gas shales in the United States (Parshall, 2008). Development of the Barnett Shale has been a proving ground for combining the technologies of horizontal drilling and large-volume hydraulic fracture treatments. Drilling operations continue expanding the play boundaries outward; at the same time, operations have turned toward infill drilling to increase the amount of gas recovered (Halliburton, 2008). Horizontal well completions in the Barnett are occurring at well spacings ranging from 60 to 160 acres per well. The Barnett Shale covers an area of more than 5000 square miles, with an approximate thickness ranging from 100 feet to more than 600 feet. The original gas-in-place estimate for the Barnett Shale is 327 Tcf, with estimated technically recoverable resources of 44 Tcf.

5.3.2 Fayetteville Shale Play

The Fayetteville Shale Play is located in the Arkoma Basin of northern Arkansas and eastern Oklahoma and ranges in depth from 1000 to 7000 feet (Table 5.3). The Fayetteville Shale is a black, organic-rich, Mississippian-age shale bounded by limestone (Pitkin Limestone) above and sandstone (Batesville Sandstone) below (Figure 5.3).

TABLE 5.3

Fayetteville Shale Gas Play Data (2005–2008)

Parameter	Data	Ref.
Estimated basin area (square miles)	9000	—
Depth (feet)	1000–7000	Halliburton (2008)
Net thickness (feet)	20–200	Hayden and Pursell (2005)
Depth to base of treatable water (feet)[a]	~500	Arkansas Oil & Gas (2008)
Rock column thickness between top of play and bottom of treatable water (feet)	500–6500	—
Total organic carbon (%)	4.0–9.8	Hayden and Pursell (2005)
Total porosity (%)	2–8	Hayden and Pursell (2005)
Gas content (scf/ton)	60–220	Hayden and Pursell (2005)
Water production (barrels/day)	NA	—
Well spacing (acres)	80–160	Hayden and Pursell (2005)
Original gas-in-place (Tcf)	52	Navigant Consulting (2008)
Technically recoverable resources (Tcf)	41.6	Navigant Consulting (2008)

Note: Information presented in this table, such as original gas-in-place and technically recoverable resources, is presented for general comparative purposes only. The numbers provided are based on the sources shown and this research did not include a resource evaluation. Rather, publicly available data was obtained from a variety of sources and is presented for general characterization and comparison with other shale gas plays. Resource estimates for any basin may vary greatly depending on individual company experience, data available at the time the estimate was performed, and other factors. Furthermore, these estimates are likely to change as production methods and technologies improve.

Abbreviations: Mcf = thousands of cubic feet; NA, data not available; scf = standard cubic feet; Tcf = trillion cubic feet.

[a] Depth to base of treatable water data are based on depth data from state oil and gas agencies and state geological survey data.

Development of the Fayetteville Shale Play began in the early 2000s, when gas companies that had experienced success in the Barnett Shale of the Fort Worth Basin identified parallels between it and the Mississippian-aged Fayetteville Shale in terms of age and geologic character (Anon., 2008). Lessons learned from horizontal drilling and hydraulic fracturing techniques employed in the Barnett Shale, when adapted to development of the Fayetteville Shale, make this play economical (Boughal, 2008). Between 2004 and 2007, the number of gas wells drilled annually in the Fayetteville Shale jumped from 13 to more than 600; gas production for the shale increased from just over 100 MMcf/yr

DID YOU KNOW?

The Mississippian subperiod is so named because rocks of this age are exposed in the Mississippi River Valley.

Period		Group/Unit	
Carboniferous	Pennsylvanian	Atoka	
		Bloyd	
		Hale	Prairie Grove
			Cane Hill
	Mississippian	(IMO)	
		Pitkin	
		Fayetteville	
		Batesville	
		Moorefield	
		Boone	

FIGURE 5.3
Stratigraphy of the Fayetteville Shale. (Adapted from HIE, *Fayetteville Shale Power*. Hillwood International Energy, Dallas, TX, 2007.)

to approximately 88.85 Bcf/yr (Boughal, 2008). With over 1000 wells in production to date, the Fayetteville Shale is currently on its way to becoming one of the most active plays in the United States (Williams, 2008).

The area of the Fayetteville Shale play is nearly double that of the Barnett Shale at 9000 square miles, with well spacing ranging from 80 to 160 acres per well and play zone thickness averaging between 20 and 200 feet. The gas content of the Fayetteville Shale has been measured at 60 to 220 scf/ton, which is less than the 300 to 350 scf/ton gas content of the Barnett Shale. The lower gas content of the Fayetteville has resulted in lower estimates of the original gas-in-place and technically recoverable resources: 52 Tcf and 41.6 Tcf, respectively.

DID YOU KNOW?

During the Mississippian subperiod, an important phase of orogeny (mountain building) occurred in the Appalachian Mountains.

5.3.3 Haynesville Shale Play

As noted earlier, the Haynesville Shale gas play has surpassed the Barnett Shale gas play as the nation's leading shale play (USEIA, 2011). Several factors contributed to this favorable positioning of the Haynesville Shale: (1) gas-directed drilling continues to increase dramatically; (2) experience gained from early horizontal programs has helped the Haynesville operators ramp up natural gas production; and (3) regional infrastructure expanded to accommodate the Haynesville's rising natural gas production.

TABLE 5.4

Haynesville Shale Gas Play Data (2005–2008)

Parameter	Data	Refs.
Estimated basin area (square miles)	9000	—
Depth (feet)	10,500–13,5000	Halliburton (2008)
Net thickness (feet)	200–300	Berman (2008); Boughal (2008)
Depth to base of treatable water (feet)[a]	~400	—
Rock column thickness between top of play and bottom of treatable water (feet)	10,100–13,100	—
Total organic carbon (%)	0.5–4.0	Berman (2008)
Total porosity (%)	8–9	Berman (2008)
Gas content (scf/ton)	100–330	Petroleum Listing Services (2005)
Water production (barrels/day)	NA	—
Well spacing (acres)	40–560	Sumi (2008)
Original gas-in-place (Tcf)	717	Navigant Consulting (2008)
Technically recoverable resources (Tcf)	251	Navigant Consulting (2008)

Note: Information presented in this table, such as original gas-in-place and technically recoverable resources, is presented for general comparative purposes only. The numbers provided are based on the sources shown and this research did not include a resource evaluation. Rather, publicly available data was obtained from a variety of sources and is presented for general characterization and comparison with other shale gas plays. Resource estimates for any basin may vary greatly depending on individual company experience, data available at the time the estimate was performed, and other factors. Furthermore, these estimates are likely to change as production methods and technologies improve.

Abbreviations: Mcf = thousands of cubic feet; NA, data not available; scf = standard cubic feet; Tcf = trillion cubic feet.

[a] Depth to base of treatable water data are based on depth data from state oil and gas agencies and state geological survey data.

The Haynesville Shale (also known as the Haynesville/Bossier) is situated in the North Louisiana Salt Basin in northern Louisiana and eastern Texas with depths ranging from 10,500 to 13,500 feet (Table 5.4). The Haynesville is an Upper Jurassic-age shale bounded by sandstone (Cotton Valley Group) above and limestone (Smackover Limestone) below (Figure 5.4).

After several years of drilling and testing, the Haynesville Shale made headlines in 2007 as a potentially significant gas reserve, although the full extent of the play will only be known after several more years of development are completed (Durham, 2008). The Haynesville Shale covers an area of approximately 9000 square miles with an average thickness of 200 to 300

DID YOU KNOW?

The Jurassic Smackover formation does not outcrop and is only encountered in subsurface penetrations located in the U.S. Gulf Coast area.

Period		Group/Unit
Cretaceous	Upper	Navarro
		Taylor
		Austin
		Eagle Ford
		Tuscaloosa
		Washita
		Fredericksburg
		Trinity Group
		Nuevo Leon
Jurassic		Cotton Valley Group
		Haynesville
		Smackover
		Norphlet
	Middle	Louann
	Lower	Werner
Triassic	Upper	Eagle Mills

FIGURE 5.4
Stratigraphy of the Haynesville Shale. (Adapted from Johnson III, J. et al., *Stratigraphic Charts of Louisiana*, Louisiana Geological Survey, Baton Rouge, 2000.)

feet. The thickness and areal extent of the Haynesville Shale have allowed operators to evaluate a wider variety of spacing intervals ranging from 40 to 560 acres per well. Gas content estimates for the play are 100 to 330 scf/ton. The Haynesville formation has the potential to become a significant shale play gas resource of the United States with an original gas-in-place estimate of 717 Tcf and technically recoverable resources estimated at 251 Tcf.

5.3.4 Marcellus Shale Play

The Marcellus Shale is the most expansive shale gas play, spanning six states in the northeastern United States. The estimated depth of production for the Marcellus is between 4000 and 8500 feet (Table 5.5). The Marcellus Shale is a Middle Devonian-age shale bounded by shale (Hamilton Group) above and limestone (Tristates Group) below (Figure 5.5).

DID YOU KNOW?

Several forms of bog iron (impure iron) were deposited near outcrops of the Marcellus. In the 19th century, iron ore from these deposits was used as a mineral paint pigment (Lesley, 1892).

TABLE 5.5

Marcellus Shale Gas Play Data (2005–2008)

Parameter	Data	Ref.
Estimated basin area (square miles)	95,000	—
Depth (feet)	4000–8500	Halliburton (2008)
Net thickness (feet)	50–200	Anon. (2000)
Depth to base of treatable water (feet)[a]	~850	—
Rock column thickness between top of play and bottom of treatable water (feet)	2125–7650	—
Total organic carbon (%)	3–12	Nyahay et al. (2007)
Total porosity (%)	10	Soeder (1986)
Gas content (scf/ton)	60–100	Soeder (1986)
Water production (barrels/day)	NA	—
Well spacing (acres)	40–160	Sumi (2008)
Original gas-in-place (Tcf)	1500	Navigant Consulting (2008)
Technically recoverable resources (Tcf)	262	Navigant Consulting (2008)

Note: Information presented in this table, such as original gas-in-place and technically recoverable resources, is presented for general comparative purposes only. The numbers provided are based on the sources shown and this research did not include a resource evaluation. Rather, publicly available data was obtained from a variety of sources and is presented for general characterization and comparison with other shale gas plays. Resource estimates for any basin may vary greatly depending on individual company experience, data available at the time the estimate was performed, and other factors. Furthermore, these estimates are likely to change as production methods and technologies improve.

Abbreviations: Mcf = thousands of cubic feet; NA, data not available; scf = standard cubic feet; Tcf = trillion cubic feet.

[a] Depth to base of treatable water data are based on depth data from state oil and gas agencies and state geological survey data.

In 1978, natural gas prices were on the increase along with increased development of Devonian shale gas because of passage of the Natural Gas Policy Act (NGPA). Later, however, falling gas prices resulted in uneconomical wells and declining production through the 1990s (West Virginia Geological and Economic Survey, 1997). In 2003, Range Resources Corporation drilled the first economically producing wells into the Marcellus Shale formation in Pennsylvania using horizontal drilling and hydraulic fracturing techniques similar to those used in the Barnett Shale formation of Texas (Harper, 2008). Range Resources began producing this formation in 2005. As of 2008, there were a total of 518 wells permitting in Pennsylvania in the Marcellus Shale, and 277 of the approved wells had been drilled (USGS, 2008).

The Marcellus Shale covers an area of 95,000 square miles at an average thickness of 50 to 200 feet. Whereas the Marcellus is lower in relative gas content at 60 to 100 scf/ton, the much larger area of this play compared to the other shale gas plays results in a higher original gas-in-place estimate of

<table>
<tr><th>Period</th><th colspan="3">Group/Unit</th></tr>
<tr><td>Pennsylvanian</td><td></td><td colspan="2">Pottsville</td></tr>
<tr><td>Mississippian</td><td></td><td colspan="2">Pocono</td></tr>
<tr><td rowspan="16">Devonian</td><td rowspan="6">Upper</td><td colspan="2">Conewango</td></tr>
<tr><td colspan="2">Conneaut</td></tr>
<tr><td colspan="2">Canadaway</td></tr>
<tr><td colspan="2">West Falls</td></tr>
<tr><td colspan="2">Sonyea</td></tr>
<tr><td colspan="2">Genesee</td></tr>
<tr><td rowspan="6">Middle</td><td colspan="2">Tully</td></tr>
<tr><td rowspan="4">Hamilton Group</td><td>Moscow</td></tr>
<tr><td>Ludlowville</td></tr>
<tr><td>Skaneateles</td></tr>
<tr><td>Marcellus</td></tr>
<tr><td colspan="2">Onandaga</td></tr>
<tr><td rowspan="2">Lower</td><td colspan="2">Tristates</td></tr>
<tr><td colspan="2">Helderberg</td></tr>
</table>

FIGURE 5.5
Stratigraphy of the Marcellus Shale. (Adapted from Arthur, J.D. et al., *An Overview of Modern Shale Gas Development in the United States*, ALL Consulting, Tulsa, OK, 2008.)

up to 1500 Tcf. The average well spacing in the Marcellus Shale is 40 to 160 acres per well. The data in Table 5.5 show technically recoverable resources for the formation to be 262 Tcf, although much like the Haynesville Shale the potential estimates for this play are frequently being revised upward due to its early stage of development.

NATURAL GAS POLICY ACT OF 1978

In 1978, the U.S. Congress passed the Natural Gas Policy Act (NGPA), which allowed the well head price of much of the nation's gas to rise to levels suggested by the then-assumed future price of oil and to be decontrolled in 1985 (CBO, 1983). The oil price assumptions that underlay the NGPA's gas price paths, however, proved to be substantially lower than the prices that materialized. This and other features of both the NGPA and natural gas markets have made the smooth transition to decontrol imagined in the act unlikely. These effects, coupled with the expiration of NGPA controls in 1985, led many in Congress to reconsider the nation's long-term pricing policy for natural gas at the well head.

TABLE 5.6

Woodford Shale Gas Play Data (2005–2008)

Parameter	Data	Ref.
Estimated basin area (square miles)	11,000	—
Depth (feet)	6000–11,000	Halliburton (2008)
Net thickness (feet)	120–220	Haines (2006)
Depth to base of treatable water (feet)[a]	~400	—
Rock column thickness between top of play and bottom of treatable water (feet)	5600–10,600	—
Total organic carbon (%)	1–14	Cardott (2004)
Total porosity (%)	3–9	Vulgamore (2007)
Gas content (scf/ton)	200–300	Jochen (2006)
Water production (barrels/day)	NA	—
Well spacing (acres)	640	Sumi (2008)
Original gas-in-place (Tcf)	23	Navigant Consulting (2008)
Technically recoverable resources (Tcf)	11.4	Navigant Consulting (2008)

Note: Information presented in this table, such as original gas-in-place and technically recoverable resources, is presented for general comparative purposes only. The numbers provided are based on the sources shown and this research did not include a resource evaluation. Rather, publicly available data was obtained from a variety of sources and is presented for general characterization and comparison with other shale gas plays. Resource estimates for any basin may vary greatly depending on individual company experience, data available at the time the estimate was performed, and other factors. Furthermore, these estimates are likely to change as production methods and technologies improve.

Abbreviations: Mcf = thousands of cubic feet; NA, data not available; scf = standard cubic feet; Tcf = trillion cubic feet.

[a] Depth to base of treatable water data are based on depth data from state oil and gas agencies and state geological survey data.

5.3.5 Woodford Shale Play

The Woodford Shale play is located in south-central Oklahoma; it ranges in depth from 6000 to 11,000 feet (Table 5.6). This formation is a Devonian-age shale bounded by limestone (Osage Limestone) above and undifferentiated strata below (Figure 5.6). Natural gas production in the Woodford Shale began in 2003 and 2004 with vertical well completions only (Cardott, 2004); however, horizontal drilling has been adopted in the Woodford Shale, as in

DID YOU KNOW?

The Devonian Period of the Paleozoic Era lasted from 417 million years ago to 354 million years ago. It is known as the Age of Fishes and came to an end in a dramatic mass extinction.

<table>
<tr><th>Period</th><th colspan="3">Group/Unit</th></tr>
<tr><td rowspan="7">Permian</td><td>Ochoan</td><td colspan="2">Cloyd Chief Formation</td></tr>
<tr><td rowspan="2">Guadalupian</td><td colspan="2">White Horse Group</td></tr>
<tr><td colspan="2">El Reno Group</td></tr>
<tr><td>Leonardian</td><td colspan="2">Enid Group</td></tr>
<tr><td rowspan="3">Wolfcampian</td><td colspan="2">Chase Group</td></tr>
<tr><td colspan="2">Council Grove Group</td></tr>
<tr><td colspan="2">Admire Group</td></tr>
<tr><td rowspan="2">Pennsylvanian</td><td>Atokan</td><td colspan="2">Atoka Group</td></tr>
<tr><td>Morrowan</td><td colspan="2">Morrow Group</td></tr>
<tr><td rowspan="4">Mississippian</td><td>Chesterian</td><td colspan="2">Chester Group</td></tr>
<tr><td>Meramecian</td><td rowspan="2">Mississippi Limestone</td><td>Meramec Limestone</td></tr>
<tr><td>Osagean</td><td>Osage Limestone</td></tr>
<tr><td>Kinderhookian</td><td colspan="2" rowspan="2">Woodford Shale</td></tr>
<tr><td rowspan="4">Devonian</td><td>Upper</td></tr>
<tr><td>Middle</td><td colspan="2">Undifferentiated</td></tr>
<tr><td rowspan="2">Lower</td><td rowspan="2">Hunton Group</td><td>Haragan Formation</td></tr>
<tr><td>Henryhouse Formation</td></tr>
</table>

FIGURE 5.6
Stratigraphy of the Woodford Shale. (Adapted from Cardott, B., *Overview of Woodford Gas-Shale Play in Oklahoma*, Oklahoma Geological Survey, Norman, OK, 2007; AAPG, *Correlation of Stratigraphic Units of North America (CASUNA) Project*, American Association of Petroleum Geologists, Tulsa, OK, 1987.)

other shale gas plays, due to its success in the Barnett Shale (Boughal, 2008). The Woodford Shale play encompasses an area of nearly 11,000 square miles, is in an early stage of development, and is occurring at a spacing interval of 640 acres per well. The average thickness of the Woodford Shale varies from 120 to 220 feet across the play. Gas content in the Woodford Shale is higher on average than some of the other shale gas plays at 200 to 300 scf/ton. The original gas-in-place estimate for the Woodford Shale is similar to the Fayetteville Shale at 23 Tcf, while the technically recoverable resources are 11.4 Tcf.

5.3.6 Antrim Shale Play

The Antrim Shale play is a brown to black, pyritic, and organic-rich shale located in the upper portion of the lower peninsula of Michigan within the Michigan Basin. This Late Devonian-age shale is bounded by shale (Bedford Shale) above and by limestone (Squaw Bay Limestone) below and occurs at

TABLE 5.7

Antrim Shale Gas Play Data (2005–2008)

Parameter	Data	Ref.
Estimated basin area (square miles)	12,000	—
Depth (feet)	600–2200	Hayden and Pursell (2005)
Net thickness (feet)	70–120	Hayden and Pursell (2005)
Depth to base of treatable water (feet)[a]	~300	—
Rock column thickness between top of play and bottom of treatable water (feet)	300–1900	—
Total organic carbon (%)	1–20	Hayden and Pursell (2005)
Total porosity (%)	9	Hayden and Pursell (2005)
Gas content (scf/ton)	40–100	Hayden and Pursell (2005)
Water production (barrels/day)	5–500	Hayden and Pursell (2005)
Well spacing (acres)	40–160	Hayden and Pursell (2005)
Original gas-in-place (Tcf)	76	Navigant Consulting (2008)
Technically recoverable resources (Tcf)	20	Navigant Consulting (2008)

Note: Information presented in this table, such as original gas-in-place and technically recoverable resources, is presented for general comparative purposes only. The numbers provided are based on the sources shown and this research did not include a resource evaluation. Rather, publicly available data was obtained from a variety of sources and is presented for general characterization and comparison with other shale gas plays. Resource estimates for any basin may vary greatly depending on individual company experience, data available at the time the estimate was performed, and other factors. Furthermore, these estimates are likely to change as production methods and technologies improve.

Abbreviations: Mcf = thousands of cubic feet; NA, data not available; scf = standard cubic feet; Tcf = trillion cubic feet.

[a] Depth to base of treatable water data are based on depth data from state oil and gas agencies and state geological survey data.

depths of 600 to 2200 feet, which is more typical of coal bed natural gas (CBNG) formations than most gas shales (Table 5.7 and Figure 5.7). Along with the Barnett Shale, the Antrim Shale has been one of the most actively developed shale gas plays, with major expansion taking place in the late 1980s (Harrison, 2006). The Antrim Shale encompasses an area of approximately 12,000 square miles and is characterized by distinct differences from other gas shales: shallow depth; small stratigraphic thickness, with average net pay of 70 to 120 feet; and greater volumes of produced water, in the range of 5 to 500 bbl/day/well (Hayden and Pursell, 2005). The gas content of the

DID YOU KNOW?

The Mississippian Age was rather uniformly warm. On land, major habitat division began between seed plants and the club mosses.

<table>
<tr><th colspan="2">Period</th><th colspan="2">Formation</th></tr>
<tr><td>Quaternary</td><td>Pleistocene</td><td colspan="2">Glacial Drift</td></tr>
<tr><td>Jurassic</td><td>Middle</td><td colspan="2">Ionia Formation</td></tr>
<tr><td rowspan="3">Pennsylvanian</td><td>Late</td><td colspan="2">Grand River Formation</td></tr>
<tr><td rowspan="2">Early</td><td colspan="2">Saginaw Formation</td></tr>
<tr><td colspan="2" rowspan="2">Parma Formation</td></tr>
<tr><td rowspan="6">Mississippian</td><td rowspan="3">Late</td></tr>
<tr><td colspan="2">Bayport Limestone</td></tr>
<tr><td colspan="2">Michigan Formation</td></tr>
<tr><td rowspan="3">Early</td><td colspan="2">Marshall Sandstone</td></tr>
<tr><td colspan="2">Coldwater Shale</td></tr>
<tr><td colspan="2">Sunbury Shale</td></tr>
<tr><td rowspan="7">Devonian</td><td rowspan="7">Late</td><td rowspan="2">Ellsworth Shale</td><td>Berea Sandstone</td></tr>
<tr><td>Bedford Shale</td></tr>
<tr><td rowspan="4">Antrim Shale</td><td>Upper Member</td></tr>
<tr><td>Lachine Member</td></tr>
<tr><td>Paxton Member</td></tr>
<tr><td>Norwood Member</td></tr>
<tr><td colspan="2">Squaw Bay Limestone</td></tr>
</table>

FIGURE 5.7
Stratigraphy of the Antrim Shale. (Adapted from Catacosinos, P.W. et al., *Stratigraphic Nomenclature for Michigan*, Michigan Department of Environmental Quality, Geological Survey Division, and Michigan Geological Survey, Kalamazoo, MI, 2000.)

Antrim Shale ranges between 40 and 100 scf/ton. The original gas-in-place for the Antrim is estimated at 76 Tcf, with technically recoverable resources estimated at 20 Tcf. Well spacing ranges from 40 to 160 acres per well.

5.3.7 New Albany Shale Play

The New Albany Shale is an organic-rich geologic formations located in the Illinois Basin in portions of southern Illinois, southwestern Indiana, and northwestern Kentucky (IGS, 1986). Similar to the Antrim Shale, the New Albany Shale occurs at depths between 500 and 2000 feet (Table 5.8) and is a shallower, water-filled shale with a more CBNG-like character than the other gas shales discussed in this text. The New Albany formation is a Devonian- to Mississippian-age Shale bounded by limestone above (Rockford Limestone) and below (North Vernon Limestone) (Figure 5.8). The New Albany Shale is one of the largest shale gas plays, encompassing an area of approximately 43,500 square miles with approximately 80-acre spacing between wells. Similar to the Antrim Shale, the New Albany play has a

TABLE 5.8

New Albany Shale Gas Play Data (2005–2008)

Parameter	Data	Ref.
Estimated basin area (square miles)	43,500	—
Depth (feet)	500–2000	Hayden and Pursell (2005)
Net thickness (feet)	50–100	Hayden and Pursell (2005)
Depth to base of treatable water (feet)[a]	~400	—
Rock column thickness between top of play and bottom of treatable water (feet)	100–1600	—
Total organic carbon (%)	1–25	Hayden and Pursell (2005)
Total porosity (%)	10–14	Hayden and Pursell (2005)
Gas content (scf/ton)	40–180	Hayden and Pursell (2005)
Water production (barrels/day)	5–500	Hayden and Pursell (2005)
Well spacing (acres)	80	Hayden and Pursell (2005)
Original gas-in-place (Tcf)	160	Navigant Consulting (2008)
Technically recoverable resources (Tcf)	19.2	Navigant Consulting (2008)

Note: Information presented in this table, such as original gas-in-place and technically recoverable resources, is presented for general comparative purposes only. The numbers provided are based on the sources shown and this research did not include a resource evaluation. Rather, publicly available data was obtained from a variety of sources and is presented for general characterization and comparison with other shale gas plays. Resource estimates for any basin may vary greatly depending on individual company experience, data available at the time the estimate was performed, and other factors. Furthermore, these estimates are likely to change as production methods and technologies improve.

Abbreviations: Mcf = thousands of cubic feet; NA, data not available; scf = standard cubic feet; Tcf = trillion cubic feet.

[a] Depth to base of treatable water data are based on depth data from state oil and gas agencies and state geological survey data.

thinner average net pay thickness of 50 to 100 feet and has wells that average 5 to 500 bbls of water per day (AAPG, 1987). The measured gas content of the New Albany Shale ranges from 40 to 80 scf/ton. The original gas-in-place for the New Albany formation is estimated at 160 Tcf with technically recoverable resources estimated at less than 20 Tcf.

5.4 Thought-Provoking Discussion Questions

1. Is horizontal shale gas drilling superior to vertical drilling?
2. Why is it important to implement reclamation practices soon after fracking operations are completed?

Period		Formation	
Pennsylvanian	Missourian	Mattoon	
		Bond	
		Patoka	
	Desmoinesian	Shelburn	
		Dugger	
		Petersburg	
		Linton	
		Staunton	
	Atokan	Brazil	
	Morrowan	Mansfield	
Mississippian	Chesterian	Tobinsport	
		Branchville	
		Tar Springs	
		Glen Dean Limestone	
		Hardinsburg	
		Haney Limestone	
		Big Clifty	
		Beech Creek Limestone	
		Cypress	Elwren
		Reelsville Limestone	
		Sample	
		Beaver Bend Limestone	
		Bethel	
		Paoli Limestone	
		Ste. Genevieve Limestone	
	Valmeyeran	St. Louis Limestone	
		Harrodsburg Limestone	
		Muldraugh	Ramp Creek
		Edwardsville	
		Spickert Knob	
		Rockford Limestone	Coldwater Shale
	Kinderhookian	New Albany Shale	Sunbury Shale
			Ellsworth Shale
Devonian	Senecan Chautauquan		
			Antrim Shale
	Erian	North Vernon Limestone	Transverse
		Jeffersonville Limestone	Detroit River

FIGURE 5.8
Stratigraphy of the New Albany Shale. (Adapted from *General Stratigraphic Column for Paleozoic Rocks in Indiana*, Indiana Geological Survey, Bloomington, IN, 1986.)

References and Recommended Reading

AAPG. (1987). *Correlation of Stratigraphic Units of North America (CASUNA) Project*, American Association of Petroleum Geologists, Tulsa, OK.

Anon. (2000). Alabama lawsuit poses threat to hydraulic fracturing across U.S. *Drilling Contractor*, January/February, pp. 42–43.

Anon. (2008). *Projecting the Economic Impact of the Fayetteville Shale Play for 2008–2012*. Center for Business and Economic Research, Sam M. Walton College of Business, Fayetteville, AR.

Arkansas Oil and Gas Commission. (2012). *General Rules and Regulations*. Arkansas Oil and Gas Commission, Little Rock.

Arthur, J.D., Bohm, B., and Lane, M. (2008). *An Overview of Modern Shale Gas Development in the United States*, ALL Consulting, Tulsa, OK.

Berman, A. (2008). The Haynesville Shale sizzles while the Barnett cools. *World Oil Magazine*, 229(9):23.

Boughal, K. (2008). Unconventional plays grow in number after Barnett Shale blazed the way. *World Oil Magazine*, 229(8):77–80.

Cardott, B. (2004). *Overview of Unconventional Energy Resources of Oklahoma*. Oklahoma Geological Survey, Norman, OK (www.ogs.ou.edu/fossilfuels/coalpdfs/UnconventionalPresentation.pdf)

Catacosinos, P., Harrison, W., Reynolds, R., Westjohn, D., and Wollensak. M. (2000). *Stratigraphic Nomenclature for Michigan*, Michigan Department of Environmental Quality, Geological Survey Division, and Michigan Basin Geological Survey, Kalamazoo, MI.

CBO. (1983). *Understanding Natural Gas Price Decontrol*. Congressional Budget Office, Congress of the United States, Washington, DC.

Durham, L.S. (2008). A spike in interest and activity: Louisiana play a "company maker"? *AAPG Explorer*, 29(7):18.

Franz, Jr., J.H. and Jochen, V. (2005b). *When Your Gas Reservoir Is Unconventional, So Is Our Solution*, Shale Gas White Paper 05-OF-299. Schlumberger, Houston, TX.

Halliburton. (2008). *U.S. Shale Gas: An Unconventional Resource. Unconventional Challenges*. Halliburton, Houston, TX, 8 pp.

Harper, J. (2008). The Marcellus Shale—an old "new" gas reservoir in Pennsylvania. *Pennsylvania Geology*, 28(1):2–13.

Harrison, W. (2006). *Production History and Reservoir Characteristics of the Antrim Shale Gas Play, Michigan Basin*. Western Michigan University, Kalamazoo.

Hayden, J. and Pursell, D. (2005). *The Barnett Shale: Visitor's Guide to the Hottest Gas Play in the US*. Pickering Energy Partners, Inc., Houston, TX.

HIE. (2007). *Fayetteville Shale Power*. Hillwood International Energy, Dallas, TX.

IGS. (1986). *General Stratigraphic Column for Paleozoic Rocks in Indiana*. Indiana Geological Survey, Indiana University, Bloomington, IN (www.usi.edu/science/geology/My%20Web%20Sites/stratcolumn.pdf).

ISGS. (2011). *Mississippian Rocks in Illinois*, GeoNote 1. Illinois State Geological Survey, Champaign, IL (www.isgs.uiuc.edu/maps-data-pub/publications/geonotes/geonote1.shtml).

Jochen, V. (2006). *New Technology Needs to Produce Unconventional Gas*. Schlumberger, Houston, TX.

Johnston III, J., Heinrich, B., Lovelace, J., McCulluh, R., and Zimmerman, R. (2000). *Stratigraphic Charts of Louisiana*, Louisiana Geological Survey, Baton Rouge, LA.

Lesley, J.P. (1892). *A Summary Description of the Geology of Pennsylvania, Vol. II*. Geological Survey of Pennsylvania, Pittsburgh.

NaturalGas.org. (2011). *Unconventional Natural Gas Resources*, www.naturalgas.org/overview/unconvent_ng_resource.asp.

Navigant Consulting. (2008). *North American Natural Gas Supply Assessment*, prepared for American Clean Skies Foundation, Washington, DC.

Nyahay, R., Leone, J., Smith, L., Martin, J., and Jarvie, D. (2007). Update on Regional Assessment of Gas Potential in the Devonian Marcellus and Ordovician Utica Shales of New York, paper presented at the American Association of Petroleum Geologists (AAPG) Eastern Section Meeting, Lexington, KY, September 16–18, 2007 (www.searchanddiscovery.com/documents/2007/07101nyahay/).

Parshall, J. (2008). Barnett Shale showcases tight-gas development. *Journal of Petroleum Technology*, 60(9):48–55.

PLS. (2008). *Other Players Reporting Haynesville Success*. Petroleum Listing Services, Houston, TX, August 15.

Railroad Commission of Texas. (2008). *Newark, East (Barnett Shale) Field*. Oil and Gas Division, Railroad Commission of Texas, Austin.

Soeder, D.J. (1986). Porosity and permeability of Eastern Devonian gas shale. *SPE Formation Evaluation*, 1(1):116–124.

Sumi, L. (2008). *Shale Gas Focus on the Marcellus Shale*. Oil and Gas Accountability Project (OGAP)/Earthworks, Washington, DC, 25 pp.

USEIA. (2011). *Today in Energy: Haynesville Surpasses Barnett as the Nation's Leading Shale Play*. U.S. Energy Information Administration, Washington, DC (205.254.135.7/todayinenergy/detail.cfm?id=570).

USGS. (2008). *National Oil and Gas Assessment*. U.S. Geological Survey, Washington, DC (http://energy.usgs.gov/OilGas/AssessmentsData/NationalOilGasAssessment.aspx).

Vulgamore, T., Clawson, T., Pope, C., Wolhart, W., Mayerhofer, M., and Waltman. C. (2007). *Applying Hydraulic Fracture Diagnostics to Optimize Stimulations in the Woodford Shale*, SPE-110029. Society of Petroleum Engineers, Allen, TX.

West Virginia Geological and Economic Survey. (1997). *Enhancement of the Appalachian Basin Devonian Shale Resource Base in the GRI Hydrocarbon Model*, GRI Contract No. 5095-890-3478, prepared for Gas Research Institute, Arlington, VA.

Williams, P. (2008). A vast ocean of natural gas. *American Clean Skies*, Summer, pp. 44–50.

6

Hydraulic Fracturing: The Process

> The Gas Era is coming, and the landscape north and west of [New York] will inevitably be transformed as a result. When the valves start opening next year, a lot of poor farm folk may become Texas rich.
>
> **—David France (*New York Magazine*, September 21, 2008)**

6.1 Poor Farm Folk to Texas Rich

While exploring the Appalachian Basin in a section of the Marcellus Shale region of northwestern Pennsylvania (as part of the research conducted for this book), I visited a hilly, outcropped farm area located outside a small village. I was lucky enough to run across an elderly gentleman named Karl who was actively using one of those brand-new, Farm Boss-type chainsaws to cut up a stack of deadfall timber on his 160-acre farm. I judged Karl to be somewhere in his 80s (he was actually 88); his was an experience-lined face covered with parchment-like skin with deep-set, penetrating blue–green eyes crowned by snow-white eyebrows, all accentuated with a wide, gap-toothed, tobacco-stained smile that radiated a warm hello when he looked up and saw me standing a few feet from him—me, just a harmless-looking guy with notebook and pencil in hand.

He stopped the buzz of the chainsaw, maintained the wide smile, and asked me if he could help me with something. I introduced myself and asked if he owned the property, including the finished directional gas drilling site a couple of hundred yards behind me. He looked off over my shoulder at the site and then at me and said, "Yep, the land is all mine … she is all mine … my personal shale-gas gold mine," he chuckled. "Them frack people, they own the drilling rig and send me a check each month. Haven't cashed most of 'em yet … well, not all of them. Sort of growing a nest egg, you could say."

While he talked and looked me over, I was silent and looked him over, and the only thing that looked new and somewhat out of place about this gentleman was his chainsaw and what appeared to be a new pair of expensive high-top boots. While we stood on the northeast corner of his property he explained, with some humor, that he had raised three wives and eleven children on his hard-scrabble acreage. He concluded his brief personal history about his lifelong struggle as poor farm folk to his transition to new Texas Rich with the following: "Yep, struggled for years to make a living here. Grew crops and raised a few livestock here and there and ate up a bunch of venison and rabbit. The work here wore on 'em … the wives, that is. Wore 'em out and they eventually passed onto the land of the angels. Tough life it was until them gas shalers showed up. They showed up one day a couple years ago and offered me a small fortune to drill on my land for gas. I could not believe my good luck. Funny thing is when I got all that money—and don't you know I still get a monthly check from them—all I ever wanted to buy was three nice headstones for them wives and a modern chainsaw and a new pair of comfortable boots." He spat out a long stream of tobacco juice, smiled at me, and said, "That old saying about good things happening to those who wait sure turned out to be true in my case."

I mentioned to Karl that I had noticed several dozen drilling operations in all stages of operation in the general area of his land and his neighbors, and that there must be a bunch of happy, wealthy farmers around here. His smile shifted to almost a grimace as he said: "Well, young fellow (I was every day and a minute past 67 years old but totally appreciative of his compliment), not all. Some are bitter and refuse to allow them drillers to drill for all that money … real sad and hard to figure. You might say those neighbors of mine just frack me up. Ha, ha, ha. Sometimes they torque my jaws, too, though. You see, young feller, there is this law called the Capture Law that allows the drillers to drill from my land into the neighbors' land 100 feet in all directions." Letting go with another long stream of yellow-brown tobacco, Karl went on to say, "Just hard to figure, since the drillers are going to take the gas anyways, why not get paid for it?" While rubbing sawdust from the front of his shirt, Karl continued: "You know, these neighbors of mine are a bit strange. They kick a gift horse in the mouth by turning down the gas shalers."

"Why do they turn down the drilling?" I asked.

"Simple—the key word is *drilling, drilling, drilling*. They just don't like all the noise, traffic, road building, chemical use, tree removal, and so forth." Shaking his head in wonder, he said, "Makes no sense to me and makes no never mind to me otherwise when they turn down the big bucks … makes more for me. Just seems strange is all."

"Have the drilling and gas shale processing hereabouts caused a lot of environmental damage?" I asked.

Rolling his chew from one side of mouth to the other, Karl answered, "Well, things are a bit different now. You know there is some stream contamination below us here and our well water tastes a bit funny, but, heck, I don't. Tastes like money to me." He laughed. Karl's final statement to me that day: "To each his own … heck, whenever we damage our environment ol' Mother Nature steps in and cleans up the mess anyway, so what's the worry? For years we have all been digging them water wells by hand or drilling 'em to get our water. What is the difference between drilling for that shale gas and drilling for water? I sure don't get it. One thing is certain, though. I'd rather be Texas Rich than dirt poor. Yes, sir, no more hard livin' for me."

As I walked back down the country road I could hear that chainsaw buzzing away again, and all I could think about were Karl's words of wisdom concerning our messing things up and Mother Nature coming to the rescue. I sure hope the old man's philosophy is correct because in my lifetime I have stumbled into one environmental mess after another; all of them could use that *motherly* touch right about now, thank you very much!

6.2 Drilling for Water Versus Drilling for Shale Gas

Up until recently, it could be said with some assurance that digging, driving, boring, or drilling for water has always been more important to humankind than drilling for natural gas. It can be said with even more assurance that water has always been an important and life-sustaining drink to humans and is essential to the survival of all organisms. It can also be said that drilling for water and drilling for shale gas are two entirely different ventures; however, if we were to construct a Venn diagram depicting drilling for shale natural gas, we would also have to include a water set somewhere within the diagram. Why? The simple answer is because it would be difficult to drill 6000 feet down through soil and rock strata without running into water in the soil itself or into a water-saturated aquifer. Water would have to be included in the diagram because hydraulic fracturing requires and uses water—millions and millions of gallons of water. In the following section, a brief description of a water well operation is provided to show the similarities and differences between drilling for water and drilling for shale natural gas.

DID YOU KNOW?

"We used to think that energy and water would be the critical issues. Now we think water will be the critical issue," Mostafa Tolba, former head of the United Nations Environmental Programme, has declared. The author's view suggests that water will always be a critical issue, but when we finish the current global warming cycle we are experiencing and eventually cycle back into another global cool down (ice age), energy will certainly move back into the forefront. The question is will we have enough energy to keep everyone on Earth warm? Or will chaos reign? Will freezing conditions cause people to raid their neighbors' homes to steal their furniture to burn to keep warm?

6.3 Water Well Systems*

As mentioned, a water well is an excavation created in the ground by digging, driving, boring, or drilling into underground aquifers to access groundwater for consumption or other uses. The most common method for withdrawing groundwater is to penetrate the aquifer with a vertical well and then pump the water up to the surface. Today, in most locations in the United States, developing a well water supply usually involves a more complicated step-by-step process. Local, state, and federal requirements specify the actual requirements for development of a well supply in this country. The standard sequence for developing a well supply generally involves a seven-step process:

1. *Application*—Depending on the location, filling out and submitting an application (to the applicable authorities) to develop a well supply is standard procedure.
2. *Well site approval*—Once the application has been made, local authorities check various local geological and other records to ensure that the siting of the proposed well coincides with mandated guidelines for approval.
3. *Well drilling*—The well is then drilled.
4. *Preliminary engineering report*—After the well is drilled and the results documented, a preliminary engineering report is made on the suitability of the site to serve as a water source. This procedure involves performing a pump test to determine if the well can supply the required amount of water. The well is generally pumped for at least six hours at a rate equal to or greater than the desired yield. A stabilized drawdown should be obtained at that rate, and the original static level should be recovered within 24 hours after pumping stops. During this test period, samples are taken and tested for bacteriological and chemical quality.
5. *Submission of documents for review and approval*—The application and test results are submitted to an authorized reviewing authority that determines if the well site meets approval criteria.
6. *Construction permit*—If the site is approved, a construction permit is issued.
7. *Operation permit*—When the well is ready for use, an operation permit is issued.

* This section is adapted from Spellman, F.R., *Handbook of Water and Wastewater Treatment Plant Operations*, 2nd ed., CRC Press, Boca Raton, FL, 2008.

6.3.1 Water Well Site Requirements

To protect the groundwater source and provide high-quality safe water, the waterworks industry has developed standards and specifications for wells. The following listing includes industry standards and practices, as well as those items included in example State Division of Environmental Compliance regulations.

Note: Check with your local regulatory authorities to determine specific well site requirements.

1. Minimum well lot requirements
   a. 50 feet from well to all property lines
   b. All-weather access road provided
   c. Lot graded to divert surface runoff
   d. Recorded well plan and dedication document
2. Minimum well location requirements
   a. At least 50 feet horizontal distance from any actual or potential sources of contamination involving sewage
   b. At least 50 feet horizontal distance from any petroleum or chemical storage tank or pipeline or similar source of contamination, except where plastic-type well casing is used the separation distance must be at least 100 feet
3. Vulnerability assessment
   a. Well head area = 1000-ft radius from the well
   b. What is the general land use of the area (residential, industrial, livestock, crops, undeveloped, other)?
   c. What are the geologic conditions (sinkholes, surface, subsurface)?

6.3.2 Types of Water Wells

Water supply wells may be characterized as shallow or deep. In addition, wells are classified as follows:

1. Class I—Cased and grouted to 100 ft
2. Class II A—Cased to a minimum of 100 ft and grouted to 20 ft
3. Class II B—Cased and grouted to 50 ft

Note: During the well development process, mud and silt forced into the aquifer during the drilling process is removed, allowing the well to produce the best-quality water at the highest rate from the aquifer.

6.3.2.1 Shallow Wells

Shallow wells are those that are less than 100 ft deep. Such wells are not particularly desirable for municipal supplies because the aquifers they tap are likely to fluctuate considerably in depth, making the yield somewhat uncertain. Municipal wells in such aquifers cause a reduction in the water table (or phreatic surface) that affects nearby private wells, which are more likely to utilize shallow strata. Such interference with private wells may result in damage suits against the community. Shallow wells may be dug, bored, or driven:

- *Dug wells*—Dug wells are the oldest type of well and date back many centuries; they are dug by hand or by a variety of unspecialized equipment. They range in size from approximately 4 to 15 ft in diameter and are usually about 20 to 40 ft deep. Such wells are usually lined or cased with concrete or brick. Dug wells are prone to failure from drought or heavy pumpage. They are vulnerable to contamination and are not acceptable as a public water supply in many locations.
- *Driven wells*—Driven wells consist of a pipe casing terminating in a point slightly greater in diameter than the casing. The pointed well screen and the lengths of pipe attached to it are pounded down or driven in the same manner as a pile, usually with a drop hammer, to the water-bearing strata. Driven wells are usually 2 to 3 inches in diameter and are used only in unconsolidated materials. This type of shallow well is not acceptable as a public water supply.
- *Bored wells*—Bored wells range from 1 to 36 inches in diameter and are constructed in unconsolidated materials. The boring is accomplished with augers (either hand or machine driven) that fill with soil and then are drawn to the surface to be emptied. The casing may be placed after the well is completed (in relatively cohesive materials), but must advance with the well in noncohesive strata. Bored wells are not acceptable as a public water supply.

6.3.2.2 Deep Wells

Deep wells are the usual source of groundwater for municipalities. Deep wells tap thick and extensive aquifers that are not subject to rapid fluctuations in water level and that provide a large and uniform yield. Deep wells typically yield water of more constant quality than shallow wells, although the quality is not necessarily better. Deep wells are constructed by a variety of techniques; we discuss two of these techniques below:

- *Jetted wells*—Jetted well construction commonly employs a jetting pipe with a cutting tool. This type of well cannot be constructed in clay or hardpan or where boulders are present. Jetted wells are not acceptable as a public water supply.

- *Drilled wells*—Drilled wells are usually the only type of well allowed for use in most public water supply systems. Several different methods of drilling are available, all of which are capable of drilling wells of extreme depth and diameter. Drilled wells are constructed using a drilling rig that creates a hole into which the casing is placed. Screens are installed at one or more levels when water-bearing formations are encountered.

6.4 Components of a Water Well

The components that make up a well system include the well itself, the building and the pump, and related piping system. In this section, we focus on the components that make up the well itself. Many of these components are shown in Figure 6.1.

6.4.1 Well Casing

A well is a hole in the ground called the *borehole*. A well casing placed inside the borehole prevents the walls of the hole from collapsing and prevents contaminants (either surface or subsurface) from entering the water source. The casing also provides a column of stored water and housing for the pump mechanisms and pipes. Well casings constructed of steel or plastic material are acceptable. The well casing must extend a minimum of 12 inches above grade.

6.4.2 Grout

To protect the aquifer from contamination, the casing is sealed to the borehole near the surface and near the bottom where it passes into the impermeable layer with grout. This sealing process keeps the well from being polluted by surface water and seals out water from water-bearing strata that have undesirable water quality. Sealing also protects the casing from external corrosion and restrains unstable soil and rock formations. Grout consists of near cement that is pumped into the annular space (it is completed within 48 hours of well construction); it is pumped under continuous pressure starting at the bottom and progressing upward in one continuous operation.

6.4.3 Well Pad

The well pad provides a ground seal around the casing. The pad is constructed of reinforced concrete 6 × 6 ft (6 inches thick) with the well head located in the middle. The well pad prevents contaminants from collecting around the well and seeping down into the ground along the casing.

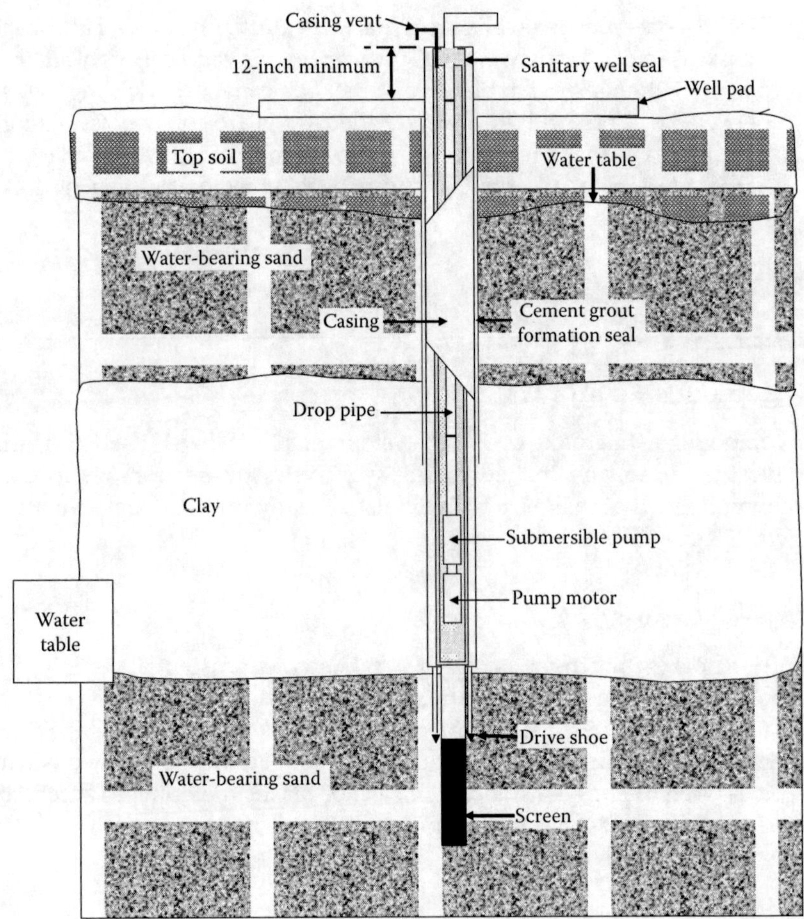

FIGURE 6.1
Components of a water well.

6.4.4 Sanitary Seal

To prevent contamination of the well, a sanitary seal is placed at the top of the casing. The type of seal varies depending on the type of pump used. The sanitary seal contains openings for power and control wires, pump support cables, a drawdown gauge, discharge piping, pump shaft, and air vent, while providing a tight seal around them.

6.4.5 Well Screen

Screens can be installed at the intake points on the end of a well casing or on the end of the inner casing on gravel-packed wells. These screens perform two functions: (1) supporting the borehole, and (2) reducing the amount of sand that enters the casing and the pump. They are sized to allow passage of the maximum amount of water while preventing passage of sand, sediment, or gravel.

6.4.6 Casing Vent

The well casing must have a vent to allow air into the casing as the water level drops. The vent terminates 18 inches above the floor with a return bend pointing downward. The opening of the vent must be screened with #24 mesh stainless steel to prevent entry of vermin and dust.

6.4.7 Drop Pipe

The drop pipe or riser is the line leading from the pump to the well head. It ensures adequate support so that an aboveground pump does not move and so that a submersible pump is not lost down the well. This pipe is either steel or PVC. Steel is the most desirable.

6.4.8 Miscellaneous Well Components

Miscellaneous well components include the following:

- *Gauges and air lines* are used to measure the water level of the well.
- *Check valves* are located immediately after the well to prevent system water from returning to the well. The check valve must be located above ground and be protected from freezing.
- *Flowmeters* are required to monitor the total amount of water withdrawn from the well, including any water blown off.
- *Control switches* control well pump operation.
- *Blow-off valves* are located between the well and storage tank and are used to flush the well of sediment or turbid or superchlorinated water.

DID YOU KNOW?

Below the hydrocarbons may be a groundwater aquifer. Water, as with all liquids, is compressible to a very small degree. When the hydrocarbons are depleted, the reduction in pressure in the reservoir causes the water to expand slightly. Although this expansion is quite small, if the aquifer is large enough this will translate into a large increase in volume, which will push up on the hydrocarbons, maintaining pressure. This effect is known as *water drive*.

- *Sample taps* include a raw water sample tap, which is located before any storage or treatment to permit sampling of the water directly from the well, and an entry point sample tap, which is located after treatment.
- *Control valves* isolate the well for testing or maintenance or to control water flow.

6.5 Shale Gas Drilling Development Technology*

When initially developed, conventional petroleum reservoirs depend on the pressure of their gas cap and oil-dissolved gas to lift the oil to the surface (i.e., *gas drive*). Water trapping the petroleum from below also exerts an upward hydraulic pressure (i.e., *water drive*). The combined pressure in petroleum reservoirs produced by the natural gas and water drives is known as the *conventional drive*. As a reservoir's production declines, lifting further petroleum to the surface, like the lifting of water, requires pumping, or *artificial lift*. In the late 1940s, drilling companies began inducing hydraulic pressure in wells to fracture the producing formation. This stimulated further production by effectively increasing the contact of a well with a formation. Advances in directional drilling technology have allowed wells to deviate from nearly vertical to extend horizontally into the reservoir formation, which further increases contact of a well with the reservoir. Directional drilling technology also enables drilling a number of wells from a single well pad, thus cutting costs while reducing environmental disturbance. Combining hydraulic fracturing with directional drilling has opened up the production of tight (less permeable) petroleum and natural gas reservoirs, particularly unconventional gas shales such as the Marcellus Shale formation.

6.6 Drilling

From the art of divining, dowsing, and witching for water (still practiced all over the world) to the use of scientific exploration and discovery methods and innovative drilling technology, locating water and petroleum products and drilling for them have come a long way, evolving from an art to a

* This section is adapted from Andrews, A., Ed., *Unconventional Gas Shales: Development, Technology, and Policy Issues*, CRS Report for Congress, Congressional Research Service, Washington, DC, 2009.

science. Originally, drillers used cable tool rigs and percussion bits. The drill operator would raise the bit and release it to pulverize the sediment. From time to time, the driller would stop to "muck out" the pulverized rock cuttings to advance the well. Though time-consuming, this method was simple and required minimal labor. Some drillers still use this method for water wells and even some shallow gas wells. The introduction of rotary drill rigs at the beginning of the 20th century marked a big advance in drilling, particularly with the development of the tricone rotary bit. (Howard Hughes, Jr., of the Hughes Tool Company, developed the modern tricone rotary bit. His father, Howard Robert Hughes, Sr., had invented the bit's ancestor, a two-cone rotary bit.) This method, as the name implies, uses a weighted rotating bit to penetrate the sediment (see Figure 6.2). Following is an explanation of the components shown in the figure:

1. *Derrick* is the support structure for the equipment used to lower and raise the drill string into and out of the wellbore.
2. *Crown block* is the stationary end of the *block and tackle.*
3. *Drill line* is thick, stranded metal cable threaded through the two blocks (traveling and crown) to raise and lower the drill string.
4. *Traveling block* is the moving end of the block and tackle. Together, they give a significant mechanical advantage for lifting.
5. *Goose-neck* is a thick metal elbow connected to the swivel and standpipe that supports the weight of and provides a downward angle for the kelly hose to hang from.
6. *Swivel* is the top end of the kelly that allows the rotation of the drill string without twisting the block.
7. *Kelly hose* is a flexible, high-pressure hose that connects the standpipe to the kelly (or, more specifically, to the gooseneck on the swivel above the kelly) and allows free vertical movement of the kelly while facilitating the flow of drilling fluid through the system and down the drill string.
8. *Standpipe* is a thick metal pipe, situated vertically along the derrick, that facilitates the flow of drilling fluid and has attached to it and supports one end of the kelly hose.
9. *Power source motor* powers the drill line.
10. *Mud pump* is a reciprocal type of pump used to circulate drilling fluid through the system.
11. *Suction line (mud pump)* is an intake line for the mud pump to draw drilling fluid from the mud tanks.
12. *Shale shaker* separates drill cuttings from the drilling fluid before it is pumped back down the wellbore.

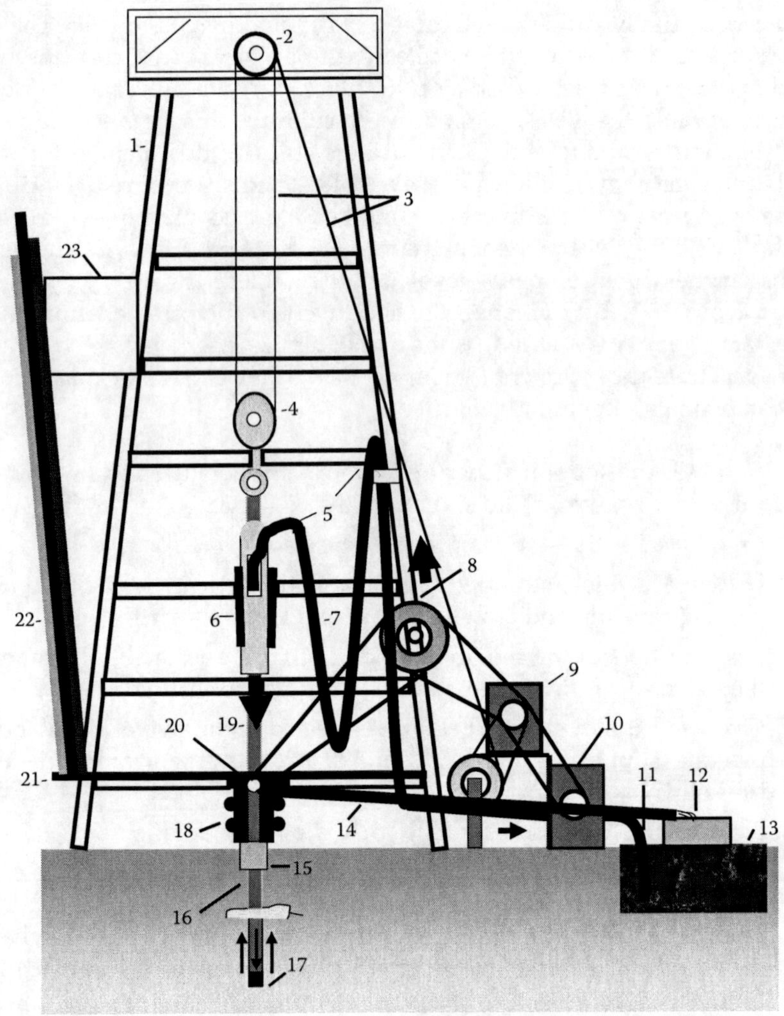

FIGURE 6.2
Simple diagram of a drilling rig and its basic components (see text for details).

13. *Mud tanks,* often called *mud pits,* store drilling fluid until it is required down the wellbore.
14. *Flow line* is a large-diameter pipe that is attached to the bell nipple and extends to the shale shakers to facilitate the flow of drilling fluid back to the mud tanks.
15. *Casing head* or *well head* is a large metal flange welded or screwed onto the top of the conductor pipe (drive pipe) or the casing; it is used to bolt surface equipment such as blowout preventers or Christmas tree assemblies.

16. *Drill string* is an assembled collection of drill pipe, heavyweight drill pipe, drill collars, and any of a whole assortment of tools, connected together and run into the wellbore to facilitate drilling a well.
17. *Drill bit* is a device attached to the end of the drill string that breaks apart the rock being drilled. It contains jets through which the drilling fluid exits.
18. *Blowout preventers* are devices installed at the well head to prevent fluids and gases from unintentionally escaping from the wellbore.
19. *Kelly drive* is square-, hexagonal-, or octagonal-shaped tubing that is inserted through and is an integral part of the rotary table that moves freely vertically while the rotary table turns it.
20. *Rotary table* rotates, along with its constituent parts, the kelly and kelly bushing, the drill string, and the attached tools and bit.
21. *Pipe rack* is part of the drill floor where the stands of drill pipe stand upright. It is typically made of a metal frame structure with large wooden beams situated within it. The wood helps to protect the end of the drill pipe.
22. *Stand* is a section of two or three joints of drill pipe connected together and standing upright in the derrick.
23. *Monkey board* is the structure used to support the top end of the stands of drill pipe vertically situated in the derrick.

The key to a rotary drill's speed is the relative ease of adding new sections of drill pipe (or drill string) while the drill bit continues turning. Drilling mud (fluid) circulates down through the center of the hollow drill pipe and

DID YOU KNOW?

The *reciprocating pump* (or *piston pump*) is one type of positive displacement pump. This pump works just like the piston in an automobile engine—on the intake stroke, the intake valve opens, filling the cylinder with liquid. As the piston reverses direction, the intake valve is pushed closed and the discharge valve is pushed open; the liquid is pushed into the discharge pipe. With the next reversal of the piston, the discharge valve is pulled closed and the intake valve is pulled open, and the cycle then repeats. A piston pump is usually equipped with an electric motor and a gear and cam system that drives a plunger connected to the piston. Just like an automobile engine piston, the piston must have packing rings to prevent leakage and must be lubricated to reduce friction. Because the piston is in contact with the liquid being pumped, only good-grade lubricants can be used when pumping materials that will be added to drinking water. The valves must be replaced periodically as well.

DID YOU KNOW?

The oil field service company individual who is charged with maintaining a drilling fluid or completion fluid system on an oil or gas drilling rig is called the *mud engineer.*

up through the wellbore to lift the drill cuttings to the surface. Modern drill bits are studded with industrial diamonds to make them abrasive enough to grind through any rock type. From time to time, the drill string must be removed (a process termed *tripping*) to replace dulled drill bits.

To function properly, drilling fluids must lubricate the drill bit, keep the wellbore from collapsing, and remove cuttings. The main functions of drilling mud are as follows (Schlumberger, 2012):

- Remove cuttings from well.
- Suspend and release cuttings.
- Control formation pressures.
- Seal permeable formations.
- Maintain wellbore stability.
- Minimize formation damage.
- Cool, lubricate, and support the bit and drilling assembly.
- Transmit hydraulic energy to tools and bit.
- Ensure adequate formation evaluation.
- Control corrosion.
- Facilitate cementing and completion.
- Minimize impact on environment.

The weight of the mud column prevents a blowout from occurring when high-pressure reservoir fluids are encountered. Drillers base the mud's composition on natural bentonite clay, a thixotropic material that is solid (viscous) when still and fluid (less viscous) when shaken, agitated, or otherwise disturbed. This essential rheological property keeps the drill cuttings suspended in the mud. The chemistry and density of the mud must be carefully monitored and adjusted as the drilling deepens (e.g., adding a barium compound increases mud density). *Mud pits,* excavated adjacent to the drill rig, provide a reservoir for mixing and holding the mud. The mud pits also serve as settling ponds for the cuttings. At the completion of drilling, the mud may be recycled at another drilling operation, but the cuttings will be disposed of into the pit. Several environmental concerns over drilling stem from the hazardous composition of the drilling mud and cuttings and from the potential for mud pits to overflow and contaminate surface water.

The most recent advance in drilling is the ability to direct the drill bit beyond the region immediately beneath the drill rig. Early directional drilling involved placing a steel wedge downhole (whipstock) that deflected the drill toward the desired target, but this approach lacked control and was time consuming. Advances such as steerable downhole drill motors that operated on the hydraulic pressure of the circulating drilling mud offered improved directional control; however, to change drilling direction the operator had to halt drill string rotation in such a position that a bend in the motor pointed in the direction of the new trajectory (referred to as the sliding mode). Newer rotary steerable systems introduced in the 1990s eliminated the need to slide a steerable downhole motor (Schlumberger, 2011). The newer tools drill directionally while continuously rotated from the surface by the drilling rig. This enables a much more complex, and thus accurate, drilling trajectory. Continuous rotation also leads to higher rates of penetration and fewer incidents of the drill string sticking (see Figure 6.3).

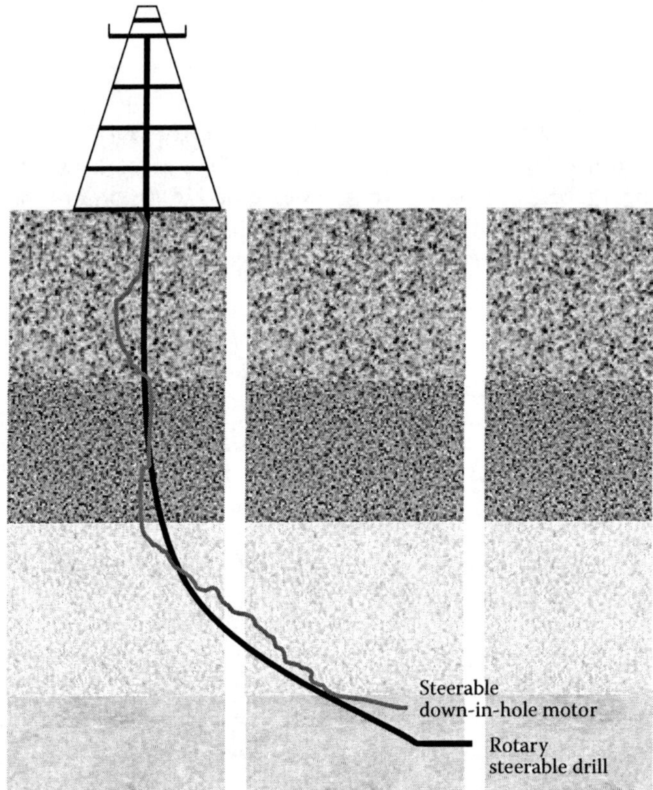

FIGURE 6.3
Directional drilling: steerable downhole motor vs. rotary steerable system. (Courtesy of Schlumberger, Houston, TX.)

Directional drilling offers another significant advantage in developing gas shales. In the case of thin or inclined shale formations, a long horizontal well increases the length of the wellbore in the gas-bearing formation and therefore increases the surface area for gas to flow into the well; however, the increased well surface (length) is often insufficient without some means of artificially stimulating flow. In some sandstone and carbonate formations, injecting dilute acid dissolves the natural cement that binds sand grains, thus increasing permeability. In tight formations such as shale, inducing fractures can increase flow by orders of magnitude; however, before stimulation or production can take place, the well must be completed and cased.

6.6.1 Well Casing Construction

Telescoping steel well casings that prevent wellbore collapse and water infiltration are commonly used in the drilling of commercial gas and oil (Figure 6.4) and municipal water-supply wells (see Figure 6.1). The casing also conducts the produced reservoir fluids to the surface. A properly designed and cemented casing also prevents reservoir fluids (gas or oil) from infiltrating the overlying groundwater aquifers.

During the first phase of drilling, termed *spudding in*, shallow casing installed underneath the platform serves to reinforce the ground surface. Drilling continues to the bottom of the water table (or the potable aquifer), at which point the drill string is tripped out (removed) in order to lower a second casing string, which is cemented in and plugged at the bottom. Drillers use special oil-well cement that expands when it sets to fill the void between the casing and the wellbore.

Surface casing and casing to the bottom of the water table prevent water from flooding the well while also protecting the groundwater from contamination by drilling fluids and reservoir fluids. (The initial drilling stages may

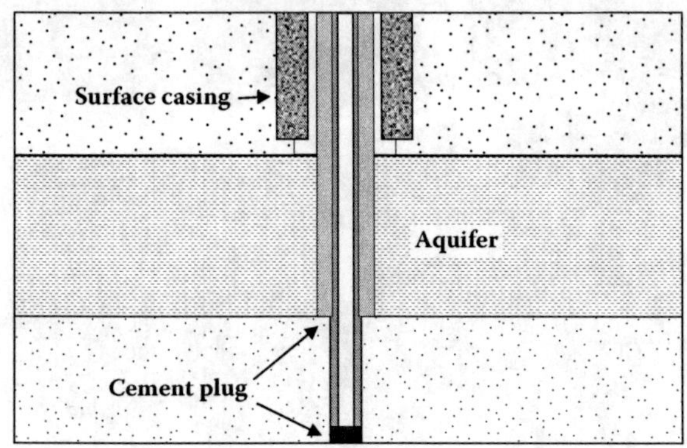

FIGURE 6.4
Theoretical well casing. (Adapted from USDOE, *Modern Shale Gas Development in the United States: A Primer*, U.S. Department of Energy, Washington, DC, 2009.)

use compressed air in place of drilling fluids to avoid contaminating the potable aquifer.) Drilling and casing then continue to the *pay zone*—the formation that produces gas or oil. The number and length of the casings, however, depend on the depth and properties of the geologic strata.

After completing the well to the target depth and cementing in the final casing, the drilling operator may hire an oil-well service company to run a *cement evaluation log*. An electrical probe lowered into the well measures the cement thickness. The cement evaluation log provides the critical confirmation that the cement will function as designed—preventing well fluids from bypassing outside the casing and infiltrating overlying formations. As mentioned earlier, additionally, state oil and gas regulatory agencies often specify the required depth of protective casings and regulate the time that is required for cement to set prior to additional drilling. These requirements are typically based on regional conditions, are established for all wildcat wells, and may be modified when field rules are designated. These requirements are instituted by state oil and gas agencies to provide protection of groundwater resources (Arkansas Oil and Gas Commission, 2012). When the casing strings have been run and cemented, there could be five or more layers or barriers between the inside of the production tubing and a water-bearing formation (salt or fresh).

Analysis of the redundant protections provided by casings and cement was presented by Michie & Associates (1988) in a series of reports and papers prepared for the American Petroleum Institute (API). These investigations evaluated the level of corrosion that occurred in Class II injection wells. Class II injection wells are used for the routine injection of water associated with oil and gas production. The research resulted in the development of a method of calculating the probability (or risk) that fluids injected into Class II injection wells could result in an impact to underground sources of drinking water (USDWs). This research began by evaluating data for oil- and gas-producing basins to determine if there were natural formation waters present that were reported to cause corrosion of well casings. The United States was divided into 50 basins, and each basin was ranked by its potential to have a casing leak resulting from such corrosion.

Detailed analysis was performed for those basins in which there was a possibility of casing corrosion. Risk probability analysis provided an upper boundary for the probability of the fracturing fluids reaching an underground source of drinking water. Based on the values calculated, a modern horizontal well completion in which 100% of the USDWs are protected by properly installed surface casings (and for geologic basins with a reasonable likelihood of corrosion), the probability that fluids injected at depth could impact a USDW would be between 2×10^{-5} (one well in 200,000) and 2×10^{-8} (one well in 200,000,000) if these wells were operated as injection wells. Other studies in the Williston Basin found that the upper-bound probability of injection water escaping the wellbore and reaching an underground source of drinking water was 7 chances in 1 million well-years where surface casings cover the drinking water aquifers (Michie and Koch, 1991). Note that these values do not account for the differences between the operation of a shale gas well and the operation of an injection well. An injection well is constantly injecting fluid under pressure and thus raises the pressure of the receiving aquifer, increasing the chance of a leak or well failure. A production well is reducing the pressure in the producing zone by giving the gas and associated fluid a way out, making it less likely that they will try to find an alternative path that could contaminate a freshwater zone. Furthermore, a producing gas well would be less likely to experience a casing leak because it is operated at a reduced pressure compared to an injection well. It would be exposed to lesser volumes of potentially corrosive water flowing through the production tubing, and it would only be exposed to the pumping of fluids into the well during fracture stimulations.

Because the API study included an analysis of wells that had been in operation for many years when the study was performed in the 1980s, it does not account for advances that have occurred in equipment and applied technologies and changes to regulations. As such, a calculation of the probability of any fluids, including hydraulic fracturing fluids, reaching a USDW from a gas well would indicate an even lower probability, perhaps by as much as two to three orders of magnitude. The API report came to another important conclusion relative to the probability of the contamination of a USDW when it stated that (Michie & Associates, 1988):

> ... for injected water to reach USDW in the 19 identified basins of concern, a number of independent events must occur at the same time and go *undetected* [emphasis added]. These events include simultaneous leaks in the [production] tubing, production casing, [intermediate casing], and the surface casing coupled with the unlikely occurrence of water moving long distances up the borehole past saltwater aquifers to reach a USDW.

As indicated by the analysis conducted by the API and others, the potential for groundwater to be impacted by injection is low. It is expected that the probability for treatable groundwater to be impacted by the pumping of fluids during hydraulic fracture treatments of newly installed, deep shale gas wells when a high level of monitoring is being performed would be even less than the 2×10^{-8} estimated by the API.

In addition to the protections provided by multiple casings and cements, there are natural barriers in the rock strata that act as seals holding the gas in the target formation. Without such seals, gas and oil would naturally migrate to the surface of the Earth. A fundamental precept of oil and gas geology is that, without an effective seal, gas and oil would not accumulate in a reservoir in the first place and so could never be tapped and produced in usable quantities. These sealing strata act as barriers to vertical migration of fluids upward toward useable groundwater zones. Most shale gas wells (outside of those completed in the New Albany and the Antrim) are expected to be drilled at depths greater than 3000 feet below the land surface.

When the cement log has been run, and absent any cement voids, the well is ready for completion. A perforating tool that uses explosive shape charges punctures the casing sidewall at the pay zone. The well may then start producing under its natural reservoir pressure or, as in the case of gas shales, may require stimulation treatment. Both domestic-use gas wells and water wells are common throughout regions experiencing recent shale gas development. In the absence of regulation, domestic-use wells (gas or water) may not meet standard practices of construction. If the well head of a water supply well is improperly sealed, for example, surface water may infiltrate down along the casing exterior and contaminate the drinking-water aquifer. Some domestic water wells have also produced natural gas, and some shallow gas wells have leaked into nearby building foundations. To avoid some of these problems, Pennsylvania has instituted regulations that require a minimum 2000-foot setback between a new gas well and an existing water well.

6.6.2 Drilling Fluids and Retention Pits

Drilling fluids are a necessary component of the drilling process: They circulate cuttings (rock chips created as the drill bit advances through rock, much like sawdust) to the surface to clear the borehole, lubricate and cool the drilling bit, stabilize the wellbore (preventing cave in), and control downhole fluid pressure (Schlumberger, 2008a). In order to maintain sufficient volumes of fluids onsite during drilling, operators typically use pits to store make-up water used as part of the drilling fluids. Storage pits are not used in every development situation. In the case of shale gas drilling practices, they should be adapted to facilitate development in both settings. Drilling with compressed air has become an increasingly popular alternative to drilling with fluids due to the increased cost savings from both reduction in mud costs and the shortened drilling times as a result of air-based drilling (Singh, 1965). The air, like drilling mud, functions to lubricate, cool the bit, and remove cuttings. Air drilling is generally limited to low-pressure formations, such as the Marcellus shale in New York (Kennedy, 2000). In rural areas, storage pits may be used to hold freshwater for drilling and hydraulic fracturing. In urban settings, due to space limitations, steel storage tanks may be used to hold drilling fluids as well as to store water and fluids for use during hydraulic fracturing. Tanks used in closed-loop drilling systems allow for the reuse of drilling fluids and the use of lesser amounts of drilling fluids (Swaco, 2006). Closed-loop drilling systems have also been used with water-based fluids in environmentally sensitive environments in combination with air-rotary drilling techniques (Oklahoma DEQ, 2008). Although closed-loop drilling has been used to address specific situations, the practice is not necessary for every well drilled. Drilling is a regulated practice managed at the state level, and although state oil and gas agencies have the ability to require operators to vary standard practices, the agencies typically do so only when it is necessary to protect the gas resources and the environment.

In rural environments, storage pits may be used to hold water. They are typically excavated containment ponds that, based on the local conditions and regulatory requirements, may be lined. Pits can also be used to store additional make-up water for drilling fluids or to store water used in the hydraulic fracturing of wells.

Water storage pits used to hold water for hydraulic fracturing purposes are typically lined to minimize the loss of water from infiltration. Water storage pits are becoming an important tool in the shale gas industry because the drilling and hydraulic fracturing of these wells often require significant volumes of water as the base fluid for both purposes (Harper, 2008).

6.7 Hydraulic Fracturing

Despite the abundant natural gas content of some shales, they do not produce gas freely. Economic production depends on some means of artificially stimulating shale to liberate gas. In the late 1940s in Texas oil fields, fluids pumped down wells under pressures high enough to fracture stimulated the producing formation. Hydraulic fracturing is a formation stimulation practice used to create additional permeability in a producing formation, thus causing gas to flow more readily toward the wellbore (Jennings and Darden, 1979; Veatch et al., 1999). Hydraulic fracturing can be used to overcome natural barriers to the flow of fluids (gas or water) to the wellbore. Such barriers may include naturally low permeability common in shale formations or reduced permeability resulting from near-wellbore damage during drilling activities (Boyer et al., 2006). Hydraulic fracture stimulation treatments have been adapted to tight gas formations such as the Barnett Shale in Texas, and more recently the Marcellus Shale. Typical fracking treatments, or frack jobs, are relatively large operations compared to some drilling operations. The oilfield service company contracted for the work may take a week to stage the job, and a convoy of trucks will be necessary to deliver the equipment and materials needed.

A company involved in developing Texas shale gas offered the following description of a frack job (Franz and Jochen, 2005):

> Shale gas wells are not hard to drill, but they are difficult to complete. In almost every case the rock around the wellbore must be hydraulically fractured before the well can produce significant amounts of gas. Fracturing involves isolating sections of the well in the producing zone, then pumping fluids and proppant (grains of sand or other material used to hold the cracks open) down the wellbore through perforations in the casing and out into the shale [see Figure 6.5]. The pumped fluid, under pressures up to 8000 psi, is enough to crack shale as much as 3000 ft in each direction from the wellbore. In the deeper high-pressure shales,

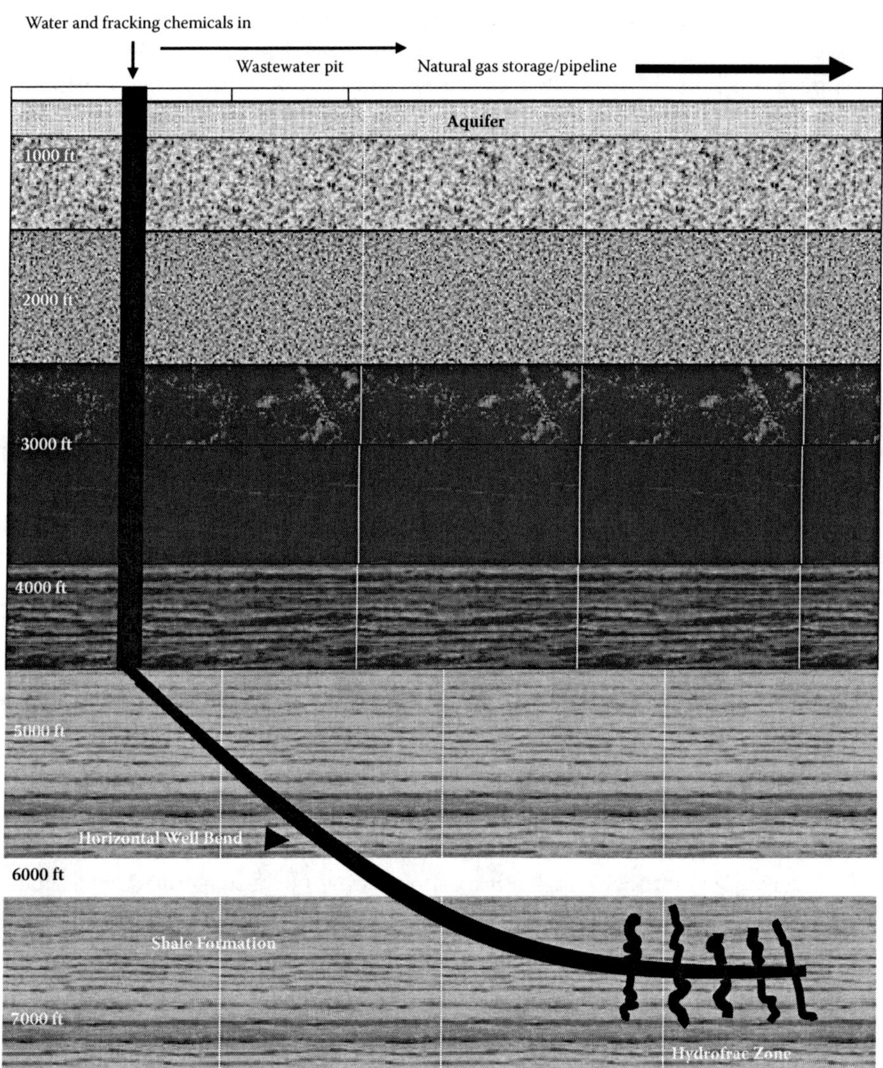

FIGURE 6.5
Diagram of the hydraulic fracturing operation.

> operators pump slickwater (a low-viscosity water-based fluid) and proppant. Nitrogen-foamed fracturing fluids are commonly pumped on shallower shales and shales with lower reservoir pressures.

As shown in Figure 6.5, hydraulic fracturing involves the pumping of fracturing fluid into a formation at a calculated, predetermined rate with enough pressure to generate fractures or cracks in the target formation. For shale gas development, fracture fluids are primarily water-based fluids mixed with additives that help the water to carry sand (or other material) proppant into the fractures. The proppant is needed to keep the fractures open when the pumping of fluid has stopped. When the fracture has initiated, additional fluids are pumped into the wellbore to continue the development of the fracture and to carry the proppant deeper into the formation. The additional fluids are needed to maintain the downhole pressure necessary to accommodate the increasing length of opened fractures in the formation. Each rock formation has inherent natural variability resulting in different fracture pressures for different formations. The process of designing hydraulic fracture treatments requires identifying properties of the target formation, including fracture pressure, and the desired length of fractures. The following discussion addresses some of the processes involved in the design of a hydraulic fracture stimulation of a shale gas formation.

6.7.1 Fracture Design

Modern formation stimulation practices are sophisticated, engineered processes designed to emplace fracture networks in specific rock strata (Boyer et al., 2006). A hydraulic fracture treatment is a controlled process designed to the specific conditions of the target formation (e.g., thickness of shale, rock fracturing characteristics). Understanding the *in situ* reservoir conditions present and their dynamics is critical to successful stimulations. Hydraulic fracturing designs are continually refined to optimize fracture networking and maximize gas production. Whereas the concepts and general practices are similar, the details of a specific fracture operation can vary substantially from basin to basin and from well to well. Fracture design can incorporate many sophisticated and state-of-the-art techniques to accomplish an effective, economic, and highly successful fracture stimulation. Some of these techniques include modeling, microseismic fracture mapping, and tilt-meter analysis.

A computer model can be used to aid in candidate selection with regard to identifying wells, fields, or formations that would be good fracture candidates; such models take into consideration complex factors such as multifractured wells, non-Darcy (nonlaminar) flow, and multiphase flow. Computer modeling can also be utilized in the treatment design process through the use of refined geologic parameters to generate the final pump schedule, in the execution and analysis of the hydraulic fracture treatment (i.e., post-frack production analysis), and in simulations of hydraulic fracturing designs (Meyer & Associates, 2012). Computer models help to maximize effectiveness and to economically design a treatment event. The modeling programs allow

DID YOU KNOW?

Stimulations are optimized to ensure that fracture development is confined to the target formation.

geologists and engineers to modify the design of a hydraulic fracture treatment and evaluate the height, length, and orientation of potential fracture development (Schlumberger, 2008b). These simulators also allow designers to use the data gathered during a fracture stimulation to evaluate the success of the fracture job performed. From these data and analyses, engineers can optimize the design of future fracture stimulations.

Additional advances in hydraulic fracturing design have targeted the analysis of hydraulic fracture treatments through technologies such as microseismic fracture mapping, which is used to pinpoint fracturing and aids in well stimulation, and tilt measurements (Meyer & Associates, 2012). These technologies can be used to define the success and orientation of the fractures created, thus providing the engineers with the ability to manage the resource through the strategic placement of additional wells, taking advantage of the natural reservoir conditions and expected fracture results in new wells.

DID YOU KNOW?

Microseismic mapping technology is rooted in the observation that when a fracture is induced into a reservoir or the bounding rock layers the *in situ* stress is disturbed, resulting in shear failure and subsequent "mini-earthquakes" (Walser, 2010).

As more formation-specific data are gathered, service companies and operators can optimize fracture patterns. Operators have strong economic incentives to ensure that fractures do not propagate beyond the target formation and into adjacent rock strata (Parshall, 2008). Allowing the fractures to extend beyond the target formation would be a waste of materials, time, and money. In some cases, fracturing outside of the target formation could potentially result in the loss of the well and the associated gas resource. Fracture growth outside of the target formation can result in excess water production from bounding strata. Having to pump and handle excess water increases production costs, negatively impacting well economics. This is a particular concern in the Barnett Shale of Texas where the underlying Ellenberger Group limestones are capable of yielding significant formation water.

6.7.2 Hydraulic Fracturing Process*

Fracture treatments of horizontal shale gas wells are carefully controlled and monitored operations that are performed in stages. Lateral lengths in horizontal wells for shale gas development may range from 1000 feet to more than 5000 feet. Before beginning a treatment, the service company will perform

* This section is based on information contained in USDOE, *Modern Shale Gas Development in the United States: A Primer*, U.S. Department of Energy, Washington, DC, 2009, pp. 55–59.

a series of tests on the well to determine if it is competent to hold up to the hydraulic pressures generated by the fracture pumps. In the initial stage, hydrochloric acid (HCl) solution is pumped down the well to clean up the residue left from cementing the well casing. Each successive stage pumps discrete volumes of fluid (slickwater) and proppant down the well to open and propagate the fracture further into the formation. The treatment may last up to an hour or more, with the final stage designed to flush the well. Some wells may receive several or more treatments to produce multiple fractures at different depths or further out into the formation in the case of horizontal wells.

DID YOU KNOW?

Because of the length of exposed wellbore, it is usually not possible to maintain a downhole pressure sufficient to stimulate the entire length of a lateral in a single stimulation event (Overbey et al., 1988). Because of the lengths of the laterals, hydraulic fracture treatments of horizontal shale gas wells are usually performed by isolating smaller portions of the lateral.

The fracturing of each portion of the lateral wellbore is called a *stage*. Stages are fractured sequentially beginning with the section at the farthest end of the wellbore, moving uphole as each stage of the treatment is completed until the entire lateral well has been stimulated (Chesapeake Energy, 2008b). Horizontal wells in the various shale gas basins may be treated using two or more stages to fracture the entire perforated interval of the well. Each stage of a horizontal well fracture treatment is similar to a fracture treatment for a vertical shale gas well. For each stage of a fracture treatment, a series of different volumes of fracturing fluids, called *substages*, with specified additives and proppant concentrations, is injected sequentially. Table 6.1 presents an example of the substages of a single-stage hydraulic fracture treatment for a well completed in the Marcellus Shale (Arthur et al., 2008). This is a single-stage treatment typical of what might be performed on a vertical shale well or for each stage of a multistage horizontal well treatment. The total volume of the substages in Table 6.1 is 578,000 gallons. If this were one stage of a four-stage horizontal well, the entire fracture operation would require approximately four times this amount, or 2.3 million gallons of water. Note that the actual *rate* of water usage is measured in gallons per minute (gal/min); 42 gal/min = 1 bbl/min, and 500 gal/min = ~12 bbl/min. In Table 6.1, with the exception of the 500-gal/min rate for diluted acid (15%), the other rates from pad through flush are 3000 gal/min.

Guided by state oil and gas regulatory agencies—to ensure that a well is protective of water resources and is safe for operation—operators or service companies perform a series of tests. These tests are designed to ensure that the well, well equipment, and hydraulic fracturing equipment are in proper

TABLE 6.1

Example of a Single Stage of a Sequenced Hydraulic Fracture Treatment

Hydraulic Fracture Treatment Substage	Volumes (gallons)
Diluted acid (15%)	5000
Pad	100,000
Prop 1	50,000
Prop 2	50,000
Prop 3	40,000
Prop 4	40,000
Prop 5	40,000
Prop 6	30,000
Prop 7	30,000
Prop 8	20,000
Prop 9	20,000
Prop 10	20,000
Prop 11	20,000
Prop 12	20,000
Prop 13	20,000
Prop 14	10,000
Prop 15	10,000
Flush	13,000

Source: Adapted from Arthur, J.D. et al., *An Overview of Modern Shale Gas Development in the United States*, ALL Consulting, Tulsa, OK, 2008.

Note: Volumes are presented in gallons (42 gallons = 1 bbl, 5000 gallons = ~120 bbl). Flush volumes are based on the total volume of open borehole; therefore, as each stage is completed, the volume of flush decreases as the volume of borehole is decreased. Total amount of proppant used is approximately 450,000 pounds.

DID YOU KNOW?

A single fracture treatment may consume more than 500,000 gallons of water (USDOE, 2009). Wells subject to multiple treatments consume several million gallons of water. For comparison, an Olympic-size swimming pool (164 × 82 × 6 ft deep) holds over 660,000 gallons of water, and the average daily per capita consumption of freshwater (roughly 1430 gallons per day) adds up to 522,000 gallons over one year (USGS, 2000).

working order and will safely withstand application of the fracture treatment pressures and pump flow rates. The tests start with the testing of well casings and cement during the drilling and well construction process. Testing continues with pressure testing of hydraulic fracturing equipment prior to the fracture treatment process (Harper, 2008). As mentioned earlier, construction requirements for wells are mandated by state oil and gas regulatory agencies to ensure that a well is protective of water resources and is safe for operation.

After the testing of equipment has been completed, the hydraulic fracture treatment process begins. The substage sequence is usually initiated with the pumping of an acid treatment. Again, this acid treatment helps to clean the near-wellbore area, which can be damaged as a result of the drilling and well installation process; for example, pores and pore throats can become plugged with drilling mud or casing cement. The next sequence after the acid treatment is a slickwater pad, which is a water-based fracturing fluid mixed with a friction-reducing agent. The pad is a volume of fracturing fluid large enough to effectively fill the wellbore and the open formation area. The slickwater pad helps to facilitate the flow and placement of the proppant further into the fracture network.

DID YOU KNOW?

Slickwater fracturing is a method or system of hydrofracturing that involves the addition of chemicals to water to increase the fluid flow. Fluid (friction reducer) can be pumped down the wellbore at a rate as high as 100 bbl/min to fracture the shale; otherwise, without the use of slickwater, the top speed of pumping is about 60 bbl/min.

After the pad is pumped, the first proppant substage combining a large volume of water with fine mesh sand is pumped. The next several substages increase the volume of fine-grained proppant while the volume of fluids pumped is decreased incrementally from 50,000 gal to 30,000 gal. This fine-grained proppant is used because the finer particle size is capable of being carried deeper in the developed fractures (Cramer, 2008). In this example, the fine proppant substages are followed by eight substages of a coarser proppant with volumes from 20,000 gal to 10,000 gal. After completion of the final substage of coarse proppant, the well and equipment are flushed with a volume of freshwater sufficient to remove excess proppants from the equipment and the wellbore.

Hydraulic fracture stimulations are overseen continuously by operators and service companies to evaluate and document the events of the treatment process. Every aspect of the fracture stimulation process is carefully monitored, from the well head and downhole pressures to pumping rates and density of the fracturing fluid slurry. The monitors also track the volumes of each additive and the water used and ensure that equipment is functioning

DID YOU KNOW?

The fracture is ideally represented by a vertical plane that intersects the well casing. It does not propagate in a random direction but opens perpendicular to the direction of least stress underground (which is nearly horizontal in orientation).

properly. For a 12,000-bbl (504,000-gal) fracture treatment of a vertical shale gas well there may between 30 and 35 people onsite to monitor the entire stimulation process.

The staging of multiple fracture treatments along the length of the lateral leg of the horizontal well allows the fracturing process to be performed in a very controlled manner. By fracturing discrete intervals of the lateral wellbore, the operator is able to make changes to each portion of the completion zone to accommodate site-specific changes in the formation. These site-specific variations may include variations in shale thickness, presence or absence of natural fractures, proximity to another wellbore fracture system, and boreholes that are not centered in the formation.

6.7.3 Fracturing Fluids

As mentioned, the current practice of hydraulic fracturing of shale gas reservoirs is to apply a sequenced pumping event in which millions of gallons of water-based fracturing fluids mixed with proppant materials are pumped in a controlled and monitored manner into the target shale formation above fracture pressure (Harper, 2008). The fracturing fluids used for gas shale stimulations consist primarily of water but also include a variety of additives. The number of chemical additives used in a typical fracture treatment varies depending on the conditions of the specific well being fractured. A typical fracture treatment will use very low concentrations of between 3 and 12 additive chemicals, depending on the characteristics of the water and shale formation being fractured. Each component serves a specific, engineered purpose (Schlumberger, 2008b). The predominant fluids currently being used for fracture treatments in the shale gas plays are water-based fracturing fluids mixed with friction-reducing additions (slickwater).

The addition of friction reducers allows fracturing fluids and proppant to be pumped to the target zone at a higher rate and reduced pressure than if water alone were used. In addition to friction reducers, other additions include biocides to prevent microorganism growth and to reduce biofouling of the fractures, oxygen scavengers and other stabilizers to prevent corrosion of metal pipes, and acids that are used to remove drilling mud damage with the near-wellbore area (Schlumberger, 2008b). These fluids function in two ways: opening the fracture and transporting the propping agent (or proppant) the length of the fracture (Economides and Nolte, 2000).

DID YOU KNOW?

As the term *propping* implies, the agent functions to prop or hold the fracture open. The fluid must have the proper viscosity and low friction pressure when pumped, it must break down and clean up rapidly when treatment is over, and it must provide good fluid-loss control (not dissipate). The fluid chemistry may be water, oil, or acid based, depending on the properties of the formation. Water-based fluids (slickwater) are the most widely used in shale formations because of their low cost, high performance, and ease of handling.

Table 6.2 lists the volumetric percentages of additives that were used for a nine-stage hydraulic fracturing treatment of a Fayetteville Shale horizontal well. The make-up of fracturing fluid varies from one geologic basin or formation to another. Evaluating the relative volumes of the components of a fracturing fluid reveals the relatively small volume of additives that are present. The additives represent less than 0.5% of the total fluid volume. Overall, the concentration of additives in most slickwater fracturing fluids is a relatively consistent 0.5% to 2%, with water making up 98% to 99.5%.

Because the make-up of each fracturing fluid varies to meet the specific needs of each area, there is no one-size-fits-all formula for the volumes for each additive. In classifying fracturing fluids and their additives, it is important to realize that service companies that provide these additives have developed a number of compounds for different recipes with similar functional

TABLE 6.2

Volumetric Composition of a Fracturing Fluid

Component	Percent (%) by Volume
Water and sand	99.51
Surfactant	0.085
KCl	0.06
Gelling agent	0.056
Scale inhibitor	0.043
pH adjusting agent	0.011
Breaker	0.01
Crosslinker	0.007
Iron control	0.004
Corrosion inhibiter	0.002
Biocide	0.001
Acid	0.123
Friction reducer	0.088

Source: Based on data (2008) from ALL Consulting for a fracture operation in the Fayetteville Shale.

DID YOU KNOW?

Some fracturing fluids may include nitrogen and carbon dioxide to help foaming. Oil-based fluids find use in hydrocarbon-bearing formations susceptible to water damage, but they are expensive and difficult to use. Acid-based fluids use hydrochloric acid to dissolve the mineral matrix of carbonate formations (limestone and dolomite) and thus improve porosity; the reaction produces inert calcium chloride salt and carbon dioxide gas.

properties to be used for the same purpose in different well environments. The difference between additive formulations may be as small as a change in concentration of a specific compound. Although the hydraulic fracturing industry may have a number of compounds that can be used in a hydraulic fracturing fluid, any single fracturing job would use only a few of the available additives. In Table 6.1, for example, 12 additives are used, covering the range of possible functions that could be built into a fracturing fluid. It is not uncommon for some fracturing recipes to omit some compound categories if their properties are not required for the specific application.

Most industrial processes use chemicals, and almost any chemical can be hazardous in large enough quantities if not handled properly. Even chemicals that go into our food or drinking water can be hazardous. Drinking water treatment plants use large quantities of chlorine. When used and handled properly, it is safe for workers and nearby residents and provides clean, safe drinking water for the community. Although the risk is low, the potential exists for an unplanned release that could have serious effects on human health and the environment. By the same token, hydraulic fracturing uses a number of chemical additives that could be hazardous but are safe when properly handled according to requirements and long-standing industry practices. In addition, many of these additives are common chemicals that people regularly encounter in everyday life.

DID YOU KNOW?

Proppants hold the fracture walls apart to create conductive paths for the natural gas to reach the wellbore. Silica sands are the most commonly used proppants. Resin coating the sand grains improves their strength.

Table 6.3 provides a summary of the additives, their main compounds, the reason the additive is used in a hydraulic fracturing fluid, and some of the other common uses for these compounds. Hydrochloric acid (HCl) is the single largest liquid component used in a fracturing fluid aside from water;

TABLE 6.3

Fracturing Fluid Additives, Main Compounds, and Common Uses

Additive Type	Main Component(s)	Purpose	Common Use of Main Compound
Diluted acid (15%)	Hydrochloric acid or muriatic acid	Helps dissolve minerals and initiates cracks in the rock	Swimming pool chemical and cleaner
Biocide	Glutaraldehyde	Eliminates bacteria in the water that produce corrosive byproducts	Disinfectant; used to sterilize medical and dental equipment
Breaker	Ammonium persulfate	Allows delayed breakdown of the gel polymer chains	Bleaching agent in detergent and hair cosmetics, manufacture of household plastics
Corrosion inhibitor	*N,n*-dimethyl formamide	Prevents corrosion of the pipe	Pharmaceuticals, acrylic fibers, plastics
Crosslinker	Borate salts	Maintain friction between the fluid and the pipe	Laundry detergents, hand soaps, cosmetics
Friction reducer	Polyacrylamide	Minimizes friction between the fluid and the pipe	Water treatment, soil conditioner
	Mineral oil		Make-up remover, laxative, candy
Gel	Guar gum or hydroxyethyl cellulose	Thickens the water in order to suspend the sand	Cosmetics, toothpaste, sauces, baked goods, ice cream
Iron control	Citric acid	Prevents precipitation of metal oxides	Food additive, flavoring in foods and beverages; lemon juice is 7% citric acid
KCl	Potassium chloride	Creates a brine carrier fluid	Low-sodium table salt substitute
Oxygen scavenger	Ammonium bisulfite	Removes oxygen from the water to protect pipes from corrosion	Cosmetics, food and beverage processing, water treatment
pH adjusting agent	Sodium or potassium carbonate	Maintains effectiveness of other components, such as crosslinkers	Washing soda, detergents, soap, water softener, glass and ceramics
Proppant	Silica, quartz sand	Allows fractures to remain open so gas can escape	Drinking water filtration, play sand, concrete, brick mortar
Scale inhibitor	Ethylene glycol	Prevents scale deposits in pipes	Automotive antifreeze, household cleansers, deicing agent
Surfactant	Isopropanol	Used to increase viscosity of the fracture fluid	Glass cleaner, hair color, antiperspirant

Source: USDOE, *Modern Shale Gas Development in the United States: A Primer*, U.S. Department of Energy, Washington, DC, 2009.

Note: Specific compounds used in a given fracturing operation will vary depending on company preference, source water quality, and site-specific characteristics of the target formation. Compounds shown above are representative of the major compounds used in hydraulic fracturing of gas shales.

although the concentration of the acid may vary, a 15% HCl mix is a typical concentration. A 15% HCl mix is composed of 85% water and 15% acid; therefore, the volume of acid is diluted by 85% with water in its stock solution before it is pumped into the formation during a fracturing treatment. Once the entire stage of fracturing fluid has been injected, the total volume of acid in an example fracturing fluid from the Fayetteville Shale was 0.123%, which indicates the fluid had been diluted by a factor of 122 times before it is pumped into the formation. The concentration of this acid will only continue to be diluted as it is further dispersed in additional volumes of water that may be present in the subsurface. Furthermore, if this acid comes into contact with carbonate minerals in the subsurface, it would be neutralized by chemical reaction with the carbonate minerals, producing water and carbon dioxide as a byproduct of the reaction.

6.8 Fracking Water Supply

> You can't frack for shale gas without water. You can't frack for shale gas without a whole bunch of water ... I'm talkin' millions of gallons, mister!
>
> **—Pennsylvania hydraulic fracturing engineer (2011)**

As the Pennsylvania engineer pointed out, the drilling and hydraulic fracturing of a horizontal shale gas well requires water—2 to 4 million gallons of water (Satterfield et al., 2008), with about 3 million gallons being most common. Note that the volume of water required may vary substantially between wells. In addition, the volume of water required per foot of wellbore appears to be decreasing as technology and methods improve over time. Table 6.4 presents data regarding estimated per-well water needs for four shale gas plays currently being developed.

TABLE 6.4

Estimated Water Needs for Drilling and Fracturing Wells in Select Shale Gas Plays

Shale Gas Play (gal)	Volume of Drilling Water per Well (gal)	Volume of Fracturing Water per Well (gal)	Total Volumes of Water per Well (gal)
Barnett Shale	400,000	2,300,000	2,700,000
Fayetteville Shale	60,000[a]	2,900,000	3,060,000
Haynesville Shale	1,000,000	2,700,000	3,700,000
Marcellus Shale	80,000[a]	3,800,000	3,880,000

Source: Based on data (2008) from ALL Consulting for a fracture operation in the Fayetteville Shale.

Note: These volumes are approximate and may vary substantially between wells.

[a] Drilling performed with an air mist and/or water-based or oil-based muds for deep horizontal well completions.

DID YOU KNOW?

For very deep wells, up to 4.5 million gallons of water may be needed to drill and fracture a shale gas well; this is equivalent to the amount of water consumed by (Chesapeake Energy, 2012):

- New York City in approximately 7 minutes
- A 1000-megawatt coal-fired power plant in 12 hours
- A golf course in 25 days
- 7.5 acres of corn in a season

Water for drilling and hydraulic fracturing of these wells frequently comes from surface water bodies such as rivers and lakes, but it can also come from groundwater, private water sources, municipal water, and reused produced water. Most of the producing shale gas basins contain large amounts of local water sources.

Even though water volumes needed to drill and stimulate shale gas wells are large, they occur in areas with moderate to high levels of annual precipitation. However, even in areas of high precipitation, due to growing populations, other industrial water demands, and seasonal variation in precipitation, it can be difficult to meet the needs of shale gas development and still satisfy regional needs for water.

Even though the water volumes required to drill and stimulate shale gas wells are large, they generally represent a small percentage of the total water resource used in the shale gas basins. Calculations indicate that water use will range from less than 0.1 to 0.8% by basin (Satterfield, 2008). This volume is small in terms of the overall surface water budget for an area; however, operators need this water when drilling activity is occurring (on demand), requiring that the water be procured over a relatively short period of time. Water withdrawals during periods of low stream flow could affect fish and other aquatic life, fishing and other recreational activities, municipal water supplies, and other industries such as power plants. To put shale gas water use in perspective, the consumptive use of freshwater for electrical generation in the Susquehanna River Basin alone is nearly 150 million gallons per day, whereas the projected total demand for peak Marcellus Shale activity in the same area is 8.4 million gallons per day (Gaudlip et al., 2008).

DID YOU KNOW?

One acre-foot (ac-ft) of water is equivalent to the volume of water required to cover one acre with one foot of water.

One alternative that states and operators are pursuing is to make use of seasonal changes in river flow to capture water when surface water flows are greatest. Utilizing seasonal flow differences allows planning of withdrawals to avoid potential impacts to municipal drinking water supplies or to aquatic or riparian communities. In the Fayetteville Shale play of Arkansas, one operator is constructing a 500-ac-ft impoundment to store water withdrawals from the Little Red River obtained during periods of high flow (storm events or hydroelectric power generation releases from Greer's Ferry dam upstream of the intake) when excess water is available (Chesapeake Energy, 2008a). The project is limited to 1550 acre-ft of water annually. As additional mitigation, the company has constructed extra pipelines and hydrants to provide portions of this rural area with water for fire protection. Also included is monitoring of in-stream water quality as well as game and non-game fish species in the reach of river surrounding the intake. This design provides a water recovery system similar in concept to what some municipal water facilities use. It will minimize the impact on local water supplies because surface water withdrawals may still be limited to times of excess flow in the Little Red River. This project was developed with input from a local chapter of Trout Unlimited, an active conservation organization in the area, and represents an innovative environmental solution that serves both the community and the gas developer.

These water needs may challenge supplies and infrastructure in new areas of shale gas development—that is, in areas where the impact of shale gas operations is new and the potential impact is unknown to local inhabitants and governing officials. As operators look to develop new shale gas plays, a failure to communicate with local officials is not a viable option. Communication with local water planning agencies can help operators and communities to coexist and effectively manage local water resources. Understanding local water needs can help operators develop a water storage or management plan that will meet with acceptance in neighboring communities. Although the water needed for drilling an individual well may represent a small volume over a large area, the withdrawals may have a cumulative impact on watersheds over the short term. This potential impact can be avoided by working with local water resource managers to develop a plan outlining when and where withdrawals will occur (i.e., avoiding headwaters, tributaries, small surface water bodies, or other sensitive sources).

Before a shale gas play hydraulic fracking operation is developed, not only is it a good idea to communicate with state and local government officials but it is also prudent to obtain information and data related to the urban water cycle (i.e., if the shale gas operation is near or has an impact on the surrounding area). Moreover, a study of the effect of shale gas hydraulic fracking operations on the indirect water reuse process is called for. In the planning stage, close scrutiny must be paid to water sources, outfalls, annual precipitation levels, drought histories, indirect water reuse, and (if located in or near an urban area) the urban water cycle.

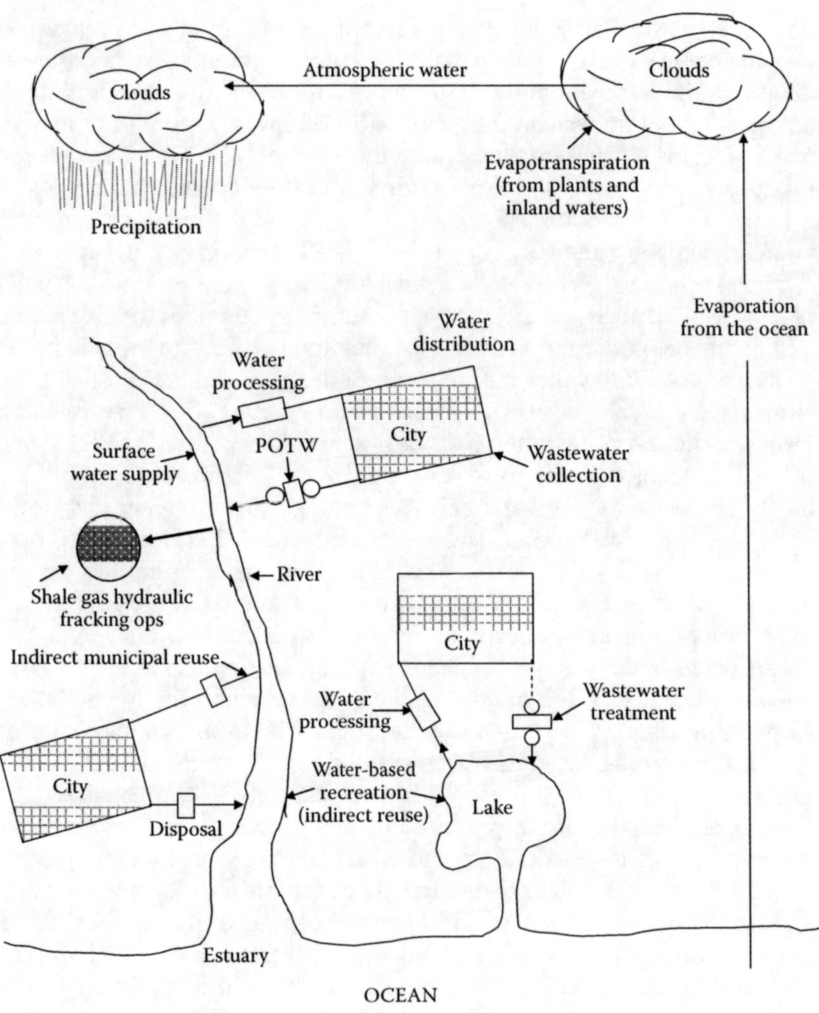

FIGURE 6.6
Urban water cycle.

Today we tend to computer model this, that, and whatever, and few can doubt the worth and advantage of computer modeling. Sometimes, though, it is best to revert back to the old adage that a picture, a sketch, or simple drawing is worth a billion words. Computer models can be complicated to conceive, depict, and understand. A simple picture is … well, it is just simple, straightforward, fundamental, eye catching, and usually understandable. The simple picture referred to is a basic (*basic* being key) drawing of the local urban surface water cycle and the regional indirect surface water reuse process (Figures 6.6 and 6.7). Note that Figure 6.6 also shows the potential

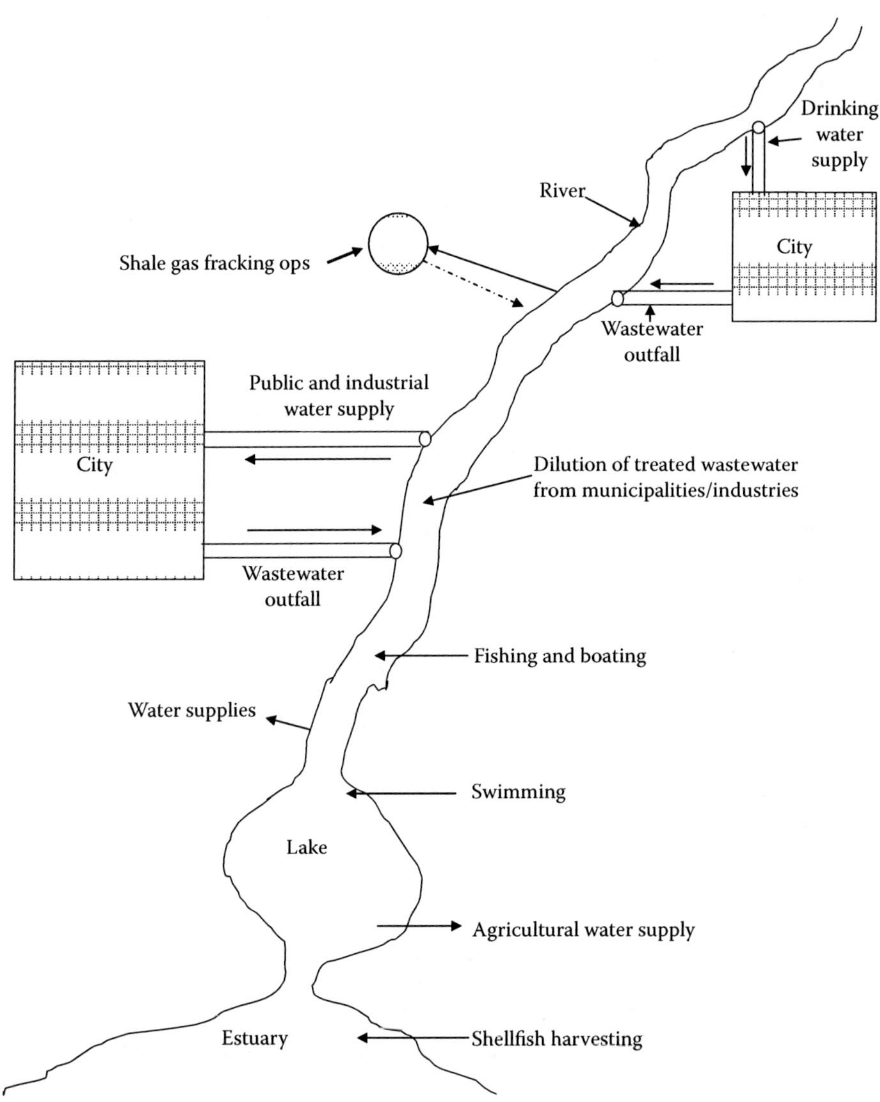

FIGURE 6.7
Indirect water reuse process.

discharge of used but "treated" frack water to the water supply for reuse. This scenario is likely only if some type of *in situ* wastestream treatment process is available to treat the frack fluid waste with, for example, an advanced oxidation process (AOP) and reverse osmosis (RO). Such a mobile treatment option could result in 75 to 80% treated clean water for discharge into the receiving body (river or lake) and the rest could be reused on site. The basic drawings represented by Figures 6.6 and 6.7, along with data pertaining to annual rainfall levels and current usage in gallons of surface water sources, would go a long way (in understandable terminology) toward showing local officials the possible impact of a nearby shale gas hydraulic fracking operation. When the shale gas fracking water supply is obtained from groundwater and not surface water sources, a simplified diagram along the lines of Figures 6.6 and 6.7 should be drawn showing the aquifer interface with the drilling and fracking operation in general and the water usage from the underground source in particular. Note that a more detailed discussion of frack water management is presented later in the text.

6.9 Thought-Provoking Discussion Questions

1. Can you relate to Karl's desire to become Texas rich as compared to dirt poor? Do you agree with his point about Mother Nature correcting any environmental damage caused by hydraulic fracturing operations? Explain.
2. How important is it to protect wildlife in a hydraulic fracturing region?
3. Do you believe that hydraulic fracturing operations damage groundwater supplies?
4. There is a lot of secrecy surrounding the chemicals used in fracking fluids. Do you have any concerns about that? Why?

References and Recommended Reading

AAPG. (1987). *Correlation of Stratigraphic Units of North America (CASUNA) Project*, American Association of Petroleum Geologists, Tulsa, OK.

Anon. (2000). Alabama lawsuit poses threat to hydraulic fracturing across U.S. *Drilling Contractor*, January/February, pp. 42–43.

Anon. (2008). *Projecting the Economic Impact of the Fayetteville Shale Play for 2008–2012*. Center for Business and Economic Research, Sam M. Walton College of Business, Fayetteville, AR.

Arkansas Oil and Gas Commission. (2012). *General Rules and Regulations*. Arkansas Oil and Gas Commission, Little Rock.

Arthur, J.D., Bohm, B., and Lane, M. (2008). *An Overview of Modern Shale Gas Development in the United States*. ALL Consulting, Tulsa, OK.

Berman, A. (2008). The Haynesville Shale sizzles while the Barnett cools. *World Oil Magazine*, 229(9):23.

Boughal, K. (2008). Unconventional plays grow in number after Barnett Shale blazed the way. *World Oil Magazine*, 229(8):77–80.

Boyer, C., Kieschnick, J., Suarez-Rivera, R., Lewis, R., and Walter, G. (2006). Producing gas from its source. *Oilfield Review*, 18(3):36–49.

Cardott, B. (2004). *Overview of Unconventional Energy Resources of Oklahoma*. Oklahoma Geological Survey, Norman, OK (www.ogs.ou.edu/fossilfuels/coalpdfs/UnconventionalPresentation.pdf).

Catacosinos, P., Harrison, W., Reynolds, R., Westjohn, D., and Wollensak. M. (2000). *Stratigraphic Nomenclature for Michigan*. Michigan Department of Environmental Quality, Geological Survey Division, and Michigan Basin Geological Survey, Kalamazoo, MI.

CBO. (1983). *Understanding Natural Gas Price Decontrol*. Congressional Budget Office, Congress of the United States, Washington, DC.

Chesapeake Energy. (2008a). Little Red River Project, paper presented to Trout Unlimited.

Chesapeake Energy. (2008b). Components of Hydraulic Fracturing, paper presented to New York Department of Environmental Conservation.

Chesapeake Energy. (2012). *Hydraulic Fracturing Facts*, http://www.hydraulicfracturing.com/Water-Usage/Pages/Information.aspxd.

Cramer, D. (2008). Stimulating unconventional reservoirs: lessons learned, successful practices, areas for improvement, in *Proceedings of SPE Unconventional Reservoirs Conference*, Keystone, CO, February 10–12, 2008. Society of Petroleum Engineers, Allen, TX.

Durham, L.S. (2008). A spike in interest and activity: Louisiana play a "company maker"? *AAPG Explorer*, 29(7):18.

Economides, M.J. and Nolte, K.G. (2000). *Reservoir Stimulation*, 3rd ed. Wiley, New York.

Franz, Jr., J.H. and Jochen, V. (2005). *When Your Gas Reservoir Is Unconventional, So Is Our Solution*, Shale Gas White Paper 05-OF-299. Schlumberger, Houston, TX.

Gaudlip, A.W., Paugh, L.O., and Hayes, T.D. (2008). Marcellus Shale water management challenges in Pennsylvania, in *Proceedings of SPE Shale Gas Production Conference*, Fort Worth, TX, November 16–18, 2008. Society of Petroleum Engineers, Allen, TX.

Halliburton. (2008). *U.S. Shale Gas: An Unconventional Resource. Unconventional Challenges*. Halliburton, Houston, TX, 8 pp.

Harper, J. (2008). The Marcellus Shale—an old "new" gas reservoir in Pennsylvania. *Pennsylvania Geology*, 28(1):2–13.

Harrison, W. (2006). *Production History and Reservoir Characteristics of the Antrim Shale Gas Play, Michigan Basin*. Western Michigan University, Kalamazoo.

Hayden, J. and Pursell, D. (2005). *The Barnett Shale: Visitor's Guide to the Hottest Gas Play in the US*. Pickering Energy Partners, Inc., Houston, TX.

HIE. (2007). *Fayetteville Shale Power*. Hillwood International Energy, Dallas, TX.

IGS. (1986). *General Stratigraphic Column for Paleozoic Rocks in Indiana*. Indiana Geological Survey, Indiana University, Bloomington, IN (www.usi.edu/science/geology/My%20Web%20Sites/stratcolumn.pdf).

ISGS. (2011). *Mississippian Rocks in Illinois*, GeoNote 1. Illinois State Geological Survey, Champaign, IL (www.isgs.uiuc.edu/maps-data-pub/publications/geonotes/geonote1.shtml).

Jennings, Jr., A.R. and Darden, W.G. (1979). Gas well stimulation in the eastern United States, in *Proceedings of Symposium on Low Permeability Gas Reservoirs*, Denver, CO, May 20–22, 1979. Society of Petroleum Engineers, Allen, TX.

Jochen, V. (2006). *New Technology Needs to Produce Unconventional Gas*. Schlumberger, Houston, TX.

Johnston III, J., Heinrich, B., Lovelace, J., McCulluh, R., and Zimmerman, R. (2000). *Stratigraphic Charts of Louisiana*, Louisiana Geological Survey, Baton Rouge, LA.

Kennedy, J. (2000). Technology limits environmental impact of drilling. *Drilling Contractor*, July/August, pp. 31–35.

Lesley, J.P. (1892). *A Summary Description of the Geology of Pennsylvania, Vol. II*. Geological Survey of Pennsylvania, Pittsburgh.

Meyer & Associates. (2012). *Meyer Fracturing Simulators User's Guide*, http://www.mfrac.com/documentation.html.

Michie & Associates. (1988). Oil and Gas Water Injection Well Corrosion, paper prepared for the American Petroleum Institute, Washington, DC.

Michie, T.W. and Koch, C.A. (1991). Evaluation of injection-well risk management in the Williston Basin. *Journal of Petroleum Technology*, 43(6):737–741.

NaturalGas.org. (2011). *Unconventional Natural Gas Resources*, www.naturalgas.org/overview/unconvent_ng_resource.asp.

Navigant Consulting. (2008). *North American Natural Gas Supply Assessment*, prepared for American Clean Skies Foundation, Washington, DC.

NRCS. (2008). *National Water & Climate Center*. Natural Resources Conservation Service, Washington, DC (www.wcc.nrcs.usda.gov/).

Nyahay, R., Leone, J., Smith, L., Martin, J., and Jarvie, D. (2007). Update on Regional Assessment of Gas Potential in the Devonian Marcellus and Ordovician Utica Shales of New York, paper presented at the American Association of Petroleum Geologists (AAPG) Eastern Section Meeting, Lexington, KY, September 16–18, 2007 (www.searchanddiscovery.com/documents/2007/07101nyahay/).

Oklahoma DEQ. (2008). *Pollution Prevention Case Study for Oxy USA, Inc*. Oklahoma Department of Environmental Quality, Oklahoma City (www.deq.state.ok.us/CSDnew/P2/Casestudy/oxyusa~1.htm).

Overbey, W.K., Yost II, A.B., and Wilkins, D.A. (1988). Inducing multiple hydraulic fractures for a horizontal wellbore, in *Proceedings of SPE Annual Technical Conference and Exhibition*, Houston, TX, October 2–5. Society of Petroleum Engineers, Allen, TX.

Parshall, J. (2008). Barnett Shale showcases tight-gas development. *Journal of Petroleum Technology*, 60(9):48–55.

PLS. (2008). *Other Players Reporting Haynesville Success*. Petroleum Listing Services, Houston, TX, August 15.

Railroad Commission of Texas. (2008). *Newark, East (Barnett Shale) Field*. Oil and Gas Division, Railroad Commission of Texas, Austin.

Satterfield, J., Mantell, M., Kathol, D., Hebert, F., Patterson, K., and Lee, R. (2008). Managing Water Resource's Challenges in Select Natural Gas Shale Plays, paper presented at the Ground Water Protection Council Annual Forum, Cincinnati, OH, September 21–24, 2008.

Schlumberger. (2008a). *The Many Roles of Drilling Fluids*. Schlumberger, Houston, TX (http://www.planetseed.com/node/15300).

Schlumberger. (2008b). *PowerSTIM Service Increases Field Production.* Schlumberger, Houston, TX (http://www.slb.com/~/media/Files/dcs/case_studies/powerstim_us_escondido.ashx)

Schlumberger. (2011). *Better Turns for Rotary Steerable Drilling*. Schlumberger, Houston, TX (http://www.slb.com/resources/publications/oilfield_review/ori/ori002.aspx).

Schlumberger. (2012). *The Oil Field Glossary: Where the Oil Field Meets the Dictionary.* Schlumberger, Houston, TX (www.glossary.oilfield.slb.com/).

Singh, Jr., M.M. (1965). Mechanism of drilling wells with air as the drilling fluid, in *Proceedings of Conference on Drilling and Rock Mechanics*, Austin, TX, January 18–19. Society of Petroleum Engineers, Allen, TX

Soeder, D.J. (1986). Porosity and permeability of Eastern Devonian gas shale. *SPE Formation Evaluation*, 1(1):116–124.

Sumi, L. (2008). *Shale Gas Focus on the Marcellus Shale.* Oil and Gas Accountability Project (OGAP)/Earthworks, Washington, DC, 25 pp.

Swaco, M. (2006). *RECLAIM Technology: The System That Extends the Life of Oil- and Synthetic-Base Drilling Fluids While Reducing Disposal and Environmental Costs.* M-I Swaco, Houston, TX.

USDOE. (2009). *Modern Shale Gas Development in the United States: A Primer*. U.S. Department of Energy, Washington, DC.

USEIA. (2011). *Today in Energy: Haynesville Surpasses Barnett as the Nation's Leading Shale Play*. U.S. Energy Information Administration, Washington, DC (205.254.135.7/todayinenergy/detail.cfm?id=570).

USGS. (2000). *Summary of Water Use in the United States*. U.S. Geological Survey, Washington, DC (http://ga.water.usgs.gov/edu/wateruse2000.html).

USGS. (2008). *National Oil and Gas Assessment*. U.S. Geological Survey, Washington, DC (http://energy.usgs.gov/OilGas/AssessmentsData/NationalOilGasAssessment.aspx).

Veatch, Jr., R.W., Meschovitis, Z.A., and Fast, C.R. (1999). An overview of hydraulic fracturing, in Gidley, J.L. et al., Eds., *Recent Advances in Hydraulic Fracturing*. Society of Petroleum Engineers, Allen, TX.

Vulgamore, T., Clawson, T., Pope, C., Wolhart, W., Mayerhofer, M., and Waltman, C. (2007). *Applying Hydraulic Fracture Diagnostics to Optimize Stimulations in the Woodford Shale*, SPE-110029. Society of Petroleum Engineers, Allen, TX.

Walser, D. (2010). Mapping hydraulic fracturing in the Permian Basin—microseismic mapping pinpoints fracturing and aids well stimulation at drilling sites. *Upstream Pumping Solutions*, Fall (www.upstreampumping.com/article/well-completion-stimulation/mapping-hydraulic-fracturing-permian-basin-geometry-and-placemen).

West Virginia Geological and Economic Survey. (1997). *Enhancement of the Appalachian Basin Devonian Shale Resource Base in the GRI Hydrocarbon Model*, GRI Contract No. 5095-890-3478, prepared for Gas Research Institute, Arlington, VA.

Williams, P. (2008). A vast ocean of natural gas. *American Clean Skies*, Summer, pp. 44–50.

7

Chemicals Used in Hydraulic Fracturing

> After sitting idle for two decades, there's steam billowing from the top of the big old steel plant in Youngstown, Ohio. This does not represent a renewal of the steel production that once created the Rust Belt. Instead, this a product of a new industry proponents say can be a game changer, not just for the depressed Youngstown–Warren area, but for the U.S. economy and the bigger energy game. It is the exploitation of oil shale.
>
> **—Ruth Ravve (2011)**

7.1 Background

In the pressing and ongoing need to produce more natural gas, technological advances have permitted the industry to drill deeper in various directions, tapping into gas reserves with greater facility and profitability. These advances have allowed the mining of vast, newly discovered gas deposits; however, the new technology depends heavily on the use of undisclosed types and amounts of toxic chemicals (Colborn et al., 2010).

7.1.1 Types of Fracturing Fluids and Additives*

Service companies have developed a number of different oil- and water-based fluids and treatments to more efficiently induce and maintain permeable and productive fractures. The composition of these fluids varies significantly, from simple water and sand to complex polymeric substances with a multitude of additives. Each type of fracturing fluid has unique characteristics, and each possesses its own positive and negative performance traits. For ideal performance, fracturing fluids should possess the following four qualities (Powell et al., 1999):

* This section is adapted from USEPA, *The Central Appalachian Coal Basin—Attachment 6: Evaluation of Impacts to Underground Sources of Drinking Water by Hydraulic Fracturing of Coalbed Methane Reserves*, EPA 816-R-04-003, U.S. Environmental Protection Agency, Washington, DC, 2004; USEPA, *Hydraulic Fracturing Study Plan*, U.S. Environmental Protection Agency, Washington, DC, 2011.

- Be viscous enough to create a fracture of adequate width.
- Measure fluid travel distance to extend fracture length.
- Be able to transport large amounts of proppant into the fracture.
- Require minimal gelling agent to allow for easier degradation or "breaking" and reduced cost.

The main fluid categories are

- Gelled fluids, including linear or cross-linked gels
- Foamed gels
- Plain water and potassium chloride (KCl) water
- Acids
- Combination treatments (any combination of two or more of the aforementioned fluids

7.1.1.1 Gelled Fluids

Water alone is not always adequate for fracturing certain formations because its low viscosity limits its ability to transport proppant. In response to this problem, the industry developed linear and cross-linked fluids, which are higher viscosity fracturing fluids. Water gellants or thickeners are used to create these gelled fluids. Gellant selection is based on formation characteristics such as pressure, temperature, permeability, porosity, and zone thickness. These gelled fluids are described in more detail below.

7.1.1.1.1 Linear Gels

A substantial number of fracturing treatments are completed using thickened, water-based linear gels. The gelling agents used in these fracturing fluids are typically guar gum, guar derivatives such as hydroxypropylguar (HPG), and carboxymethylhydroxypropylguar (CMHPG), or cellulose derivatives such as carboxymethylguar or hydroxethylcelluslose (HEC). In general, these products are biodegradable. Guar is a polymeric substance derived from the ground endosperm of seeds of the guar plant (Ely, 1994). Guar gum, also called *guaran,* on its own, is nontoxic and, in fact, is a food-grade product commonly used to increase the viscosity and elasticity of foods such as ice cream, baked goods, pastry fillings, dairy products, meat,

DID YOU KNOW?

Jeff Gelski, of *Good Business News,* has stated that, "The oil industry's increased use of guar gum has driven up its price, leaving food and beverage processors struggling to find supply of the ingredient used in such applications as ice cream and tortillas" (Rohrlich, 2011).

and condiments, in addition to being used in dry soups, instant oatmeal, sweet desserts, and frozen food and animal feed. Its industrial applications include uses in textiles, paper, explosives, pharmaceutical, mining, oil and gas drilling, and other products.

To formulate a viscous fracturing gel, guar powder or concentrate is dissolved in a carrier fluid such as water or diesel fuel. Increased viscosity improves the ability of the fracturing fluid to transport proppant and decreases the need for more turbulent flow. Concentrations of guar gelling agents within fracturing fluids have decreased over the past several years, as it was determined that reduced concentrations provide better and more complete fractures (Powell et al., 1999); this decreased use may make the food industry happier.

Diesel fuel has frequently been used in lieu of water to dissolve the guar powder because its carrying capacity per unit volume is much higher (Haliburton, 2002). Diesel is a common solvent additive, especially in liquid gel concentrates, that is used by many service companies for continuous delivery of gelling agents in fracturing treatments (Penny and Conway, 1996). Diesel does not enhance the efficiency of the fracturing fluid—it is merely a component of the delivery system, and using diesel instead of water minimizes the number of transport vehicles required to carry the liquid gel to the site (Haliburton, 2002). Based on typical practice and observation, the percentage of diesel fuel in the slurried thickener can range between 30% and almost 100%. Diesel fuel is a petroleum distillate that may contain known carcinogens. One such component of diesel fuel is benzene, which, according to literature sources, can make up anywhere between 0.003 and 0.1% by weight of diesel fuel (Clark and Brown, 1977; Morrison & Associates, 2001). Slurried diesel and gel are diluted with water prior to injection into the subsurface. The dilution is approximately 4 to 10 gal of concentrated liquid gel (guar slurried in diesel) per 1000 gal of make-up water to produce an adequate polymer slurry (CIS, 2001; USEPA, 2004).

7.1.1.1.2 Cross-Linked Gels

The development of cross-linked gels in 1968 was one of the major advances in fracturing fluid technology (Ely, 1994). When cross-linking agents are added to linear gels, the result is a complex, high-viscosity fracturing fluid that provides higher proppant transport performance than do linear gels (Ely, 1994; Messina, 2001; USEPA, 2004). Cross-linking reduces the need for fluid thickener and extends the viscous life of the fluid indefinitely. The fracturing fluid remains viscous until a breaking agent is introduced to break the cross-linker and eventually the polymer. Although cross-linkers make the fluid more expensive, they can considerably improve hydraulic fracturing performance. Cross-linked gels are typically metal ion–cross-linked guar (Ely, 1994). Service companies have used metal ions such as chromium, aluminum, titanium, and other metal ions to achieve cross-linking, and low-residue (cleaner) forms of cross-linked gels, such as cross-linked hydroxypropylguar, have been developed (Ely, 1994). Cross-linked gels may contain

DID YOU KNOW?

The final concentration of cross-linkers is typically 1 to 2 gal of cross-linker per 1000 gal of gel (USEPA, 2004).

boric acid, sodium tetraborate decahydrate, ethylene glycol, and monoethylamine. These constituents are hazardous in their undiluted form and can cause kidney, liver, heart, blood, and brain damage through prolonged or repeated exposure. According to a Bureau of Land Management environmental impact statement, cross-linkers may contain hazardous constituents such as ammonium chloride, potassium hydroxide, zirconium nitrate, and zirconium sulfate (USDOI, 1998).

DID YOU KNOW?

Using foam under high pressure in gas reservoirs has an advantage over high-pressure water injection because it does not create as much damage to the formation, and well clean-up operations are less costly.

7.1.1.2 Foamed Gels

Foam fracturing technology uses foam bubbles to transport and place proppant into fractures. The most widely used foam fracturing fluids employ nitrogen or carbon dioxide as their base gas. Incorporating inert gases with foaming agents and water reduces the amount of fracturing liquid required. Foamed gels use fracturing fluids with higher proppant concentrations to achieve highly effective fracturing. The gas bubbles in the foam fill voids that would otherwise be filled by fracturing fluid. The high concentrations of proppant allow for an approximately 75% reduction in the overall amount of fluid that would be necessary using a conventional linear or cross-linked gel (Ely, 1994; USEPA, 2004). Foaming agents can be used in conjunction with gelled fluids to achieve an extremely effective fracturing fluid. Foam emulsions experience high leakoff; therefore, typical protocol involves the addition of fluid-loss agents, such as fine sands (Ely, 1994; USEPA, 2004). Foaming agents suspend air, nitrogen, or carbon dioxide within the aqueous phase of a fracturing treatment. The gas/liquid ratio determines if a fluid will be true foam or simply a gas-energized liquid (Ely, 1994). Carbon dioxide can be injected as a liquid, whereas nitrogen must be injected as a gas to prevent freezing (USEPA, 2004). Foaming agents can contain diethanolamine and alcohols such as isopropanol, ethanol, and 2-butoxyethanol. They can also contain such hazardous substances as glycol ethers (USDOI, 1998). One type of foaming agent can cause negative liver and kidney effects. The final concentration is typically 3 gal of foamer per 1000 gal of gel (USEPA, 2004).

DID YOU KNOW?

Similar to plain water, another fracturing fluid uses water with potassium chloride (KCl), which is harmless if ingested at lower concentrations, in addition to small quantities of gelling agents, polymers, and surfactants (Ely, 1994).

7.1.1.3 Water and Potassium Chloride Water Treatments

Many shale gas service companies use groundwater pumped directly from the formation or treated water for their fracturing jobs. In some well stimulations, proppants are not needed to prop fractures open, so simple water or slightly thickened water can be a cost-effective substitute for an expensive polymer of foam-based fracturing fluid with proppant (Ely, 1994). Hydraulic fracturing performance is not exceptional with plain water, but, in some cases, the production rates achieved are adequate. Plain water has a lower viscosity than gelled water, which reduces proppant transport capacity.

7.1.1.4 Acids

Acids are used in limestone formations that overlay or are interbedded within shale gas formations to dissolve the rock and create a conduit through which formation water and shale gas can travel. Typically, the acidic stimulation fluid is hydrochloric acid or a combination of hydrochloric and acetic or formic acid. For acid fracturing to be successful, thousands of gallons of acid must be pumped far into the formation to etch the face of the fracture; some of the cellulose derivatives used as gelling agents in water and water/methanol fluids can be used in acidic fluids to increase treatment distance (Ely, 1994). Note that acids may also be used as a component of breaker fluids. In addition, acid can be used to clean perforations of the cement surrounding the well casing prior to fracturing fluid injection. The cement is perforated at the zone of injection to ease fracturing fluid flow into the formation (Halliburton, 2002; USEPA, 2004). Acids, such as formic and hydrochloric acids, are corrosive and can be extremely hazardous in concentrated form. Acids are substantially diluted with water-based or water- and gas-based fluids prior to injection into the subsurface. The injected concentration is typically 1000 times weaker than the concentrated versions (USEPA, 2004).

7.1.1.5 Use of Chemical Additives

Several fluid additives have been developed to enhance the efficiency and increase the success of fracturing fluid treatments. Chemicals are used not only in drilling for shale gas but also in fracking fluids for the purposes listed in Table 7.1.

TABLE 7.1

Fracking Fluid Additions and Function

Addition	Function
Acids	Acids, typically hydrochloric acid, are pumped into the formation to dissolve some of the rock material (minerals and clays) to clean out pores and to allow gas and fluid to flow more readily into the well.
Biocides	Because the watery fracking fluid used to fracture rocks gets hotter when pumped into the ground at high pressure and high speed, bacteria and mold multiply. When the bacteria grow, they secrete enzymes that break down the gelling agent, which causes a black slime or ooze to form in the lines and reduces viscosity. Reduced viscosity translates into poor proppant placement and poor fracturing performance. Biocides work to reduce the formation of bacteria and mold.
Breakers	Water-based gels are used in the gas industry to viscosify fluids used in the fracking of production wells, where they serve to increase the force applied to the rock and to improve the transport of proppants used to maintain the fracture after formation. The various types of breakers include time-release breakers and temperature-dependent breakers. Most breakers are typically acids, oxidizers, or enzymes (Messina, 2001) According to a Bureau of Land Management environmental impact statement, breakers may contain hazardous constituents, including ammonium persulfate, ammonium sulfate, copper compounds, ethylene glycol, and glycol ethers (USDOI, 1998). After fracturing, the gel must be degraded to a low viscosity with enzymes or gel breakers (Barati et al., 2011).
Clay stabilizers	These additives prevent clay swelling in the shale rocks and minimize migration of clay fines.
Corrosion inhibitors	These additives reduce the potential for rusting in pipes and casings. Corrosion inhibitors are required in acid fluid mixtures because acids will corrode steel tubing, well casings, tools, and tanks. The solvent acetone is a common additive in corrosion inhibitors (Penny and Conway, 1996). Corrosion inhibitors are quite hazardous in their undiluted form. These products are diluted to a concentration of 1 gal per 1000 gal of make-up water and acid mixture (USEPA, 2004). Acids and acid corrosion inhibitors are used in very small quantities in fracturing operations (500 to 2000 gal per treatment).
Cross-linkers	Cross-linkers are used to thicken fluids, often with metallic salts, to increase viscosity and improve proppant transport.

Defoamers	These additives are used to reduce foaming after it is no longer needed in order to lower surface tension and allow trapped gas to escape.
Foamers	Foamers are used to increase carrying capacity while transporting proppants and decreasing the overall volume of fluid required.
Fluid-loss additives	These additives restrict leakoff of the fracturing fluid into the exposed rock at the fracture face. Because the additives prevent excessive leakoff, fracturing fluid effectiveness and integrity are maintained. Fluid-loss additives of the past and present include bridging materials such as 100-mesh sand, 100-mesh soluble resin, and silica flour, or plastering materials such as starch blends, talc silica flour, and clay (Ely, 1994).
Friction reducers	These additives make water slick, minimize the friction created under high pressure, and increase the rate and efficiency of moving the fracking fluid. Friction reducers are typically latex polymers or copolymers of acrylamides. They are added to slickwater treatments (water with solvent) at concentrations of 0.25 to 2.0 lb per 1000 gal (Ely, 1994). Some examples of friction reducers are oil-soluble anionic liquid, cationic polyacrilate liquid, and cationic friction reducer (Messina, 2001).
Gellants	Gellants are used to increase viscosity and suspend sand during proppant transport.
pH control	The addition of buffers maintains pH and ensures maximum effectiveness of various additions.
Proppants	Proppants, usually composed of sand and occasionally glass beads, prop or hold fissures open, allowing gas to flow out of the cracked formation. An ideal proppant should produce maximum permeability in a fracture. Fracture permeability is a function of proppant grain roundness, proppant purity, and crush strength. Larger proppant volumes allow for wider fractures, which facilitate more rapid flowback to the production well. Over a period of 30 minutes, 4500 to 15,000 gal of fracturing fluid will typically transport and place approximately 11,000 to 25,000 lb of proppant into the fracture (Powell et al., 1999).
Scale control	Scale control prevents the buildup of mineral scale that can block fluid and gas passage through the pipes.
Surfactants	These additives are used to decrease liquid surface tension and improve fluid passage through pipes in either direction.

DID YOU KNOW?

As a result of the growing use of hydraulic fracturing, natural gas production in the United States reached 21,577 billion cubic feet in 2010, a level not achieved since a period of high natural gas production between 1970 and 1974 (USEIA, 2012a).

DID YOU KNOW?

The U.S Energy Information Administration (USEIA) projects that the United States possesses 2552 trillion cubic feet of potential natural gas resources, enough to supply the country for approximately 110 years. Natural gas from shale resources accounts for 827 trillion cubic feet of this total, which is more than double what the USEIA estimated in 2010 (USEIA, 2012a).

Hydraulic fracturing creates access to more natural gas supplies, but the process requires the use of large quantities of water and fracturing fluids, which are injected underground at high volumes and pressure. Oil and gas service companies design fracturing fluids to create fractures and transport sand or other granular substances to properly open the fractures. The composition of these fluids varies by formation, ranging from a simple mixture of water and sand to more complex mixtures with amplitude of chemical additives. Fracking companies may use these chemical additives (see Table 7.1) to thicken or thin the fluids, improve the flow of the fluid, or kill bacteria that can reduce fracturing performance (USEPA, 2004). Some of these chemicals, if not disposed of safely or if allowed to leach into the drinking water supply, could damage the environment or pose a risk to human health. During hydraulic fracturing, fluids containing chemicals are injected deep underground, where their migration is not entirely predictable. Well failures (such as those due to the use of insufficient well casing) could lead to the release of these fluids at shallower depths, closer to drinking water supplies. Although some fracturing fluids are removed from the well at the end of the fracturing process, a substantial amount remains underground (Veil, 2010).

DID YOU KNOW?

Pennsylvania's Department of Environmental Protection has cited Cabot Oil & Gas Corporation for contamination of drinking water wells with seepage caused by weak casing or improper cementing of a natural gas well (Lustgarten, 2009).

Although most underground injections of chemicals are subject to the protections of the Safe Drinking Water Act (SDWA), Congress in 2005 modified the law to exclude "the underground injection of fluids or propping agents (other than diesel fuels) pursuant to hydraulic fracturing operations related to oil, gas, or geothermal production activities" from the Act's protection (42 USC §300h(d)). Unless oil and gas service companies use diesel in the hydraulic fracturing process, the permanent underground injection of chemicals used for hydraulic fracturing is not regulated by the USEPA.

DID YOU KNOW?

Many have dubbed 42 USC §300h(d) the *Halliburton Loophole* because of Halliburton's ties to then Vice President Cheney and its role as one of the largest providers of hydraulic fracturing services.

Concerns also have been raised about the ultimate outcome of chemicals that are recovered and disposed of as wastewater. This wastewater is stored in tanks or pits at the well site, where spills are possible (Urbina, 2011; USEPA, 2012b). For final disposition, well operators must recycle the fluids for use in future fracturing jobs, inject the fluids into underground storage wells (which, unlike the fracturing process itself, are subject to the Safe Drinking Water Act), discharge them to nearby surface water, or transport them to wastewater treatment facilities (Veil, 2010). A recent article raised questions about the safety of surface water discharge and the ability of water treatment facilities to process wastewater from natural gas drilling operations (Urbina, 2011).

DID YOU KNOW?

Wyoming recently enacted relatively strong disclosure regulations, requiring disclosure on a well-by-well basis and, for each stage of the well stimulation program, the chemical additives, compounds, and concentrations or rates proposed to be mixed and injected (WCWR 055-000-003 Sec. 45). Similar regulations recently became effective in Arkansas (Arkansas Oil and Gas Commission Rule B-19). In Wyoming, much of this information, after an initial period of review, is available to the public. Other states, however, do not insist on such robust disclosure. West Virginia, for example, has no disclosure requirements for hydraulic fracturing and expressly exempts fluids used during fracking from the disclosure requirements applicable to the underground injection of fluids for purposes of waste storage.

Any risk or impact to the environment and human health posed by fracking fluids depends in large part on their contents. Federal law, however, contains no public disclosure requirements for oil and gas produces or service companies involved in hydraulic fracturing, and state disclosure requirements vary greatly. Although the industry has recently announced that it will soon create a public database of fluid components, reporting to this database is strictly voluntary, disclosure will not include the chemical identity of products labels as proprietary, and there is no way to determine if companies are accurately reporting information for all wells.

The absence of a minimum national baseline for disclosure of fluids injected during the hydraulic fracturing process and the exemption of most hydraulic fracturing injections from regulation under the Safe Drinking Water Act has left an information void concerning the contents, chemical concentrations, and volumes of fluids that go into the ground during fracturing operations and return to the surface in the form of wastewater. As a result, regulators and the public arc unable to effectively assess any impact the use of these fluids may have on the environment or public health.

7.1.2 Naturally Occurring Radioactive Material*

Before presenting a detailed discussion of the major chemical constituents that make up hydraulic fracking fluids currently in use it is important to briefly discuss naturally occurring radioactive material (NORM) that could be involved in the fracking process. Some soils and geologic formations contain low levels of radioactive material. This naturally occurring radioactive material emits low levels of radiation to which everyone is exposed on a daily basis. Radiation from natural sources is also referred to as *background radiation*. Other sources of background radiation include radiation for space and sources that occur naturally in the human body. This background radiation accounts for about 50% of the total exposures for Americans. Most of this background exposure is from radon gas encountered in homes (35% of the total exposure). The average person in the United States is exposed to about 360 millirem (mrem) of radiation from natural sources each year (a mrem, or 1/1000 of a rem, is a measure of radiation exposure) (RRC, 2012). The other 50% of exposures for Americans comes primarily from medical sources. Consumer products and industrial and occupational sources contribute less than 3% of the total exposure (NCRP, 2009).

In addition to the background radiation normally found at the surface of the Earth, NORM can also be brought to the surface in the natural gas production process. When NORM is associated with oil and natural gas production, it begins as small amounts of uranium and thorium within the rock. These elements, along with some of their decay elements, notably radium-226 and

* This section is adapted from USDOE, *Modern Shale Gas Development in the United States: A Primer*, U.S. Department of Energy, Washington, DC, 2009.

radium-228 (USGS, 1999), can be brought to the surface in drill cuttings and produced water. Radon-222, a gaseous decay element of radium, can come to the surface along with the shale gas.

When NORM is brought to the surface, it remains in the rock pieces of the drill cuttings, remains in solution with produced water, or, under certain conditions, precipitates out in scales or sludge. The radiation from this NORM is weak and cannot penetrate dense materials such as the steel used in pipes and tanks (Smith et al., 1996). The principal concern for NORM in the oil and gas industry is that, over time, it can become concentrated in field production equipment (API, 2004) and as sludge or sediment inside tanks and process vessels that have an extended history of contact with formation water (BSEEC, 2012). Because the general public does not come into contact with oilfield equipment for extended periods, there is little exposure risk from oilfield NORM. Studies have shown that exposure risks for workers and the public are low for conventional oil and gas operations (BSEEC, 2012; Smith et al., 1996).

If measured NORM levels exceed state regulatory levels or U.S. Occupational Safety and Health Administration (OSHA) exposure dose risks (29 CFR 1910.1096), the material is taken to licensed facilities for proper disposal. In all cases, OSHA requires employers to evaluate radiation hazards, post caution signs, and provide personal protection equipment for workers when radiation doses could exceed 5 mrem in 1 hour or 100 mrem in any 5 consecutive days. In addition to these federal worker protections, states have regulations that require operators to protect the safety and health of both workers and the public.

Currently, no existing federal regulations specifically address the handling and disposal of NORM wastes. Instead, states producing oil and gas are responsible for promulgating and administering regulations to control the reuse and disposal of NORM-contaminated equipment, produced water, and oilfield wastes. Although regulations vary by state, generally, if NORM concentrations are less than regulatory standards, operators are allowed to dispose of the material by methods approved for standard oilfield waste. Conversely, if NORM concentrations are above regulatory limits, then the material must be disposed of at a licensed facility. These regulations, standards, and practices ensure that oil and gas operations present negligible risk to the general public with respect to potential NORM exposure. They also present negligible risk to workers when proper controls are implemented (Smith et al., 1996).

7.2 Fracking Fluids and Their Constituents

In 2011, the U.S. Congressional Committee on Energy and Commerce published a report, *Chemicals Used in Hydraulic Fracturing*, which lauded hydraulic fracturing as a new technological device in the ongoing pursuit of oil

DID YOU KNOW?

Each hydraulic fracturing product is a mixture of chemicals or other components designed to achieve a certain performance goal, such as increasing the viscosity of water. Some oil and gas service companies create their own products, but most purchase these products from chemical vendors. The service companies then mix these products together at the well site to formulate the hydraulic fracturing fluids that they pump underground.

and natural gas products. Moreover, the report pointed out that hydraulic fracturing has opened access to vast domestic reserves of natural gas that could provide an important stepping stone to a clean energy future. Yet, the Committee also observed that questions about the safety of hydraulic fracturing persist and are compounded by the secrecy surround the chemicals used in fracking fluids. The report indicated that between 2005 and 2009, the 14 leading hydraulic fracturing companies in the United States (Basic Energy Services, BJ Services, Calfrac Well Services, Complete Production Services, Frac Tech Services, Halliburton, Key Energy Services, RPC, Sanjel Corporation, Schlumberger, Superior Well Services, Trican Well Service, Universal Well Services, and Weatherford) used over 2500 hydraulic fracturing products containing 750 compounds. More that 650 of these products contained chemicals that are known or possible human carcinogens, are regulated under the Safe Drinking Water Act for their risks to human health, or are listed as hazardous air pollutants under the Clean Air Act. Overall, these companies used 780 million gal of hydraulic fracturing products in their fluids in this period of time. This volume does not include water that the companies added to the fluids at well sites before injection.

Hydraulic fracturing products are comprised of a wide range of chemicals. Some are seemingly harmless, such as sodium chloride (salt), gelatin, walnut hulls, instant coffee, and citric acid. Others, though, could pose severe risks to human health or the environment.

DID YOU KNOW?

Along with walnut hulls and instant coffee, the amazing list of fracturing fluid additives includes coffee grinds, salt, ceramic balls, lead petroleum distillates, methanol, benzene, toluene, xylene, and millions of gallons of diesel. The problem is that, in the absence of a minimum U.S national baseline for disclosure of fracking fluids combined with a special industry exemption from U.S. water safety standards, it's nearly impossible to assess any impact the use of these fluids may have on the environment or public health. It's frackin' impossible (Nikiforuk, 2011).

Some of the components are surprising. One company told the Congressional Committee that it used instant coffee as one of the components of a fluid designed to inhibit acid corrosion. Two companies reported using walnut hulls as part of a breaker (recall that a breaker is a product used to degrade the fracturing fluid viscosity, which helps to enhance post-fracturing fluid recovery). Another company reported using carbohydrates as a breaker. One company used tallow soap (soap made from beef, sheep, or other animals) to reduce the loss of fracturing fluid into the exposed rock.

7.2.1 Commonly Used Chemical Components

The most widely used chemical in hydraulic fracturing from 2005 to 2009, as measured by the number of products containing the chemical, was methanol. Methanol is a hazardous air pollutant and a candidate for regulation under the Safe Drinking Water Act. It was a component in 342 hydraulic fracturing products. Some of the other most widely used chemicals include isopropyl alcohol, which was used in 274 products, and ethylene glycol, which was in 119 products. Crystalline silica (silicon dioxide) appeared in 207 products, generally proppants used to hold open fractures. Table 7.2 provides a list of the most commonly used compounds in hydraulic fracturing fluids.

Hydraulic fracturing companies used 2-butoxyethanol (2-BE) as a foaming agent or surfactant in 126 products. According to USEPA scientists, 2-BE is easily absorbed and rapidly distributed in humans following inhalation, ingestion, or dermal exposure. Studies have shown that exposure to 2-BE can cause hemolysis (destruction of red blood cells) and damage to the spleen, liver, and bone marrow (USEPA, 2010a). The hydraulic fracturing companies injected 21.9 million gal of products containing 2-BE between 2005 and 2009. The highest volume of products containing 2-BE was found in Texas, which accounted for more than half of the volume used. The USEPA recently found this chemical in drinking water wells tested in Pavillion, Wyoming (USEPA, 2010b). Table 7.3 shows the use of 2-BE by state.

TABLE 7.2

Chemical Components Appearing Most Often in Hydraulic Fracturing Products (2005–2009)

Chemical Component	No. of Products Containing Chemical
Methanol (methyl alcohol)	342
Isopropanol (isopropyl alcohol, propan-2-ol)	274
Crystalline silica (quartz) (SiO_2)	207
Ethylene glycol monobutyl ether (2-butoxyethanol)	126
Ethylene glycol (1,2-ethanediol)	119
Hydrotreated light petroleum distillates	89
Sodium hydroxide (caustic soda)	80

TABLE 7.3
States with the Highest Volume of Hydraulic Fracturing Fluids Containing 2-Butoxyethanol (2005–2009)

State	Fluid Volume (gal)
Texas	12,031,734
Oklahoma	2,186,613
New Mexico	1,871,501
Colorado	1,147,614
Louisiana	890,068
Pennsylvania	747,416
West Virginia	464,231
Utah	382,874
Montana	362,497
Arkansas	348,959

DID YOU KNOW?

2-Butoxyethanol is an organic solvent that is produced by monoethoxylation of butanol:

$$C_2H_4O + BuOH \rightarrow BuOC_2H_4OH$$

7.2.2 Toxic Chemicals

The oil and gas service companies used hydraulic fracturing products containing 29 chemicals that are (1) known or possible human carcinogens, (2) regulated under the Safe Drinking Water Act for their risks to human health, or (3) listed as hazardous air pollutants (HAPs) under the Clean Air Act. These 29 chemicals were components of 652 different products used in hydraulic fracturing. Table 7.4 lists these toxic chemicals and their frequency of use.

DID YOU KNOW?

Diesel contains benzene, toluene, ethylbenzene, and xylene (USEPA, 2004).

7.2.2.1 Carcinogens

Between 2005 and 2009, the hydraulic fracturing companies used 95 products containing 13 different carcinogens. These included naphthalene (a possible human carcinogen), benzene (a known human carcinogen), and acrylamide

TABLE 7.4

Chemicals Components of Concern: Carcinogens, SDWA-Regulated Chemicals, and Hazardous Air Pollutants

Chemical Component	Chemical Category	No. of Products
Methanol (methyl alcohol)	HAP	342
Ethylene glycol (1,2-ethanediol)	HAP	119
Diesel	Carcinogen, SDWA, HAP	51
Naphthalene	Carcinogen, HAP	44
Xylene	SDWA, HAP	44
Hydrogen chloride (hydrochloric acid)	HAP	42
Toluene	SDWA, HAP	29
Ethylbenzene	SDWA, HAP	28
Diethanolamine (2,2-iminodiethanol)	HAP	14
Formaldehyde	Carcinogen, HAP	12
Sulfuric acid	Carcinogen	9
Thiourea	Carcinogen	9
Benzyl chloride	Carcinogen, HAP	8
Cumene	HAP	6
Nitrilotriacetic acid	Carcinogen	6
Dimethyl formamide	HAP	5
Phenol	HAP	5
Benzene	Carcinogen, SDWA, HAP	3
Di(2-ethylhxyl)phthalate	Carcinogen, SDWA, HAP	3
Acrylamide	Carcinogen, SDWA, HAP	2
Hydrogen fluoride (hydrofluoric acid)	HAP	2
Phthalic anhydride	HAP	2
Acetaldehyde	Carcinogen, HAP	1
Acetophenone	HAP	1
Copper	SDWA	1
Ethylene oxide	Carcinogen, HAP	1
Lead	Carcinogen, SDWA, HAP	1
Propylene oxide	Carcinogen, HAP	1
p-Xylene	HAP	1
Number of products containing a component of concern		652

Note: HAP, hazardous air pollutant; SDWA, Safe Drinking Water Act.

DID YOU KNOW?

Here, a chemical is considered a carcinogen if it is on one of two lists: (1) substances identified by the National Toxicology Program as "known to be human carcinogens" or as "reasonably anticipated to be human carcinogens"; and (2) substances identified by the International Agency for Research on Cancer, part of the World Health Organization, as "carcinogenic" or "probably carcinogenic" to humans (IARC, 2011; NTP, 2005).

TABLE 7.5
States with at Least 100,000 gal of Hydraulic Fracturing Fluids Containing a Carcinogen (2005–2009)

State	Fluid Volume (gal)
Texas	3,877,273
Colorado	1,544,388
Oklahoma	1,098,746
Louisiana	777,945
Wyoming	759,898
North Dakota	557,519
New Mexico	511,186
Montana	394,873
Utah	382,338

(a probable human carcinogen). Overall, these companies injected 102 million gal of fracturing products containing at least one carcinogen. The companies used the highest volume of fluids containing one or more carcinogens in Texas, Colorado, and Oklahoma. Table 7.5 shows the use of these chemicals by state.

7.2.2.2 Safe Drinking Water Act Chemicals

Under the Safe Drinking Water Act, the USEPA regulates 53 chemicals that may have an adverse effect on human health and are known to or are likely to occur in public drinking water systems at levels of public health concern. Between 2005 and 2009, the hydraulic fracturing companies used 67 products containing at least one of eight SDWA-regulated chemicals. Overall, these companies injected 11.7 million gal of fracturing products containing at least one chemical regulated under SDWA. Most of these chemicals were injected in Texas. Table 7.6 shows the use of these chemicals by state.

TABLE 7.6
States with at Least 100,000 gal of Hydraulic Fracturing Fluids Containing a SDWA-Regulated Chemical (2005–2009)

State	Final Volume (gal)
Texas	9,474,631
New Mexico	1,157,721
Colorado	375,817
Oklahoma	202,562
Mississippi	108,809
North Dakota	100,479

The vast majority of these SDWA-regulated chemicals were the BTEX compounds—benzene, toluene, xylene, and ethylbenzene. The BTEX compounds appeared in 60 hydraulic fracturing products used between 2005 and 2009 and were used in 11.4 million gal of hydraulic fracturing fluids. The Department of Health and Human Services, the International Agency for Research on Cancer, and the USEPA have determined that benzene is a human carcinogen (ATSDR, 2007). Chronic exposure to toluene, ethylbenzene, or xylene can also damage the central nervous system, liver, and kidneys (USEPA, 2012b).

In addition, it is common to use diesel in hydraulic fracturing fluids; for example, the hydraulic fracturing companies injected more than 32 million gal of diesel fuel or hydraulic fracturing fluids containing diesel fuel in wells in 19 states. The USEPA has stated that the "use of diesel fuel in fracturing fluids poses the greatest threat" to underground sources of drinking water (USEPA, 2004). Diesel fuel contains toxic constituents, including BTEX compounds; thus, the use of diesel fuel should be avoided. According to the company Halliburton, "Diesel does not enhance the efficiency of the fracturing fluid; it is merely a component of the delivery system." According to the USEPA, it is technologically feasible to replace diesel with nontoxic delivery systems, such as plain water (Earthworks, 2012). The USEPA has created candidate contaminant lists of contaminants that are currently not subject to national primary drinking water regulations but are known or anticipated to occur in public water systems and may require regulation under the Safe Drinking Water Act in the future (USEPA, 2012a). These lists include, among others, pesticides, disinfection byproducts, chemicals used in commerce, waterborne pathogens, pharmaceuticals, and biological toxins. Nine listed chemicals were pertinent to or used in hydraulic fracking between 2005 and 2009: 1-butanol, acetaldehyde, benzyl chloride, ethylene glycol, ethylene oxide, formaldehyde, methanol, *n*-methyl-2-pyrrolidone, and propylene oxide.

7.2.2.3 Hazardous Air Pollutants

The Clean Air Act (CAA) requires the USEPA to control the emission of 187 hazardous air pollutants, which are pollutants that cause or may cause cancer or other serious health effects, such as reproductive effects or birth defects, or adverse environmental and ecological effects (CAA Section 112(b), 42 USC §7412). Between 2005 and 2009, the hydraulic fracturing companies used 595 products containing 24 different hazardous air pollutants.

Hydrogen fluoride is a hazardous air pollutant that is a highly corrosive and systemic poison that causes severe and sometimes delayed health effects due to deep tissue penetration. Absorption of substantial amounts of hydrogen fluoride by any route may be fatal (ASTDR, 2011). One of the hydraulic fracturing companies used 67,222 gal of two products containing hydrogen fluoride in 2008 and 2009. Lead, a heavy metal, is a hazardous air pollutant that is particularly harmful to children's neurological development. It can also cause health problems in adults, including reproductive problems, high blood pressure, and nerve disorders (USEPA, 2012b). Methanol was the hazardous air pollutant that appeared most often in hydraulic fracturing products. Other hazardous air pollutants used in hydraulic fracturing fluids included formaldehyde, hydrogen chloride, and ethylene glycol.

7.3 Trade Secret and Proprietary Chemicals

Many chemical components of hydraulic fracturing fluids used by fracking service companies are listed on the Material Safety Data Sheets (MSDSs) as "proprietary" or "trade secret." OSHA's 29 CFR 1910.1200 (Hazard Communication Standard) section (i) states:

> (i)(1) The chemical manufacturer, importer, or employer may withhold the specific chemical identity, including the chemical name and other specific identification of a hazardous chemical, from the Material Safety Data Sheet (MSDS), provided that:
>
> (i)(1)(i) The claim that the information withheld is a trade secret can be supported;
>
> (i)(1)(ii) Information contained in the Material Safety Data Sheet concerning the properties and effects of the hazardous chemical is disclosed;
>
> (i)(1)(iii) The Material Safety Data Sheet indicates that the specific chemical identity is being withheld as a trade secret; and,
>
> (i)(1)(iv) The specific chemical identity is made available to health professionals, employees, and designated representatives in accordance with the applicable provisions of this paragraph.

Information on chemicals used during oil and gas development can also be obtained from Tier II reports and from websites such as Frac Focus or state agency sites; however, Colborn et al. (2010) of the Endocrine Disruption Exchange have enumerated several problems with the information in MSDS and Tier II reports. MSDSs and Tier II reports are fraught with gaps in information about the formulation of the products. OSHA provides only general guidelines for the format and content of MSDSs, and manufacturers of the products are left to determine what information is revealed on their MSDSs. The forms are not submitted to OSHA for review unless they are part of an inspection under the Hazard Communication Standard. Some MSDSs report little or no information about the chemical composition of a product. Those MSDSs that do may report only a fraction of the total composition, sometimes less than 0.1%. Some MSDSs provide only a general description of the content, such as "plasticizer" or "polymer," while others describe the ingredients as "proprietary" or just a chemical class. Under the existing regulatory system, all of the above identifiers are permissible; consequently, it is not surprising that a study by the U.S. General Accounting Office revealed that MSDSs could easily be inaccurate and incomplete. Tier II reports can be similarly uninformative, as reporting requirements vary from state to state, country to country, and company to company. Some Tier II forms include only a functional category name (e.g., "weight materials" or "biocides") with no product name. The percent of the total composition of the product is rarely reported on these forms.

7.4 Thought-Provoking Discussion Questions

1. Should chemical manufacturers be allowed to claim "trade secret" instead of revealing constituents of chemicals used for fracking?
2. We do not know what we do not know about the damage caused by fracking fluids. Does this statement make sense? Is it realistic? Explain.
3. Are OSHA's safety standards, if properly followed, adequate to protect hydraulic fracturing workers?
4. How should fracturing fluids be disposed of?
5. Should fracturing fluids be treated in conventional wastewater treatment plants?

References and Recommended Reading

API. (2004). *Naturally Occurring Radioactive Material in North American Oilfields.* American Petroleum Institute, Washington, DC.

ATSDR. (2007). *Public Health Statement for Benzene.* Agency for Toxic Substances & Disease Registry, Washington, DC.

ATSDR. (2011). *Medical Management Guidelines for Hydrogen Fluoride.* Agency for Toxic Substances and Disease Registry, Atlanta, GA (www.atsdr.cdc.gov/mmg/mmg.asp?id=1142&tid=250).

Barati, R., Jonson, S.J., McCool, S., Green, D.W., Willhite, G.P., and Liang, J.T. (2011). Fracturing fluid cleanup by controlled release of enzymes for polyelectrolyte complex nanoparticles. *Journal of Applied Polymer Science*, 121:1292–1298.

BSEEC. (2012). *Environment*. Barnett Shale Energy Education Council, Fort Worth, TX (www.Bseec.org/index.php/content/facts/environment).

CIS. (2001). *Hydraulic Fracturing Site Visit Notes: Western Interior Coal Region, State of Kansas*. Consolidated Industrial Services, Signal Hill, CA.

Clark, R.C. and Brown, D.W. (1977) Petroleum: properties and analysis in biotic and abiotic systems, in Malins, D.C., Ed., *Effects of Petroleum on Arctic and Subartic Environments and Organisms*. Vol. 1. *Nature and Fate of Petroleum*. Academic Press, New York, pp. 1–89.

Colborn, T., Kwiatkowski, C., Schultz, K., and Bachran, M. (2010). Natural gas operations from a public health perspective. *International Journal of Human and Ecological Risk Assessment*, 17(5):1039–1056.

Earthworks. (2012). *Hydraulic Fracturing 101*. Earthworks, Washington, DC (www.earthworksaction.org/issues/detail/hydraulic_fracturing_101).

Ely, J.W. (1994). *Stimulation Engineering Handbook*. PennWell, Tulsa, OK, 357 pp.

Ely, J.W., Zbitowski, R.I., and Zuber, M.D. (1990). How to develop a coalbed methane prospect: a case study of an exploratory five-spot well pattern in the Warrior basin, Alabama, in *Proceedings of 1990 SPE Annual Technical Conference and Exhibition*, New Orleans, LA, September 23–26, SPE Paper 206666. Society of Plastics Engineers, Newtown, CT, pp. 487–496.

Halliburton. (2002). Personal communication with Halliburton staff, fracturing fluid experts Joe Sandy, Pat Finley, and Steve Almond.

IARC. (2011). *Agents Classified by the IARC Monographs*, Vols. 1–104. International Agency for Research on Cancer, Lyon, France.

Lustharten, A. (2009). Officials in three states pin water woes on gas drilling. *Publica*, April 26 (www.Propublica.org/article/officials-in-three-states-pin-water-woes-on-gas-drilling-426).

Messina. (2001). *Fracturing Chemicals*. Messina, Dallas, TX (www.messinachemicals.com/).

Morrison, R. & Associates. (2001). Does diesel #2 fuel oil contain benzene? *Environmental Tool Box*, Fall.

NCRP. (2009). *Ionizing Radiation Exposure of the Populations of the United States*, Report No. 160. National Council on Radiation Protection & Measurements, Bethesda, MD.

Nikiforuk, A. (2011). Truth comes out on "fracking" toxins. *The Tyee*, April 20 (http://thetyee.ca/Opinion/2011/04/20/FrackingToxins/).

NTP. (2005). *The Report on Carcinogens*, 11th ed. Public Health Service, National Toxicology Program, Washington, DC.

Penny, G.S. and Conway, M.W. (1996). *Coordinated Studies in Support of Hydraulic Fracturing of Coalbed Methane*, Final Report #GRI-95/0283, prepared by STIM-Lab. Gas Research Institute, Chicago, IL.

Powell, R.J., McCabe, M.A., Salbaugh, B.F., Terracina, J.M., Yaritz, J.G., and Ferrer, D. (1999). Applications of a new, efficient hydraulic fracturing fluid system. *SPE Production and Facilities*, 14(2):139–143.

Ravve, R. (2011). Shale oil in America: economy fix or dangerous fantasy? *Foxnews.com*, www.foxnews.com/us/2011/12/27/shale-oil-in-america-economy-fix-or-dangerous-fantasy/.

Rohrlich, J. (2011). Food industry blames fracking for guar gum bubble. *Minyanville*, www.minyanville.com/dailyfeed/2011/09/02/food-industry-blames-fracking-for/.

RRC. (2012). *NORM—Naturally Occurring Radioactive Material*. Railroad Commission of Texas, Austin (www.rrc.state.tx.us/environmental/publications/norm/index.php).

Smith, K.P., Blunt, D.L., Williams, G.P., and Tebes, C.L. (1996). *Radiological Dose Assessment Related to Management of Naturally Occurring Radioactive Materials Generated by the Petroleum Industry*. Environmental Assessment Division, Argonne National Laboratory, Argonne, IL.

Urbina, I. (2011). Regulation lax as gas wells' tainted water hits rivers. *The New York Times*, February 26, p. A1.

USDOE. (2009). *Modern Shale Gas Development in the United States: A Primer*. U.S. Department of Energy, Washington, DC.

USDOI. (1998). *Glenwood Springs Resource Area: Oil & Gas Leasing & Development—Draft Supplemental Environmental Impact Statement*. Bureau of Land Management, Colorado State Office, Lakewood.

USEIA. (2012a). *Natural Gas: U.S. Dry Natural Gas Production*. U.S. Energy Information Administration, Washington, DC (www.eia.gov/dnav/ng/hist/n9070us1A.htm).

USEIA. (2012b). *What Is Shale Gas and Why Is It Important?* U.S. Energy Information Administration, Washington, DC (www.eia.doe.gov/energy_in_brief/about_shale_gas.cfm).

USEPA. (2004). *The Central Appalachian Coal Basin—Attachment 6: Evaluation of Impacts to Underground Sources of Drinking Water by Hydraulic Fracturing of Coalbed Methane Reserves*, EPA 816-R-04-003. U.S. Environmental Protection Agency, Washington, DC, 26 pp.

USEPA. (2010a). *Toxicological Review of Ethylene Glycol Monobutyl Ether*. U.S. Environmental Protection Agency, Washington, DC (www.epa.gov/iris/toxreviews/0500tr.pdf).

USEPA. (2010b). *Pavillion, Wyoming, Groundwater Investigation: January 2010 Sampling Results and Site Update*. U.S. Environmental Protection Agency, Washington, DC.

USEPA. (2011). *Hydraulic Fracturing Study Plan*. U.S. Environmental Protection Agency, Washington, DC.

USEPA. (2012a). *Water: Contaminant Candidate List 3*. U.S. Environmental Protection Agency, Washington, DC (http://water.epa.gov/scitech/drinkingwater/dws/ccl/ccl3.cfm).

USEPA. (2012b). *Basic Information about Toluene in Drinking Water* (http://water.epa.gov/drink/contaminants/basicinformation/toluene.cfm); *Basic Information about Ethylbenzene in Drinking Water* (http://water.epa.gov/drink/contaminants/basicinformation/ethylbenzene.cfm); *Basic Information about Xylenes in Drinking Water* (http://water.epa.gov/drink/contaminants/basicinformation/xylenes.cfm); *Basic Information about Lead in Drinking Water* (http://water.epa.gov/drink/contaminants/basicinformation/lead.cfm). U.S. Environmental Protection Agency, Washington, DC.

USGS. (1999). *Naturally Occurring Radioactive Materials (NORM) in Produced Water and Oil-Field Equipment—An Issue for the Energy Industry*. U.S. Geological Survey, Washington, DC (pubs.usgs.gov/fs/fs-0142-99/fs-0142-99.pdf).

Veil, J.A. (2010). *Water Management Technologies Used by Marcellus Shale Gas Producers*, ANL/EVS/R-10/3, prepared by the Environmental Science Division, Argonne National Laboratory, for the U.S. Department of Energy, Office of Fossil Energy, National Energy Technology Laboratory, Washington, DC.

8

Environmental Considerations

> The country is flush with natural gas as a result of new drilling techniques that have enabled energy companies to tap vast supplies that were out of reach not so long ago. The country's natural gas surplus has been growing even as the country burns record amounts.
>
> **—Fahey (2012)**

> The Rule of Capture generally permits a landowner to drain or "capture" oil and natural gas from a neighboring property without liability or recourse.
>
> **—The Clark Law Firm, Fort Worth, TX**

8.1 Introduction

As described throughout this text, natural gas is an important part of the nation's energy supply. As a clean-burning, affordable, and reliable source of energy, natural gas will continue to play a significant role in the energy supply picture for years to come. Unconventional sources of natural gas (e.g., shale gas) have become a major component of that future supply, and shale gas is rapidly emerging as a critical part of that resource. There exists an extensive framework of federal, state, and local requirements designed to manage virtually every aspect of the natural gas development process (see Chapter 9). These regulatory efforts are primarily led by state agencies and include such things as ensuring conservation of gas resources, prevention of waste, and protection of the rights of both surface and mineral owners while protecting the environment (Arkansas Oil and Gas Commission, 2008). As part of their environmental protection mission, state agencies are responsible for safeguarding public and private water supplies, preserving air quality, addressing safety, and ensuring that wastes from drilling and production are properly contained and disposed of (Ohio Revised Code, Chapter 1509: Division of Oil and Gas Resources Management—Oil and Gas). At the federal level, the U.S. Environmental Protection Agency (USEPA) is involved with many of the same environmental concerns as local and state governments. At the present time, the USEPA is conducting various hydraulic fracturing case studies (listed and described in Appendix B) to determine potential impacts of fracking on drinking water resources. In order to make, and comprehend the reasoning behind, the decisions involved in developing shale gas operations, it is important to understand the process of drilling and producing shale gas wells (Figure 8.1) and the attendant environmental considerations. A key element in the emergence of shale gas production has been the refinement of cost-effective horizontal drilling and hydraulic fracturing technologies. These two processes, along with the implementation of protective best management practices (BMPs), have allowed shale gas development to move into areas that previously would have been inaccessible. Accordingly, it is important to understand the technologies and practices employed by the industry and their ability to prevent or minimize the potential effects of shale gas development on human health and the environment and on the quality of life in the communities in which shale gas production is located.

Many of the human and environmental considerations associated with shale gas production are common to all oil and gas development; however, the horizontal drilling and hydraulic fracturing that have become the standard for modern shale gas development bring with them new considerations as well as new ways to reduce impacts. New challenges have been encountered as shale gas development has spread into more densely populated areas, and new technologies and practices have been developed to meet these challenges. In addition, collaborations among the industry, regulators, and the public have created innovative environmental solutions to problems that at first seemed insurmountable.

8.2 Access Roads, Well Pads, and Mineral Rights

One consideration associated with traditional gas development has been the surface disturbance required for access roads and well pads. As described in greater detail below, horizontal drilling provides a means to significantly reduce surface disturbance and a host of related concerns. Other considerations associated with traditional oil and gas development are the conflicts that arise from split estates. In some instances, mineral rights and surface rights are not owned by the same party, a condition referred to as *split estate* or *severed minerals*. Split estate is more prevalent in western states, where the federal government owns much of the mineral rights (ALL Consulting and MBOGC, 2004). In the Midwest and in eastern states, where shale gas development resources are more prevalent, only 4% of the lands are associated with a federal split estate (USDOI, 2006); instead, private–private split estate scenarios frequently occur, where the surface owner differs from the mineral estate owner. In these cases, the mineral owner may be another individual or

Mineral Leasing

Companies negotiate a private contract or lease that allows mineral development and compensates the mineral owners. Lease terms vary and can contain stipulations or mitigation measures pertinent to protect various resources. (*several weeks to years*)

Permits

The operator must obtain a permit authorizing the drilling of a new well. Surveys, drilling plans, and other technical information are frequently required for a permit application. The approved permit may require site-specific environmental protection measures. Other permits such as water withdrawal or injection permits may also be required. (*several weeks to months*)

Road and Pad Construction

Once permits are received, roads are constructed to access the wellsite. Well pads are constructed to safely locate the drilling rig and associated equipment during the drilling process. Pits may be excavated to contain drilling fluids. (*several days to weeks*)

Drilling and Completion

A drilling rig drills the well and multiple layers of steel pipe (called casing) are put into the hole and cemented in place to protect fresh water formation. (*weeks or months*)

Hydraulic Fracturing

A specially designed fracturing fluid is pumped under high pressure into the shale formation. The fluid consists primarily of water along with a proppant (usually sand) and about 2% or less of chemical additives. This process creates fractures in rock deep underground that are "propped" open by the sand, which allows the natural gas to flow into the well. (*days*)

Production

Once the well is placed in production, parts of the wellpad that are no longer needed for future operations are reclaimed. The gas is brought up the well, treated to a useable condition, and sent to market. (*interim reclamation, days; production, years*)

Workovers

Gas production usually declines over the years. Operators may perform a workover, which is an operation to clean, repair, and maintain the well for the purposes of increasing or restoring production. Mulitiple workovers may be performed over the life of a well. (*several days to weeks*)

Plugging and Abandonment/Reclamation

Once a well reaches its economic limit, it is plugged and abandoned according to shale standards. The disturbed areas, including well pads and access roads, are reclaimed back to the native vegetation and contours or to conditions requested by the surface owner. (*reclamation activity, days; full restoration, years*)

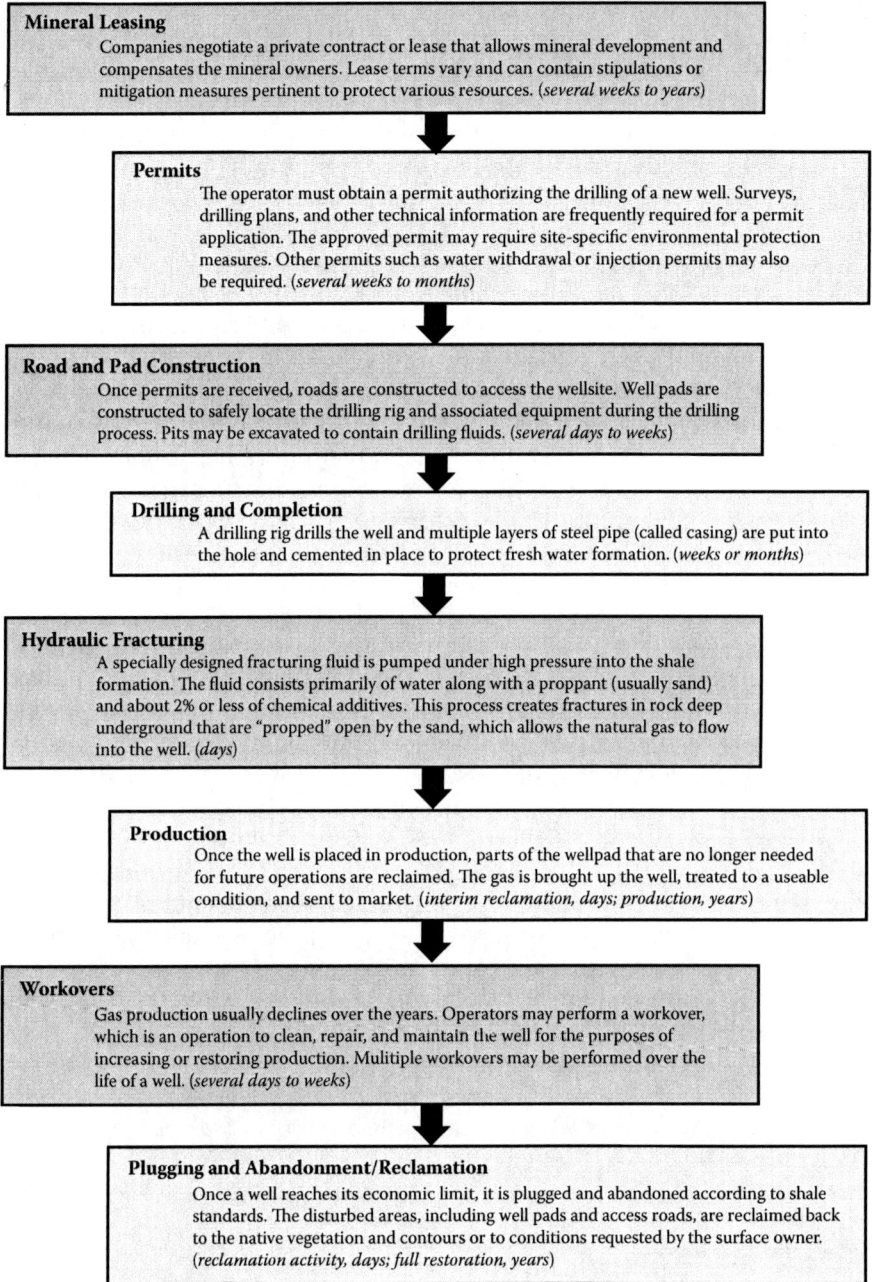

FIGURE 8.1
Process of shale gas development.

a business enterprise such as a coal company. A split estate situation, regardless of its nature, can result in conflicts—especially in areas where active mineral resource development is not commonplace. Landowners can be surprised to find that mineral lease holders are entitled to reasonable use of the land surface even though they do not own the surface. It is important, however, to understand that surface owners who do not own mineral rights are still afforded certain protections. If the mineral owner does not own the surface where drilling will occur, a separate agreement may be negotiated (in some states it is required) with the landowner to ensure that he or she is compensated for the use of the land and to set requirements for reclaiming the land when operations are complete (Bonner and Willer, 2005).

Shale gas development within or near existing communities has created challenges for production companies, but new technologies have generally allowed these challenges to be met successfully. In some cases, a combination of modern shale gas technologies and the innovative use of BMPs has been required to allow development to continue without compromising highly valued community resources. In one instance, Chesapeake Energy Corporation constructed a well pad to develop natural gas from the Barnett Shale play near a popular Fort Worth community area known as Trinity Trails. Located on private land, Trinity Trails consists of a 35-mile network of paved and natural surface pathways. The drilling pad was constructed approximately 200 feet from one portion of the trail. During the initial planning stages, proposed use of this land for development of natural gas was met with significant opposition by the public. Maintaining healthy populations of upland hardwood forest habitat was important to the community because such woodlots are rare in urban settings. To address the concerns of the community, the company sponsored public meetings and opinion surveys, provided landscape plans, planted trees and shrubs, and enhanced the general area by improving irrigation and lowering maintenance requirements. The well pad was specifically designed to be as small as possible in order to reduce the footprint of the well. Special construction practices were used to help preserve many of the existing trees. The construction zone was isolated from view using a 16-foot barrier fence with sound baffling. This approach benefited both partners: The company was able to produce the shale gas, important community resources were protected, and at no point in the process was any portion of the trail closed (Lantz, 2008).

8.3 Horizontal Wells

To adequately describe the environmental considerations that accompany shale gas development and the technologies and practices that are in place to prevent, minimize, or mitigate impacts, it is important to describe the general process of development with an emphasis on horizontal drilling and shale gas hydraulic fracturing technologies. Thus, in the following discussions, a brief review of emerging shale gas technologies and practices is provided with a focus on environmental mitigation procedures.

Modern shale gas development is a technologically driven process for the production of natural gas resources. Currently, shale gas wells include both vertical and horizontal wells. The emerging shale gas basins are expected to follow a trend similar to the Barnett Shale play, with increasing numbers of horizontal wells as plays mature (Franz and Jochen, 2005; Halliburton, 2008; Harper, 2008). The technologies used by operators to drill shale gas wells are similar to the drilling techniques that have been industry standards for drilling conventional gas wells. Both horizontal drilling and hydraulic fracturing are established technologies with significant track records; horizontal drilling dates back to the 1930s, and hydraulic fracturing has a history dating back to the 1950s (Harper, 2008). The key difference between a shale gas well and a conventional gas well is the reservoir stimulation (large-scale hydraulic fracturing) performed on shale gas wells (Halliburton, 2008).

The evolution of the Barnett Shale play toward favoring horizontal wells resulted from improvements in technology combined with the economic benefits of greater reservoir exposures that a horizontal well provides compared to a vertical well. Although both well types may be used to recover the resource, shale gas operators are increasingly relying on horizontal well completions to optimize recovery and well economics (Harper, 2008). Horizontal drilling provides more exposure to a formation than does a vertical well. In the Marcellus Shale in Pennsylvania, for example, a vertical well may be exposed to as little as 50 feet of formation, while a horizontal well may have a lateral wellbore extending in length from 2000 to 6000 feet within the 50- to 300-foot-thick formation (Harper, 2008). This increase in reservoir exposure creates a number of advantages over vertical well drilling.

A wide range of factors influence choosing between a vertical or horizontal well. Vertical wells may require a smaller capital investment on a per-well basis, but production is often less economical. A vertical well may cost as much as $800,000 (excluding pad and infrastructure) to drill compared to a horizontal well that can cost $2.5 million or more (excluding pad and infrastructures) (Marshall Miller & Associates, 2008).

8.3.1 Reducing Surface Disturbance

It is too early in shale gas drilling and production process history to determine the final well spacing that will most efficiently recover the gas resource in all basins; however, current experience indicates that the use of horizontal well technology will significantly decrease the total environmental disturbance. As a case in point, consider that complete development of a 1-square mile section could require 16 vertical wells, each located on a separate well pad. Alternatively, six to eight horizontal wells (potentially more), drilled from only

DID YOU KNOW?

The low natural permeability of shale requires vertical wells to be developed at closer spacing intervals than conventional gas reservoirs in order to effectively manage the resource. This can result in initial development of vertical wells at spacing intervals of 40 acres per well, or less, to efficiently drain the gas resources from the tight shale reservoirs.

one well pad, could access the same reservoir volume, or even more (Parshall, 2008). In addition, horizontal drilling can significantly reduce the overall number of well pads, access roads, pipeline routes, and production facilities required, thus minimizing habitat fragmentation, impacts to the public, and the overall environmental footprint. Personal observations of an active drilling site in the Barnett Shale play and discussions with onsite operators revealed that the use of horizontal wells has allowed the service companies involved to replace three, four, or more vertical wells with a single horizontal well.

The spacing interval for vertical wells in the shale gas plays averages 40 acres per well for initial development, and the average spacing interval for horizontal wells is approximately 160 acres per well. A 640-acre section of land, therefore, could be developed with a total of 16 vertical wells, each on its own individual well pad, or by as few as 4 horizontal well all drilled from a single multiple-well drilling pad. Analysis performed in 2008 for the U.S. Department of Interior estimated that a shallow vertical gas well completed in the Fayetteville Shale in Arkansas would require a 2-acre well pad, 0.10 miles of road, and 0.55 miles of utility corridor, resulting in a total of 4.8 acres of disturbance per well (USDOI, 2008). A horizontal well pad in Arkansas could occupy approximately 3.5 acres plus roads and utilities, resulting in a total of 6.9 acres. Slightly larger multiple-well pads are often used for horizontal wells to minimize the environmental footprint, resulting in a 0.5-acre increase in the size of the area disturbed. It has been estimated that a horizontal 4-well pad and necessary roads and utilities would disturb approximately 7.4 acres, whereas 16 vertical wells would disturb approximately 77 acres. In this example, 16 vertical wells would disturb more than 10 times the area of 4 horizontal wells to produce the same resource volume. This difference in development footprint when considered in terms of both rural and urban development scenarios highlights the desire for operators to move toward horizontal development of shale gas plays.

8.3.2 Reducing Wildlife Impacts

Research has documented that activities associated with gas development can affect wildlife and its habitat during the exploration, development, operations, and abandonment phases (Bromley, 1985). The development of shale gas production utilizing horizontal wells and multiple-well pads not only reduces surface area disturbances by reducing the total number of wells drilled and well pad sites constructed, but also results in fewer roadways and utility corridors. This overall reduction in the footprint of a project results in significantly less habitat disturbance while allowing for more operation flexibility. Furthermore, by drilling underneath sensitive areas such as wetlands areas near streams and rivers and wilderness habitats, gas can be produced without disturbing some of these resources.

This ability to reduce surface disturbance is especially important in certain critical habitats; for example, certain portions of New York (e.g., Catskill Park, Shawangunk Ridge, Hudson Highlands, Poconos) are dominated by hardwood forests, which are important wildlife habitats susceptible to fragmentation (Catskill Mountainkeeper, 2008). In addition, state regulations and, in some cases, local ordinances include operational restrictions to provide added protection for wildlife or sensitive resources. In the city of Flower Mound, Texas, ordinances have been adopted to protect the surface resources and allow for future growth of the community without detracting from the land value or sense of community. These ordinances prevent construction in or near streams or rivers, floodplains, and sensitive upland forests to protect wildlife species and their associated habitats.

At the state level, special plans or waivers are required when surface use actions may affect threatened or endangered species. Such waivers must demonstrate that contemplated disturbances will not adversely impact the species in question. In Pennsylvania, wildlife is further protected on state lands by the Pennsylvania Game Commission through the use of lease agreements that require, whenever feasible, the use of existing timber and maintenance roads to access wells and avoidance of areas such as wetlands and unique and critical habitats for threatened or endangered species (Venesky, 2008). When disturbances to wildlife habitat are unavoidable, energy companies mitigate land disturbances by implementing land reclamation practices to restore disturbed land to original conditions. In general, reclamation practices (or mitigation measures) designed to protect and maintain wildlife will depend on project features, regional characteristics, and the potentially affected species; however, because technologies associated with modern shale gas development can reduce impacts in the first place, the need for additional protective restoration measures may also be reduced. Regardless of the situation, the timely reclamation of disturbed lands (e.g., reseeding, land contouring, re-vegetating) can minimize short- and long-term disturbances to natural habitat (USDOI and USDA, 2007).

8.3.3 Reducing Community Impacts

In the initial planning phase of development, states, local governments, and industry can work together to minimize long-term effects and to address citizen concerns such as traffic congestion, damage to roads, dust, and noise

(ALL Consulting and MBOGC, 2004). The process of shale gas development, especially drilling and hydraulic fracturing, can create short-term increases in traffic volume, dust, and noise. These nuisance impacts are usually limited to the initial 20- to 30-day drilling and completion period (Perryman Group, 2007). Along with increases in traffic volume, damage to road surfaces can occur if design parameters for traffic volume and weight loads are exceeded. Where these effects are an issue, developers have worked with authorities to adjust work schedules to help alleviate congestion, to water unpaved roads to reduce dust, and to adjust the timing of some operations and install special sound barriers to reduce noise for nearby residents. When feasible, developers can also use avoidance practices to help minimize traffic congestion on heavily traveled roads. In the Barnett Shale play around the Dallas–Fort Worth International Airport, operators have constructed permanent pipelines to transfer produced water from well sites to disposal facilities, thereby reducing traffic and potential damage to roads (Satterfield et al., 2008). When these practices are coupled with the benefits of multiple directional wells from fewer pads, the number of access roads and associated traffic can be further reduced.

In many cases, developers have negotiated to compensate local municipalities for road damage that does occur as a result of their activities. Alternatively, they may negotiate road maintenance and repair agreements to ensure that damage to roadways is repaired and that the cost is absorbed by the drilling enterprises (Douglas, 2008). In their 2007 study of the Barnett Shale play, The Perryman Group noted that traffic volume is a legitimate concern in the area, but developers have been effectively addressing the issue through maintenance agreements so road repairs are not adversely affecting local taxpayers.

From a traffic perspective, members of the public or local municipalities often have the ability to limit traffic volume in residential areas by developing restrictions in neighborhood lease agreements or by developing ordinances that prevent road construction in certain areas, respectively. In urban areas, these agreements can be used to coordinate local traffic patterns to minimize congestion, control speed limits to address safety concerns, and specify weight zones to reduce road damage. With continued advances in technologies, modern developers are afforded a higher level of drilling flexibility than in the past. This provides producers with the ability to adjust their operations plans, allowing them to access drilling locations that would otherwise be inaccessible. Although drilling circumstances vary by geologic region and well locations, in many cases shale gas plays are being developed with both vertically and horizontally drilled wells. Based on the current development activities of active shale gas basins, horizontal drilling has become the preferred method of drilling in most shale gas plays. Horizontal wells have also been used in many areas of the country to remotely access natural gas resources beneath existing infrastructure, buildings, environmentally sensitive areas, or other features that would prevent the use of vertical wells.

The development of the Barnett Shale near the Dallas–Fort Worth International Airport is a prime example of how development of urban areas is possible with horizontal wellbores (Parshall, 2008). Development practices there have been altered to suit local ordinances implemented to lessen community impacts and protect environmental resources. These ordinances include detailed setbacks from residences, roadways, churches, and schools, as well as the use of BMPs to control visual and noise impacts, including the required use of directional lighting. Typically, drilling operations in rural gas development areas continue around the clock until the well is completed. When these same operations moved into urban areas around the Texas cities of Arlington, Burleson, Cleburne, Fort Worth, Joshua, and North Richland Hills, specific ordinances were developed requiring additional permitting, well setbacks from properties, daytime and nighttime limits, and directional lighting. Well sites must be illuminated for worker safety, but directing the light downward and shielding the surrounding areas prevents illuminating neighboring residences, roads, or other buildings (Chesapeake Energy Corporation, 2008).

In a similar concept, drilling rigs are also being outfitted with blanket-like enclosures that act as acoustic barriers to reduce engine noise. Sound-deadening technology is a BMP that is also being applied to reduce noises from compressor facilities in both rural and urban settings (ALL Consulting and MBOGC, 2002); for example, sound barriers may be manufactured from alternative building materials with integral sound-absorbing properties. Such BMPs are not appropriate for all operations, though, and they must be applied on a case-by-case basis. In some cases, a given BMP may actually be counterproductive. In other cases, a particular BMP may create other environmental, safety, or operational problems that must be weighed against each other. While BMPs have certain benefits in certain situations, they cannot be universally applied or required.

8.3.4 Protecting Groundwater

State oil and gas regulatory programs place great emphasis on protecting groundwater. Current well construction requirements consist of installing multiple layers of protective steel casing and cement that are specifically designed and installed to protect freshwater aquifers and to ensure that the producing zone is isolated from overlying formations. During the drilling process, a conductor casing and surface casing string are set in the borehole and cemented in place (Figure 8.2). In some instances, additional casing strings, known as intermediate casings, may be installed (ALL Consulting and MBOGC, 2008). After each string casing is set and prior to drilling any deeper into the borehole, the casing is cemented to ensure a good seal between the casing and formation or between two strings of casings (Bellabarba et al., 2008). The multiple strings of casing, layers of cement, and the production tubing prevent contamination of freshwater zones and ensure that the gas

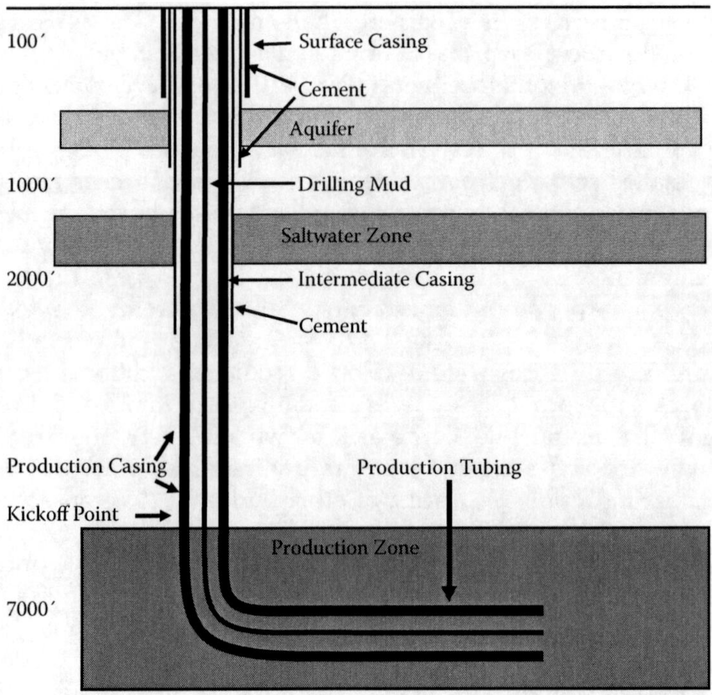

FIGURE 8.2
Casing and cementing (not to scale). (Adapted from Arthur, J.D. et al., *An Overview of Modern Shale Gas Development in the United States*, ALL Consulting, Tulsa, OK, 2008.)

resource does not flow into other, lower pressure zones around the outside of the casing rather than flowing up the well (Railroad Commission of Texas, 2008). The conductor casing serves as a foundation for the well construction and prevents caving of surface soils. The surface casing is installed to seal off potential freshwater-bearing zones; this isolation is necessary to protect aquifers from drilling mud and produced fluids.

As a further protection of freshwater zones, air-rotary drilling is often used when drilling through this portion of the wellbore to ensure that no drilling mud comes into contact with the freshwater zone. Intermediate casings, when installed, are used to isolate non-freshwater-bearing zones from the producing wellbore. Intermediate casings may be necessary because of a naturally overpressured zone or because of a saltwater zone located at depth. The borehole area below an intermediate casing may be uncemented until just above the kickoff point for the horizontal leg. This area of wellbore is typically filled with drilling muds.

Each string of casing serves as a layer of protection, separating the fluids inside and outside of the casing and preventing them from coming into contact. Operators perform a variety of checks to ensure that the desired isolation of each zone is occurring, including verifying that the casing used has sufficient strength and that the cement has properly bonded to the casing (Bellabarba et al., 2008). These checks may include acoustic cement bond logs and pressure testing to ensure the mechanical integrity of casings. Additionally, state oil and gas regulatory agencies often specify the required depth of protective casings and regulate the time that is required for cement to set prior to additional drilling. These requirements are typically based on regional conditions and are established for all wildcat wells and may be modified when field rules are designated.

8.4 Thought-Provoking Discussion Questions

1. Are the casing and cementing procedures used in sealing the wellbore from contamination of groundwater adequate?
2. Was Vice President Dick Cheney's exclusion of hydraulic fracturing from the 2005 Safe Drinking Water Act an acceptable move?
3. The USEPA is conducting several case studies to determine the environmental impact of hydraulic fracturing. What do you think they will find?
4. If the USEPA case studies determine that hydraulic fracturing is causing harm to groundwater sources, what action(s) should be taken?

References and Recommended Reading

ALL Consulting and Montana Board of Oil and Gas Conservation (MBOGC). (2002). *Handbook on Best Management Practices and Mitigation Strategies for Coal Bed Methane in the Montana Portion of the Powder River Basin*, prepared for National Energy Technology Laboratory, National Petroleum Technology Office, U.S. Department of Energy, Tulsa, OK, 44 pp.

ALL Consulting and Montana Board of Oil and Gas Conservation (MBOGC). (2004). *Coal Bed Natural Gas Handbook: Resources for the Preparation and Review of Project Planning Elements and Environmental Documents*, prepared for National Petroleum Technology Office, U.S. Department of Energy, Washington, DC, 182 pp.

Arkansas Oil and Gas Commission. (2008). *Mission Statement*. Arkansas Oil and Gas Commission, Little Rock (http://www.aogc.state.ar.us/mission.pdf).

Arthur, J.D., Bohm, B., and Lane, M. (2008). *An Overview of Modern Shale Gas Development in the United States*, ALL Consulting, Tulsa, OK.

Bellabarba, M., Bulte-Loyer, H., Froelich, B., Roy-Delage, S., Van Kuijk, R. et al. (2008). Ensuring zonal isolation beyond the life of the well. *Oilfield Review*, 20(1):18–31.

Bonner, T. and Willer, L. (2005). Royalty management 101: basics for beginners and refresher course for everyone, in *Proceedings of the 25th Anniversary Convention in National Association of Royalty Owners* (NARO), Oklahoma City, OK, November 3–5.

Bromley, M. (1985). *Wildlife Management Implications of Petroleum Exploration and Development in Wildland Environments*, General Technical Report INT-199. Forest Service, Intermountain Research Station, U.S. Department of Agriculture, Ogden, UT.

Catskill Mountainkeeper. (2008). *The Marcellus Shale—American's Next Super Giant*. Catskill Mountainkeeper, Youngsville, NY (www.catskillmountainkeeper.org/our-programs/fracking/marcellus-shale/).

Chesapeake Energy Corporation. (2008). Drilling 101, paper presented to New York Department of Environmental Conservation.

Douglas, D. (2008). *Anger over Road Damage Caused by Barnett Shale Development*. WFAA-TV, Dallas–Fort Worth, TX (www.wfaa.com/news/local/64791902.html).

Fahey, J. (2012). Natural gas price plunge aids families, businesses. *Associated Press*, January 15.

Franz, Jr., J.H. and Jochen, V. (2005). *When Your Gas Reservoir Is Unconventional, So Is Our Solution*, Shale Gas White Paper 05-OF-299. Schlumberger, Houston, TX.

Halliburton. (2008). *U.S. Shale Gas: An Unconventional Resource. Unconventional Challenges*. Halliburton, Houston, TX, 8 pp.

Harper, J. (2008). The Marcellus Shale—an old "new" gas reservoir in Pennsylvania. *Pennsylvania Geology*, 28(1):2–13.

Lantz, G. (2008). Drilling green along Trinity Trails. *The Barnett Shale Magazine*, Summer.

Marshall Miller & Associates. (2008). Marcellus Shale, paper presented to Fireside Pumpers, Bradford, PA.

Parshall, J. (2008). Barnett Shale showcases tight-gas development. *Journal of Petroleum Technology*, 60(9):48–55.

Perryman Group, The. (2007). *Bounty from Below: The Impact of Developing Natural gas Resources Associated with the Barnett Shale on Business Activity in Fort Worth and the Surrounding 14-Country Area*. The Perryman Group, Waco, TX.

Railroad Commission of Texas. (2008). *Newark, East (Barnett Shale) Field*. Oil and Gas Division, Railroad Commission of Texas, Austin.

Satterfield, J., Mantell, M., Kathol, D., Hebert, F., Patterson, K., and Lee, R. (2008). Managing Water Resource's Challenges in Select Natural Gas Shale Plays, paper presented at the Ground Water Protection Council Annual Forum, Cincinnati, OH, September 21–24.

USDOI. (2006). *Scientific Inventory of Onshore Federal Lands' Oil and Gas Resources and the Extent and Nature of Restrictions or Impediments to Their Development*, prepared by Bureau of Land Management, U.S. Department of the Interior, Washington, DC.

USDOI. (2008). *Arkansas: Reasonably Foreseeable Development Scenario for Fluid Minerals*. Eastern States Field Office, Bureau of Land Management, U.S. Department of the Interior, Jackson, MS.

USDOI and USDA. (2007). *Surface Operating Standards and Guidelines for Oil and gas Exploration and Development: The Gold Book*, 4th ed., BLM/WO/ST-06/021+3071/REV07. Bureau of Land Management, U.S. Department of the Interior, Denver, CO.

Venesky, T. (2008). State-owned parcels eyed for gas deposits. *Times Leader*, March 4 (www.timesleader.com/stories/State-owned-parcels-eyed-for-gas-deposits,110359).

9

Laws and Regulations Affecting Shale Gas Development and Operations

> Natural gas doesn't run our lives: As a fossil fuel used to generate electricity, it's in a distant second place behind coal. But with new domestic gas sources—much of it made available thanks to hydraulic fracturing—that all could change.
>
> **—Svoboda (2010)**

> Today the term "regulation" is apt to conjure up images of every manager's headache: a network of confusing and constraining rules and standards; costly modification of existing installations to meet new legal demands; inspections, fines, or time-consuming legal hearings; and, above all, an increasingly burdensome task of record keeping and paperwork.
>
> **—Ferry (1990)**

9.1 Introduction*

There is one word in our current vernacular that is responsible for a high level of consternation and trepidation among many industrial organizations (this is especially true for the oil and gas drilling industry—the so-called privileged upper 1%) while at the same time is a blessing, a God-sent protective device against those bad-boy industrialists who dare to foul the air we breathe and the water we drink (shame on them!). That Earth-shaking word? *Regulation*, of course.

Awhile back, Ferry (1990) voiced a familiar refrain: "We do not like rules and regulations. We are free, we have choices, and we intend to make them—we don't like others telling us what we can or cannot do." We can extend this idea further: "Yes, nobody should be able to tell us what to do. We have choices, and we should be able to make them." The problem is that we often give little thought to their repercussions when we make choices that satisfy our immediate wants and desires. Sound familiar? Most of us know we need

* This section is adapted from Spellman, F.R. and Whiting, N., *Safety Engineering: Principles and Practices*, 2nd ed., Government Institutes Press, Latham, MD, 2005.

rules and regulations to live with other people in our society, but the fact is we don't like rules—and we often don't abide by them. (Been on an interstate lately? The speed limit may be 65 mph, but you'll find that most people are going 70, 75, 85, or faster. The statistics say we are far less safe on the road at that speed, but rules are made to be broken—right?)

From management's point of view, there are good rules and bad rules. Management likes rules requiring workers to show up on time, to put in an honest day's work, to maintain good order and discipline in the workplace, to focus on company goals. Rules that are good for the company are obviously good rules—right?

So what are the bad rules? Typically, a business manager views any rule, law, regulation, or other requirement placed upon his or her company by an outside regulatory agency as a bad rule. Why? Primarily for the "headache-making" problems pointed out by Ferry in the chapter's opening statement. For the purposes of illustration, let's consider the federal Occupational Safety and Health (OSH) Act and break these problems down one by one. After reading the following sections, you can make your own judgment as to what the real problem is. Keep in mind that the examples listed below could just as well be applied to the Clean Water Act, Clean Air Act, Safe Drinking Water Act, National Environmental Policy Act, and others.

9.1.1 A Network of Confusing and Constraining Rules and Standards

Anyone who has attempted to read and then to comply with 29 CFR 1910, Occupational Safety and Health Standards for General Industry, will probably agree with Terry's statement. Much of the material contained within this Occupational Safety and Health Administration (OSHA) bible is indeed difficult to comprehend, a problem that is greatly exacerbated for those who have very limited safety experience. The problem is compounded by the ambiguity and vagary that contribute to the warp and woof of the twisted fabric whose tightness depends almost entirely on how the material is interpreted by the reader. More importantly, how might this material that has been interpreted and in use by a company be interpreted by OSHA auditors? Note that the OSHA auditor usually has the final word on interpretation.

9.1.2 Costly Modifications of Existing Installations to Meet New Legal Demands

This item not only is a headache-generator for any site manager but can also be very costly to the company in terms of both money and workers' time. Companies are in business, obviously, to make money—to hold or improve their bottom line. The last thing any manager who is fighting competition and other costly impediments to making his or her company profitable wants is to have "those briefcase-carrying so and so's coming into my site and telling *me* I have to have this and I have to have that. Not only do they waste my time, but they also make me spend money on things that cut profits—things that don't contribute to the bottom line." Have you heard anything similar before?

9.1.3 Inspections, Fines, or Time-Consuming Legal Hearings

A typical OSHA workplace inspection (commonly called an *audit*) and to a degree the citations that can lead to fines being assessed on the employer for noncompliance are, for the manager, a trying experience. How trying?

> You're in a business under OSHA's regulatory supervision. You've heard the industry horror stories about auditors called in on employee complaints, who ask if you mind if they have a "look around"—and later, thousands of dollars worth of fines and hours of fear, pain, and/or aggravation later, they leave again (hopefully not to return). But maybe you don't think that can happen to you. Believe me, it can. If you are at all casual, careless, or haphazard about required compliance, you're running an OSHA risk. Even facilities that make the strongest possible attempts to comply—that are, in fact, in compliance down to the dotting of the i's and crossing of the t's—will be cited by OSHA for an interpretation of the smallest detail on 3rd or 4th level instructions. Better that than having OSHA come down heavy—but a headache generator at the very least (Spelllman, 1998).

This rather pointed assessment of OSHA and its auditing process may seem silly or ridiculous to many. But have you been there? If not, then the reality is beyond imagination. If you have been there, then you may feel that this description is a rather mild one, an understatement.

The authority of OSHA is nothing to ignore. Formal regulatory inspections and fines that result from any noncompliance findings are not only costly but also a major contributor to every manager's headache—the one that begins at the base of your skull and makes even your eyebrows ache, the one that results from dealing with regulatory requirements. But, much more contributes to management's dilemma in dealing with regulatory requirements; for example, consider the legal ramifications of regulatory noncompliance. In addition to the civil or criminal penalties that might result from employer violations of or noncompliance with OSHA regulations, possible legal actions may result from noncompliance (not uncommon in this age of "let me sue you before you sue me"). Employees can sue an employer for making them work in an unsafe workplace, for making them perform unsafe work actions, or for injuries incurred while working on the job.

In addition to regulatory penalties, an employer may be exposed to workers' compensation liability for employee injuries. Another potential headache generator can be product liability. Most managers need not be told that the company can be held liable if a product it manufactures or sells causes personal injury or property damage to buyers or third parties; however, it does sometimes surprise managers (but usually not for long) that this liability is not lessened if the firm produces finished products from components manufactured by someone else. The simple fact is that if a component causes injury, the firm that assembled and sold the product can be held liable. This is, of course, a major concern and consideration for any company that produces a product to be sold to any consumer. One of an employer's worst nightmares (and sometimes its most significant headache generator) can be employee complaints—especially when these complaints are made to legal counsel and eventually in a court of law. Actually, whenever an employee decides (for whatever reason) that he or she is going to take legal action against an employer (whether the employer is in the right or wrong) in a court of law, the headache soon turns into the migraine variety.

In some instances, of course, employees should take legal action against their employers. Some employers have absolutely no regard for the safety, health, and wellbeing of their employees. In these cases, the court house door is wide open for litigants and their lawyers. How bad can this situation be in the real world? Just check your local newspaper articles and television news stories to get an answer to this question. Almost daily, an employee or group of employees sues their employer or employers for some infraction of safety and health regulations. To be fair to both sides, many of the suits brought by employees against their employer are frivolous. Some employees make false claims against their employers because they don't like the employer, a supervisor, company policies, or the results of last night's football game, or they feel they have been improperly disciplined or terminated—for whatever reason. Lawsuits generated by disgruntled employees are fairly common, with or without cause, but one cannot overlook those cases when employees have just cause to sue their employers.

9.1.4 Above All, an Increasingly Burdensome Task of Recordkeeping and Paperwork

Safety professionals must learn to avoid certain perceptions, certain so-called "rules of thumb," as well as other gobbledygook commonly accepted as fact. We commonly hear, for example, that "some people are accident prone." This statement should never be accepted as fact. The implication is, of course, that accidents just seem to follow some individuals, that no matter what they do they just seem to be plagued with bad luck. In reality, of course, no one is truly accident prone. Workers have accidents simply because they are careless, are indifferent to safety regulations and safe work practices, or are required to work under unsafe conditions.

Safety officials soon find that safety is not something you read about or practice only on occasion. It has to be observed constantly. Most industries place a high premium on safety—they simply cannot afford to ignore it. When on-the-job injuries occur, some type of causal factor is always involved. Typically workers suffer the pain of injury because they have failed to use good judgment. What is the solution to this problem? Experience has shown that well-written safety programs can aid in solving this problem; however, experience also indicates that if a well-written safety program or safe work practice is not followed by workers, then it is worth less than the paper it is written on. The safety official who views his or her job primarily as occupying a desk to write safe work practices and other policies—without ensuring that such written procedures and practices are properly disseminated through training—is likely to fail in his primary mission: to protect the safety and health of all employees.

Shortly after being hired, new safety officials discover that the training and recordkeeping function is vital, necessary, costly—and frequently overwhelming. It cannot be avoided. The new safety official also soon discovers that drawing the line between what is required by regulation and what is simply required for efficiency is difficult. Literally hundreds of records are required, for a variety of reasons and purposes. The importance of keeping and maintaining up-to-date and accurate records cannot be stressed enough. Safety officials soon discover that, although they may be tedious and time consuming to prepare, written records are their first, second, and sometimes third line of defense. Line of defense? Absolutely. Remember, safety officials hold a precarious position, constantly walking a very fine line. The seasoned safety professional will instantly understand this last statement. Accurate and complete recordkeeping is essential to the safety official's job, professional standing, and personal wellbeing.

Exactly what kind of records is the safety official responsible for? Good question. The safety official is primarily responsible for complying with recordkeeping under the OSH Act. Also, such recordkeeping is concerned with workplace safety, health, the workplace environment, and other administrative functions. Here is an important point that you should remember—one that you should learn to live by if you are going to become a safety official, if you are going to survive as a safety official (Ferry, 1990):

> Wisdom in record keeping is too often a matter of hindsight. As company safety engineer you must be able to anticipate what records will be needed by knowing what is required. The excuse of not recognizing a needed record-keeping function is unacceptable not only to OSHA but also to a court of law.

Simply put, whatever the safety official does and says as part of his or her job should be covered by a piece of paper. Notwithstanding the trend toward a paperless working environment, do not get caught in the trap of not having a piece of paper that is acceptable in a court of law, a stockholders' meeting, or for whatever other purpose an auditor, for example, can come up with. "In this CYA world, I want it on paper, please!"

9.2 Environmental Regulations for Shale Gas Operations

Again, why do we need regulations? Actually, based on personal experience, a better question would also seek to answer when we need regulations and how they are enforced. In dealing with the implications of various regulations for many years, I have come to realize that whether a regulation is a good regulation or a bad regulation is nothing more than a judgment call—a personal judgment call. This is especially the case with regulations that affect the environment and are designed to reduce, limit, or do away with pollution or contamination of the three environmental media of air, water, and soil. Preventing pollution and contamination of Earth's environmental media is, of course, a focus of this book. Again, on a personal note, in regard to shale gas hydraulic fracturing and its environmental consequences, I am reminded of one of those homemade signs in bold letters and carried by an Occupier (part of the 99%) which stated the following:

Fracking jobs are grave-digging jobs!

On a different, more sensical note, to gain understanding of the various viewpoints related to regulations, let's look at a few definitions of pollution.* Pollution has been defined as "a substance that is in the wrong place in the environment, in the wrong concentrations, or at the wrong time, such that it is damaging to living organisms or disrupts the normal functioning of the environment" (Keller, 1988, p. 496). This definition seems incomplete, although it makes the important point that often pollutants are or were useful—in the right place, in the right concentrations, at the right time. Let's take a look at some of the definitions for pollution that have been offered over the years:

1. Pollution is the impairment of the quality of some portion of the environment by the addition of harmful impurities.
2. Pollution is something people produce in large enough quantities that it interferes with our health or wellbeing.
3. Pollution is any change in the physical, chemical, or biological characteristics of the air, water, or soil that can affect the health, survival, or activities of human beings or other forms of life in an undesirable way. Pollution does not have to produce physical harm; pollutants such as noise and heat may cause injury but more often cause psychological distress and aesthetic pollution, such as foul odors and unpleasant sights that affect the senses.

* This section is adapted from Spellman, F.R., *The Science of Environmental Pollution*, 2nd ed., CRC Press, Boca Raton, FL, 2010.

Pollution that initially affects one medium frequently migrates into the other media: Air pollution falls to Earth, contaminating soil and water; soil pollutants migrate into groundwater; and acid precipitation carried by air falls to Earth as rain or snow, altering the delicate ecological balance in surface waters.

In my quest for the definitive definition, the source of last resort was consulted: the common dictionary. According to one dictionary, pollution is a synonym for contamination. A *contaminant* is a pollutant—a substance present in greater than natural concentrations as a result of human activity and having a net detrimental effect upon its environment or upon something of value in the environment. Every pollutant originates from a *source*. A *receptor* is anything that is affected by a pollutant. A *sink* is a long-time repository of a pollutant. What is actually gained from the dictionary definition is that, because pollution is a synonym for contamination, contaminants are things that contaminate the three environmental media (air, water, and soil) in some manner.

The bottom line is that we have come full circle: "Pollution is a judgment call." Why? Because people's opinions differ regarding what they consider to be a pollutant on the basis of their assessment of benefits and risks to their health and economic wellbeing. Visible and invisible chemical spewed into the air or water by an industrial facility might be harmful to people and other forms of life living nearby; however, if the facility is required to install expensive pollution controls, the industrial facility may be forced to shut down or move away. Workers who would lose their jobs and merchants who would lose their livelihoods might feel that the risks from polluted air and water are minor when weighed against the benefits of profitable employment. The same level of pollution can also affect two people quite differently. Some forms of air pollution, for example, might cause a slight eye or respiratory irritation for a healthy person but could be life threatening to someone with chronic obstructive pulmonary disease (COPD) such as severe emphysema. Differing priorities lead to differing perceptions of pollution (e.g., concern about the level of pesticides in foodstuffs generating the need for wholesale banning of insecticides is unlikely to help the starving). No one wants to hear that cleaning up the environment is going to have a negative impact on them. The fact is that public perception lags behind reality because the reality is sometimes unbearable. This perception lag is clearly demonstrated in Case Study 9.1.

Case Study 9.1. Eau de Paper Mill

With regard to certain unbearable facets of reality, consider, for example, the residents of Franklin, Virginia, and their reeking paper mill. For those of us who live close to Franklin—it is 50 miles from Norfolk/Virginia Beach—there is no need to read the road signs. The nose knows when it's close to Franklin. The uninitiated, after a stream of phew-eees courtesy of Eau de paper mill, ask that same old question: How can anyone stand to live in a town that smells like a cocktail mixture of swamp, marsh, sulfur mine, and rotten eggs? Among those who live inside the city limits and, in particular, the 1100 who work at the paper mill, few seem to appreciate the question. When the question is asked, smiles fade; attitudes get defensive. The eventual response is "What smell?" Then, waiting for that quizzical look to appear on the face of the questioner, the local's eyes will twinkle and with a chuckle he will say, "Oh, you must mean that smell of money."

So, again, what is pollution? Our best answer? Pollution is a judgment call. And preventing pollution demands continuous judgment.

9.3 Pollution: Effects Often Easy to See, Feel, Taste, or Smell

Although pollution is difficult to define, its adverse effects are often relatively easy to see; for example, some rivers are visibly polluted or have an unpleasant odor or apparent biotic population problems (such as fish kill). The infamous Cuyahoga River in Ohio became so polluted it twice caught on fire from oil floating on its surface. Air pollution from automobiles and unregulated industrial facilities is obvious. In industrial cities, soot often drifts onto buildings and clothing and into homes. Air pollution episodes can increase hospital admissions and kill people sensitive to the toxins. Fish and birds are killed by unregulated pesticide use. Trash is discarded in open dumps and burned, releasing impurities into the air. Traffic fumes in city traffic plague commuters daily. Ozone levels irritate the eyes and lungs. Sulfate hazards obscure the view.

WHY DOES A PAPER MILL STINK?

Ever take a whiff of a swamp or marsh? That distinctive scent comes from a gas known as TRS—total reduced sulfur—that's released when plants break down. Now, multiply that by the tide of trees reduced to pulp at a paper mill, and you have a stench. TRS seeps out in the steam that billows from a mill's stacks. It smells like rotten eggs, boiling cabbage, or burned matches. Most sources say it's not hazardous to your health, at least not in the concentrations emitted by a paper mill. When the odor gets strong enough, however, some people complain of nausea and headaches. TRS has been cited as a threat to the environment. It's one of the culprits behind acid rain.

—Kimberlin (2009)

Even if you are not in a position to see pollution, you are still made aware of it through the media. How about the 1984 Bhopal incident, the 1986 Chernobyl nuclear plant disaster, the 1991 pesticide spill into the Sacramento River, the *Exxon Valdez*, or the 1994 oil spill in Russia's Far North? Most of us do remember some of these, even though most of us did not directly witness any of these travesties. Events, whether manmade (e.g., Bhopal) or natural (e.g., Mount St. Helens erupting) disasters, sometimes impact us directly, but if not directly they still get our attention. Worldwide, we see constant reminders of the less dramatic, more insidious, continued, and increasing pollution of our environment. We see or hear reports of dead fish in stream beds, litter in national parks, decaying buildings and bridges, leaking landfills, and dying lakes and forests. On the local scale, air quality alerts may have been issued in your community.

Some people experience pollution more directly, firsthand—what we call "in your face," "in your nose," "in your mouth," "in your skin" type of pollution. Consider the train and truck accidents that release toxic pollutants that force us to evacuate our homes (see Case Study 1.2). We become ill after drinking contaminated water or breathing contaminated air or eating contaminated (*Salmonella*-laced) peanut butter products. We can no longer swim at favorite swimming holes because of sewage contamination. We restrict fish, shellfish, and meat consumption because of the presence of harmful chemicals, cancer-causing substances, and hormone residues. We are exposed to nuclear contaminants released to the air and water from uranium-processing plants and other industrial activities.

Science and technology notwithstanding, we damage the environment through use, misuse, and abuse of technology. This can't be avoided. We do not live sterile existences. We pollute when we breathe, use water for various purposes, and bury our dead in the soil. Frequently, we use technological advances before we fully understand their long-term effects on the environment. We weigh the advantages of a technological advance against its effects on the environment and discount the importance of the environment, whether out of greed, hubris, insanity, lack of knowledge, or stupidity, or simply for the hell of it. We often only examine short-term plans without fully developing solutions to problems that may have to be addressed years later. We assume that when the situation becomes critical, technology will be there to fix it. The scientists will figure it out, we believe; thus, we ignore the immediate consequences of our technological abuse. Although technological advances have provided us with nuclear power, electricity and the light bulb, plastics, internal combustion engines, MP3 players, laptop computers, air conditioning, refrigeration, and scores of other advances that make our modern lives pleasant and comfortable, these advances have affected the Earth's environment in ways we did not expect, in ways we deplore, and in ways we may not be able to live with. The environmentalists make these same arguments and a few more when addressing potential environmental problems caused by hydraulic fracking for shale gas; for example, the standard line environmentalists put forward goes something like this: The shale gas fracking industries put a false positive spin on their practices when they state that gas industries harp and profess the benefits of jobs, jobs, and more jobs, when the reality is they come into villages or communities and contaminate the water and air with their processes and destroy the land and people's properties—and then they pick up stakes and leave. One environmental activist in Ohio told me that she and her brothers and sisters (like-minded people) were against any form of hydraulic fracking for any purpose because the process causes long-term problems, such as earthquakes and destruction of the environment—despite several current U.S. Environmental Protection Agency (USEPA) studies and personal communications with USEPA professionals indicating that the risk to groundwater is minimal and that no earthquake has been definitively liked to fracking.

9.4 Regulatory Structure

Every aspect of the exploration, development, operation, and production of natural gas in the United States, including shale gas, is regulated by a complex set of federal, state, and laws. All of the laws, regulations, and permits that apply to conventional oil and gas exploration and production activities also apply to shale gas development. The USEPA administers most of the federal laws, although development on federally owned land is managed primarily by the Bureau of Land Management (BLM), which is part of the Department of the Interior, and the U.S. Forest Service, which is part of the Department of Agriculture. In addition, each state in which oil and gas are produced has one or more regulatory agencies that permit wells, including their design, location, spacing, operation, and abandonment, as well as environmental activities and discharges, including water management and disposal, waste management and disposal, air emissions, underground injection, wildlife impacts, surface disturbance, and worker health and safety as prescribed by the Occupational Safety and Health Act (OSH Act). Many of the federal laws are implemented by the states under agreements and plans approved by the appropriate federal agencies. Those laws and their delegation are discussed in this chapter.

9.4.1 Federal Environmental Laws and Shale Gas Development

Most environmental aspects of shale gas development are governed by a series of federal laws. The Clean Water Act (CWA) regulates surface discharges of water associated with shale gas drilling and production, as well as stormwater runoff from production sites. The Safe Drinking Water Act (SDWA) regulates the underground injection of fluids from shale gas activities. The Clean Air Act (CAA) limits air emissions from gas processing

DID YOU KNOW?

By statute, states may adopt their own standards; however, these must be at least as protective as the federal standards they replace, and they may be more protective in order to address local conditions.

equipment and other sources associated with drilling and production. The National Environmental Policy Act (NEPA) requires that exploration and production on federal lands be thoroughly analyzed for environmental impacts.

Unfortunately, federal agencies do not have the resources to administer all of these environmental programs for all of the oil and gas sites around the country. Moreover, as explained below, one set of nationwide regulations may not always be the most effective way of ensuring the desired level of environmental protection. Accordingly, most of these federal laws have provisions for granting "primacy" to states (i.e., state agencies implement the programs with federal oversight). When these state programs have been approved by the relevant federal agency (usually the USEPA), the state then has primacy jurisdiction.

9.4.2 State Regulations

State regulations related to environmental practices and the safety and health concerns of operators, usually with federal oversight, can more effectively address the regional and state-specific character of shale gas development and operation compared to cookie-cutter, one-size-fits-all regulations at the federal level (IOGCC, 2008). Some of these specific aspects include geology, hydrology, climate, topography, industry characteristics, development history, state legal structures, state worker safety and health regulations, population density, and local economics. The state agencies that permit these practices and monitor and enforce their laws and regulations may be located in the state Department of Natural Resources (e.g., Ohio) or in the Department of Environmental Protection (e.g., Pennsylvania). The Texas Railroad Commission regulates oil and gas activity in the nation's largest oil- and gas-producing state, home to the Barnett Shale. The names and organizational structures vary, but the functions are very similar. Often, multiple agencies are involved, having jurisdiction over different activities and aspects of development.

Not only do these state agencies implement and enforce federal laws, but they also have their own sets of state laws to administer. These state laws often add additional levels of environmental protection and requirements. Also, several states have their own versions of the federal NEPA law, requiring environmental assessments and reviews at the state level and extending those reviews beyond federal lands to state and private lands.

States have many tools at their disposal to ensure that shale gas operations do not adversely impact the environment. The regulation of shale gas drilling and production is a cradle-to-grave approach (referred to as *life-cycle assessment*). The states have broad powers to regulate, permit, and enforce all activities—the drilling and fracture of the well, production operations, management and disposal of wastes, and abandonment and plugging of the well. Different states take different approaches to this regulation and enforcement, but state laws generally give the state oil and gas director or agency the discretion to require whatever is necessary to protect human health and the environment.* In addition to the general protection regulations, most states have a general prohibition against pollution from oil and gas drilling and production.† Most of the state requirements are written into rules or regulations, while some are added to permits on a case-by-case basis as a result of environmental review, on–the-ground inspections, public comments, or commission hearings. All states require a permit before an operator can drill and operate a gas well. The application for this permit includes all the information about a well's location, construction, operation, and reclamation. Agency staff reviews the application for compliance with regulations and to ensure adequate environmental safeguards. The state agency responsible for regulating shale gas operations also, if deemed necessary for any reason, can conduct a site inspection before permit approval. Moreover, most states require operators to post a bond or other financial security when getting a drilling permit to ensure compliance with state regulations and to make sure that there are funds to properly plug the well once production ceases. Another safeguard is that producers generally must notify the state agencies of any significant new activity through a "sundry notice" or a new permit application so the agency is aware of that activity and can review it (BLM, 2001).

States have implemented voluntary review processes to help ensure that the state programs are as effective as possible. The Ground Water Protection Council (GWPC) has a program to review state implementation of the Underground Injection Control (UIC) program. In addition, state oil and gas environmental programs can also be periodically reviewed against a set of guidelines developed by an independent body of state, industry, and environmental stakeholders, known as STRONGER (State Review of Oil and Natural Gas Environmental Regulation). STRONGER is a nonprofit, multistakeholder organization whose purpose is to assist states in documenting the environmental regulations associated with the exploration,

* An example of this type of provision is the following from Pennsylvania's statute: "The department shall have the authority to issue such orders as are necessary to aid in the enforcement of the provisions of [the oil and gas] act" (58 P.S. Section 601.503).

† An example of such language can be found in New York's rules, which state: "The drilling, casing and completion program adopted for any well shall be such as to prevent pollution. Pollution of the land and/or surface or ground fresh water resulting from exploration or drilling is prohibited" (6 NYCRR Part 554).

development, and production of crude oil and natural gas. Periodic evaluations of state exploration and production waste management programs have proven useful in improving the effectiveness of those programs and increasing cooperation between federal and state regulatory agencies. To date, 18 states have been reviewed under the state review guidelines, and several have been reviewed more than once. The STRONGER program has documented the effectiveness of improvements in these state oil and gas environmental programs (STRONGER, 2008a). Prior to the formation of STRONGER, the Interstate Oil and Gas Compact Commission (IOGCC) also completed state reviews using earlier versions of the guidelines.

The organization of regulatory agencies within the various oil- and gas-producing states varies considerably. Some states have several agencies that may oversee some facet of oil and gas operations, especially environmental requirements. These agencies may be located within various departments or divisions within each state's organizations. These various approaches have developed over time within each state, and each state tries to create a structure that best serves its citizenry and all of the industries that it must oversee. The one constant is that each oil- and gas-producing state has one agency with primary responsibility for permitting wells and overseeing general operations. This agency may work with other agencies in the regulatory process, but they can serve as a good source of information about the various agencies that may have jurisdiction over oil and gas activities. Table 9.1 provides a list of the agencies with primary responsibility for oil and gas regulation in each of the states that have or are likely to have shale gas production.

9.4.3 Local Regulation

Along with state and federal requirements, additional requirements regarding oil and gas operations may be imposed by other levels of government in specific locations. Entities such as cities, counties, tribes, and regional water authorities (state water control boards) may each set operational requirements that affect the location and operation of wells or require permits and approvals in addition to those at the federal or state level.

When operations occur in or near populated areas, local governments may establish ordinances to protect the environment and the general welfare of its citizens. These local ordinances frequently require additional permits for issues such as well placement in flood zones, noise level, setbacks from residences or other protected sites, site housekeeping, and traffic. One example is ordinances that set limits on noise levels that may be generated during both daytime and nighttime operations (e.g., Haltom City, Texas, Ordinance No. 0-2004-026-15; Richard Hills, Texas, Ordinance No. 996-04; Southlake, Texas, Gas Well Ordinance, Article IV, Gas and Oil Well Drilling and Production). In some cases, regional water-permitting authorities that have jurisdiction in multiple states have also been established. These federally established

TABLE 9.1

Oil and Gas Regulatory Agencies in Shale Gas States

State	Agency
Alabama	Geological Survey of Alabama, State Oil and Gas Board
Arkansas	Arkansas Oil and Gas Commissions
Colorado	Colorado Department of Natural Resources, Oil and Gas Conservation Commission
Illinois	Illinois Department of Natural Resources, Division of Oil and Gas
Indiana	Indiana Department of Natural Resources, Division of Oil and Gas
Kentucky	Kentucky Department for Energy Development and Independence, Division of Oil and Gas Conservation
Louisiana	Louisiana Department of Natural Resources, Office of Conservation
Michigan	Michigan Department of Environmental Quality, Office of Geological Survey
Mississippi	Mississippi State Oil and Gas Board
Montana	Montana Department of Natural Resources and Conservation, Board of Oil and Gas
New Mexico	New Mexico Energy, Minerals and National Resources Department, Oil Conservation Division
New York	New York Department of Environmental Conservation, Division of Mineral Resources
North Dakota	North Dakota Industrial Commission, Department of Mineral Resources Oil and Gas Division
Ohio	Ohio Department of Natural Resources, Division of Mineral Resources Management
Oklahoma	Oklahoma Corporation Commission, Oil and Gas Conservation Division
Pennsylvania	Pennsylvania Department of Environmental Protection, Bureau of Oil and Gas Management
Tennessee	Tennessee Department of Environment and Conservation, State Oil and Gas Board
Texas	The Railroad Commission of Texas
West Virginia	West Virginia Department of Environmental Protection, Office of Oil and Gas

authorities have been created to protect the water quality of the entire river basin and to govern uses of the water. Additional approvals and permits may be required for operations in these river basins; for example, the Delaware River Basin Commission (DRBC) covers parts of New York, Pennsylvania, New Jersey, and Delaware (DRBC, 2010). Natural gas operators wishing to withdraw water for consumptive use in the basin must first receive a permit from the DRBC, which has the legal authority to fine violators of their rules and regulations.

Because of the wide variety of laws governing shale gas exploration and production and the multitude of federal and state agencies that implement them, the shale gas laws are sometimes confusing. Accordingly, the following discussion has been organized based on the various environmental media (water, air, and soil) that are affected by these activities. The major laws and programs affecting each of these are discussed, along with some of the federal and state programs that cut across these environmental media. Finally, various Occupational Safety and Health Administration (OSHA) requirements and standards designed to protect shale gas workers from injury and illness are also covered.

9.5 Water Quality Regulations

Potential impacts to water quality are primarily regulated under several federal statutes and the accompanying state programs. The primary federal statutes governing water quality issues related to shale gas development are the Clean Water Act (CWA), the Safe Drinking Water Act (SDWA), and the Oil Pollution Act. These statutes and their relationships to shale gas development are discussed below, but first we present a short history of clean water reform.

9.5.1 Genesis of Clean Water Reform

To help the reader better understand the history of the reform movement to clean up our water supplies, including shale gas water (fracking fluid or gas-related produced water and blowback), we trace a chronology of some of the significant events precipitated by environmental organizations and citizens' groups that have occurred since the mid-1960s:*

1. Americans came face-to-face with the grim condition of the nation's waterways in 1969, when the industrial-waste-laden Cuyahoga River caught on fire. That same year, waste from food-processing plants killed almost 30 million fish in Lake Thonotosassa, Florida.
2. In 1972, Congress enacted the Clean Water Act (after having overridden President Nixon's veto). The passage of the Clean Water Act has been called "literally a life-or-death proposition for the Nation." The Act set the goals of achieving water quality levels that are "fishable and swimmable" by 1983, receiving zero discharges of pollutants by 1985, and prohibiting the discharge of toxic pollutants in toxic amounts.

* This chronology is adapted from Sierra Club, Clean water timeline, *The Planet*, 4(8), 1997.

3. In 1974, the Safe Drinking Water Act (SDWA) passed, requiring the USEPA to establish national standards for contaminants in drinking water systems, underground wells, and sole-source aquifers, as well as several other requirements.
4. In 1984, an alliance of the Natural Resources Defense Council, the Sierra Club, and others successfully sued Phillips ECG, a New York industrial polluter that had dumped waste into the Seneca River. According to the Sierra Club's water committee chair, Samuel Sage, the case "tested the muscles of citizens against polluters under the Clean Water Act." During this same time frame, the Clean Water Act reauthorization bill drew the wrath of environmental groups, who dubbed it the "Dirty Water Act" after lawmakers added last-minute pork and weakened wetland protection and industrial pretreatment provisions. Because of grassroots action, most of these pork provisions were dropped. That same year also saw, as a result of a suit filed by the Sierra Club, the highest environmental penalty to date ($70,000), which was imposed against Alcoa Aluminum in Messina, New York (for polluting the St. Lawrence River).
5. In 1986, Tip O'Neill, Speaker of the House of Representatives, stated that he would not let a Clean Water Act reauthorization bill on the floor without the blessing of environmental groups. Later, after the bill was crafted and passed by Congress, President Reagan vetoed the bill. Also in 1986, amendments to the Safe Drinking Water Act directed the USEPA to publish a list of drinking water contaminants that require legislation.
6. In 1987, the Clean Water Act reauthorization bill was reintroduced. It became law after Congress overrode President Reagan's veto. A new provision established the National Estuary Program.
7. From 1995 to 1996, the House passed H.R. 961 (again dubbed the "Dirty Water Act"), which in some cases eliminated standards for water quality, wetlands protection, sewage treatment, and agricultural and urban runoff. The Sierra Club collected over 1 million signatures supporting the Environmental Bill of Rights and released "Danger on Tap," a report that showed polluter contributions to friends in Congress who wanted to gut the Clean Water Act. Due in part to these efforts, the bill was stopped in the Senate.
8. In 1997, the USEPA reported that more than one third of the country's rivers and half of its lakes were still unfit for swimming or fishing. The Sierra Club successfully sued the USEPA to enforce Clean Water Act regulations in Georgia. The state was required to identify polluted waters and establish their pollution-load capacity. Similar suits were filed in other states. In Virginia, for example, Smithfield Foods was assessed a penalty of more than

DID YOU KNOW?

There are approximately 155,000 public water systems in the United States. The USEPA classifies these water systems according to the number of people they serve, the source of their water, and whether they serve the same customers year-round or on an occasional basis.

$12 million—the highest ever—for violating the Clean Water Act by discharging phosphorus and other hog waste products into a tributary of the Chesapeake Bay.

This chronology of events presents only a handful of the significant actions taken by Congress (with helpful prodding and guidance provided by the Sierra Club and the National Resources Defense Council, as well as others) in enacting legislation and regulations to protect our nation's waters. No law has been more important to furthering this effort than the Clean Water Act, which we discuss in the following section.

9.6 Clean Water Act*

Concern with the disease-causing pathogens residing in many of our natural waterways and otherwise filthy water was not the initial lightning rod that got Joe or Nancy Citizen's attention regarding the condition and health of our country's waterways. Instead, it was their aesthetic qualities. Americans in general have a strong emotional response to the beauty of nature and have acted to prevent the pollution and degradation of our nation's waterways simply because many of us expect rivers, waterfalls, and mountain lakes to be natural and naturally beautiful—in the state they were intended to be, pure and clean.

Much of this emotional attachment to the environment was generated from sentimentality borne by popular literature and art of early 19th-century American writers and painters. From Longfellow's *Song of Hiawatha* to *Huckleberry Finn* to the vistas of the Hudson River School of Winslow Homer, American culture abounds with expressions of this singularly strong attachment. As the saying goes: "Once attached, detachment is never easy."

* This section is adapted from USEPA, *Summary of the Clean Water Act*, U.S. Environmental Protection Agency, Washington, DC, 2012 (www.epa.gov/lawsregs/laws/cwa.html); Spellman, F.R., *The Science of Water*, 2nd ed., CRC Press, Boca Raton, FL, 2007.

Federal water pollution legislation dates back to the turn of the century, to the Rivers and Harbors Act of 1899, though the Clean Water Act (CWA) stems from the Federal Water Pollution Control Act, which was originally enacted in 1948 to protect surface waters such as lakes, rivers, and coastal areas. The Clean Water Act is the primary federal law in the United States governing pollution of surface water. Established to protect water quality, the Act includes regulation of pollutant limits on the discharge of oil- and gas-related produced water. Regulation is achieved through the National Pollutant Discharge Elimination System (NPDES) permitting process.

The Clean Water Act was significantly expanded and strengthened in 1972 in response to growing public concern for serious and widespread water pollution problems. This 1972 legislation provided the foundation for subsequent dramatic progress in reducing water pollution. Amendments to the 1972 Clean Water Act were made in 1977, 1981, and 1987.

The Clean Water Act focuses on improving water quality by maintaining and restoring the physical, chemical, and biological integrity of the nation's waters. It provides a comprehensive framework of standards, technical tools, and financial assistance to address the many stressors that can cause pollution and adversely affect water quality, including municipal and industrial wastewater discharges, polluted runoff from urban and rural areas, and habitat destruction.

The Clean Water Act requires national performance standards for major industries (such as iron and steel manufacturing and petroleum refining) that provide a minimum level of pollution control based on the best technologies available. These national standards result in the removal of over a billion pounds of toxic pollution from our waters every year.

The Clean Water Act also establishes a framework whereby states and Indian tribes survey their waters, determine an appropriate use (such as recreation or water supply), then set specific water quality criteria for various pollutants to protect those uses. These criteria, together with the national industry standards, are the basis for permits that limit the amount of pollution that can be discharged to a water body. Under the National Pollutant Discharge Elimination System, sewage treatment plants and industries that discharge wastewater are required to obtain permits and to meet the specified limits in those permits.

Note: The Clean Water Act requires the USEPA to set effluent limitations. All dischargers of wastewaters to surface waters are required to obtain NPDES permits, which require regular monitoring and reporting.

The Clean Water Act also provides federal funding to help states and communities meet their clean water infrastructure needs. Since 1972, federal funding has provided more than $66 billion in grants and loans, primarily for building or upgrading sewage treatment plants. Funding is also provided to address another major water quality problem—polluted runoff from urban and rural areas.

DID YOU KNOW?

The Clean Water Act made it unlawful to discharge any pollutant from a point source into the navigable waters of the United States, unless done in accordance with a specific approved permit. The NPDES permit program controls discharges form point sources that are discrete conveyances, such as pipes or manmade ditches. Industrial, municipal, and other facilities such as shale gas production sites or commercial facilities that handle the disposal or treatment of shale gas produced water must obtain permits if they intend to discharge directly into surface water (USEPA, 2008a,b). Large facilities usually have individual NPDES permits. Discharge from some smaller facilities may be eligible for inclusion under general permits that authorize a category of discharge under the CWA within a geographic area. A general permit is not specifically tailored to an individual discharger. Most oil and gas production facilities with related discharges are authorized under general permits because there are typically numerous sites with common discharges in a geographic area.

Protecting valuable aquatic habitat—wetlands, for example—is another important component of this law. American waterways have suffered loss and degradation of biological habitat, a widespread cause of the decline in the health of aquatic resources. When Europeans colonized this continent, North America held approximately 221 million acres of wetlands. Today, most wetlands are lost. Roughly 22 states have lost 50% or more of their original acreage of wetlands, and 10 states have lost about 70% of their wetlands. The Clean Water Act sections dealing with wetlands have become extremely controversial. Although wetlands are among our nation's most fragile ecosystems and provide a valuable role in maintaining regional ecology and preventing flooding, while serving as home to numerous species of insects, birds, and animals, wetlands also possess potential expandable monetary value in the eyes of private landowners and developers. Herein lies the major problem. Many property owners feel they are being unfairly penalized by a Draconian regulation that restricts their right to develop their own property. Alternative methods that do not involve destroying the wetlands do exist. These methods include wetlands mitigation and mitigation banking. Since 1972, when the Clean Water Act was passed, permits from the Army Corps of Engineers have been required to work in wetland areas. To obtain these permits, builders must agree to restore, enhance, or create an equal number of wetland acres (generally in the same watershed) as those damaged or destroyed in the construction project. Landowners are given the opportunity to balance the adverse affects by replacing environmental values that are lost. This concept is known as *wetlands mitigation*.

Mitigation banking allows developers or public bodies that seek to build on wetlands to make payments to a "bank" for use in the enhancement of other wetlands at a designated location. The development entity purchases credits from the bank and transfers full mitigation responsibility to an agency or environmental organization that runs the bank. Environmental professionals design, construct, and maintain a specific natural area using these funds.

Note that a state that meets the federal primacy requirements is allowed to set more stringent state-specific standards for this program. Because individual states can acquire primacy over their respective programs, it is not uncommon to have varying requirements from state to state. This variation is important to oil and gas industry managers because it can affect how they manage produced water within a drainage basin located within two or more states, such as the Marcellus Shale in the Appalachian Basin. Effluent limitations serve as the primary mechanism under NPDES permits for controlling discharges of pollutants to receiving waters. When developing effluent limitations for NPDES permits, the permit writers must consider limitations based on both the technology available to control the pollutants (i.e., technology-based effluent standards) and the regulations that protect the water quality standards of the receiving water (i.e., water quality-based effluent standards).

The intent of technology-based effluent limits in NPDES permits is to require treatment of effluent concentrations to less than a maximum allowable standard for point source discharges to the specific surface water today. This is based on available treatment technologies, while allowing the discharger to use any available control technique to meet the limits. For industrial (and other non-municipal) facilities, technology-based effluent limits are derived by (1) using national effluent limitations guidelines and standards established by USEPA, or (2) using best professional judgment (BPJ) on a case-by-case basis in the absence of national guidelines and standards.

Prior to the granting of a permit, the authorizing agency must consider the potential impact of every proposed surface water discharge on the quality of the receiving water, not just individual dischargers. If the authorizing agency determines that technology-based effluent limits are not sufficient to ensure that water quality standards will be attained in the receiving water, the CWA (Section 303(b)(1)(c)) and NPDES (40 CFR 122.44(d)) regulations require that more stringent limits be imposed as part of the permit (USEPA, 2008b). USEPA establishes effluent limitation guidelines (ELGs) and standards for different non-municipal (i.e., industrial) categories. These guidelines are developed based on the degree of pollutant reduction attainable by an industrial category through the application of pollution control technologies.

The CWA requires the USEPA to develop specific effluent guidelines that represent the following:

1. *Best conventional technology* (BCT) for control of conventional pollutants and applicable to existing dischargers
2. *Best practicable technology* (BPT) currently available for control of conventional, toxic, and nonconventional pollutants and applicable to existing dischargers
3. *Best available technology* (BAT) economically achievable for control of toxic and nonconventional pollutants and applicable to existing dischargers
4. *New source performance standards* (NSPS) for conventional pollutants and applicable to new sources

To date, the USEPA has established national guidelines and standards for wastewater discharges to surface waters and publicly owned treatment works. At present, more than 50 different industrial categories are listed (USEPA, 2008c) (Table 9.2). The ELGs for oil and gas extraction, which were published in 1979, can be found at 40 CFR 435 (see Table 9.2). The onshore subcategory, Subpart C, is applicable to discharges associated with shale gas development and production. The Clean Water Act also includes a program to control stormwater discharges. The 1987 Water Quality Act (WQA) added Section 402(p) to the CWA that required the USEPA to develop and implement a stormwater permitting program. The USEPA developed this program in two phases (in 1990 and 1999). The regulations establish NPDES permit requirements for municipal, industrial, and construction site stormwater runoff. The WQA also added Section 402(1)(2) to the CWA which specified that the USEPA and states shall not require NPDES permits for unconventional stormwater discharges from oil and gas exploration, production, processing, or treatment operations or from transmission facilities. This exemption applies where the runoff is not contaminated by contact with raw materials or wastes.

The USEPA had previously interpreted the 402(1)(2) exemption as not applying to construction activities of oil and gas development, such as building roads and pads (i.e., an NPDES permit was required) (USEPA, 2008d); however, the Energy Policy Act of 2005 modified Section 402(1)(2) by defining the excluded oil and gas sector operations as including all oil and gas field activities and operations, including those necessary to prepare a site for drilling and for the movement and placement of drilling equipment. The USEPA promulgated a rule to implement this exemption. On May 23, 2008, the U.S. Court of Appeals for the Ninth Circuit released a decision vacating the permitting exemption for discharges of sediment from oil and gas construction activities that contribute to violations of the CWA (*NRDC v. USEPA*, 9th Cir. P. 5947, 2008).

The court based its decision on the fact that the new rule exempted runoff contaminate with sediment, while the CWA does not exempt such runoff. As a result of the court's decision, stormwater discharges contaminated with sediment resulting in a water quality violation require permit coverage under the NPDES stormwater permitting program. Although the USEPA

TABLE 9.2
Existing Effluent Guidelines

Industry Category	40 CFR Part	Industry Category	40 CFR Part
Aluminum Forming	467	Meat and Poultry Products	432
Asbestos Manufacture	427	Metal Finishing	433
Battery Manufacturing	461	Metal Molding and Casting (Foundries)	464
Canned and Preserved Fruits and Vegetable Processing	407	Metal Products and Machinery	438
Canned and Preserved Seafood (Seafood Processing)	408	Mineral Mining and Processing	436
Carbon Black Manufacturing	458	Nonferrous Metals Forming and Metal Powders	471
Cement Manufacturing	411	Nonferrous Metals Manufacturing	421
Centralized Waste Treatment	437	Oil and Gas Extraction	435
Coal Mining	434	Ore Mining and Dressing (Hard Rock Mining)	440
Coal Coating	465	Organic Chemicals, Plastics, and Synthetic Fibers (OCPSF)	414
Concentrated Animal Feeding Operations (CAFO)	412	Paint Formulating	446
Concentrated Aquatic Animal Production	451	Paving and Roofing Materials (Tars and Asphalt)	443
Copper Forming	468	Pesticide Chemicals Manufacturing, Formulating, and Packaging	455
Dairy Products Processing	405	Petroleum Refining	419
Electrical and Electronic Components	469	Pharmaceutical Manufacturing	439
Electroplating	413	Phosphate Manufacturing	422
Explosives Manufacturing	457	Photography	459
Ferroalloy Manufacturing	424	Porcelain Enameling	466
Fertilizer Manufacturing	414	Pulp, Paper, and Paperboard	430
Glass Manufacturing	426	Rubber Manufacturing	428
Grain Mills Manufacturing	406	Soap and Detergent Manufacturing	417
Gum and Wood Chemicals	454	Steam Electric Power Generating	423
Hospitals	460	Sugar Processing	409
Ink Formulating	447	Textile Mills	410
Inorganic Chemicals	415	Timber Products Processing	429
Iron and Steel Manufacturing	420	Transportation Equipment Cleaning	442
Landfills	445	Waste Combustors	444
Leather Tanning and Finishing	425		

stormwater permitting rule now contains a broad exclusion of oil and gas sector construction activities, it is important to note that individual states and Indian tribes may still regulate stormwater associated with these activities. The USEPA has clarified its position that states and tribes may not regulate such stormwater discharges under their CWA authority but are free to regulate under their own independent authorities: "This final rule is not intended to interfere with the ability of states, tribes, or local governments to regulate any discharges through a non-NPDES permit program" (71 FR 33635). In addition to state and tribal regulation, the industry has a voluntary program of Reasonable and Prudent Practices for Stabilization (RAPPS) of oil and gas construction sites (IPAA, 2004). Producers use RAPPS to control erosion and sedimentation associated with stormwater runoff from areas disturbed by clearing, grading, and excavating activities related to site preparation.

The history of the CWA is much like that of the environmental movement itself. Once widely supported and buoyed by its initial success, the CWA has since encountered increasingly difficult problems—polluted stormwater runoff, for example, and non-point-source pollution, as well as unforeseen legal challenges, such as the debate on wetlands and property rights. Unfortunately, the CWA is only part of the way toward achieving its goal. At least a third of U.S. rivers, half of U.S. estuaries, and more than half of U.S. lakes are still not safe for such uses as swimming or fishing. Thirty-one states reported toxins in fish exceeding the action levels set by the Food and Drug Administration (FDA). Every pollutant in a USEPA study on chemicals in fish showed up in at least one location. Water quality is seen as deteriorated and viewed as the cause of the decreasing number of shellfish in the waters.

9.7 Safe Drinking Water Act*

When we get the opportunity to travel the world, one of the first things we learn to ask is whether or not the water is safe to drink. Unfortunately, in most of the places in the world, the answer is "no." As much as 80% of all sickness in the world is attributable to inadequate water or sanitation (Masters, 2007). In a speech in Racine, Wisconsin, in 1998, environmentalist William C. Clark probably summed it up best: "If you could tomorrow morning make water clean in the world, you would have done, in one fell sweep, the best thing you could have done for improving human health by improving environmental quality." An estimated three fourths of the population in Asia, Africa, and Latin America lack a safe supply of water for drinking, washing, and sanitation (Morrison, 1983). Money, technology, education, and attention

* This section is adapted from USEPA, *Understanding the Safe Drinking Water Act*, EPA 816-F-04-030, U.S. Environmental Protection Agency, Washington, DC, 2004.

DID YOU KNOW?

Point-source water pollution is any discernible, defined, and discrete conveyance, including but not limited to any pipe, ditch, channel, tunnel, conduit, well, discrete fissure, container, rolling stock, vessel, or other floating craft, from which pollutants are or may be discharged into a river or other surface water body. *Non-point-source pollution* consists of runoff from irrigated agricultural land.

to the problem are essential for improving these statistics, to solving the problem this West African proverb succinctly states: "Filthy water cannot be washed." Left alone, Nature provides for us. Left alone, Nature feeds us. Left alone, Nature refreshes and sustains us with untainted air. Left alone, Nature provides and cleans the water we need to drink to survive. As Norse (1985) put it, "In every glass of water we drink, some of the water has already passed through fishes, trees, bacteria, worms in the soil, and many other organisms, including people. …Living systems cleanse water and make it fit, among other things, for human consumption." Left alone, Nature performs at a level of efficiency and perfection we cannot imagine. The problem, of course, is that our human populations have grown too large, demanding, and intrusive to allow Nature to be left alone.

Our egos allow us to think that humans are the real reason why Nature exists at all. In our eyes, our infinite need for water is why Nature works its hydrologic cycle—to provide the constant supply of drinking water we need to sustain life. But the hydrologic cycle itself is unstoppable, human activity or not. Bangs and Kallen (1985) summed it up best: "Of all our planet's activities—geological movements, the reproduction and decay of biota and even the disruptive propensities of certain species (elephants and humans come to mind)—no force is greater than the hydrologic cycle."

Nature, through the hydrologic cycle, provides us with an endless (we hope) resupply of water; however, we find that developing and maintaining an adequate supply of safe drinking water requires the coordinated efforts of scientists, technologists, engineers, planners, water treatment plant operators, and regulatory officials. In this section, we concentrate on the regulations that have been put into place in the United States to ensure that the water supplies developed are protected and are kept safe, fresh, and palatable.

Legislation to protect drinking water quality in the United States (the nation's first water quality standards) began with the Public Health Service Act of 1912. With time, the Act evolved, but not until the passage of the Safe Drinking Water Act (SDWA) in 1974 was federal responsibility extended beyond intestate carriers to include all community water systems serving 15 or more outlets, or 25 or more customers. Prompted by public concern over findings of harmful chemicals in drinking water supplies, the law established the basic federal–state partnership for drinking water used today. It focuses on ensuring safe water from public water supplies and on protecting the nation's aquifers from contamination. The law was amended in 1986 and 1996 and requires many actions to protect drinking water and its sources, including rivers, lakes, reservoirs, springs, and groundwater wells. Before we examine the basic tenets of the SWDA, we must define several of the terms used in the Act.

9.7.1 SDWA Definitions*

Action level (AL)—The amount required to trigger treatment or other action.

Best management practices (BMPs)—Schedules of activities, prohibitions of practices, maintenance procedures, and other management practices to prevent or reduce the pollution of waters of the United States.

Contaminant—Any physical, chemical, biological, or radiological substance or matter in water.

Consumer Confidence Report (CCR)—Annual water quality report that a community water system is required to provide to its customers. The CCR helps people make informed choices about the water they drink. They let people know what contaminants, if any, are in their drinking water and how these contaminants may affect their health. CCRs also give the system a chance to tell customers what it takes to deliver safe drinking water.

Discharge of a pollutant—Any addition of any pollutant to navigable waters from any point source.

Exemption—A document for water systems having technical and financial difficulty meeting the National Primary Drinking Water Regulations; it is effective for one year and is granted by the USEPA due to compelling factors.

Likely source—Where a contaminant could come from.

Maximum contaminant level (MCL)—The maximum permissible level of a contaminant in water that is delivered to any user of a public water system.

Maximum contaminant level goal (MCLG)—The level at which no known or anticipated adverse effects on the health of persons occur and which allows an adequate margin of safety.

Maximum residual disinfectant level (MRDL)—The highest level of a disinfectant allowed in drinking water.

Maximum residual disinfectant level goal (MRDLG)—The level of a drinking water disinfectant below which there is no known or expected risk to health.

* This section is adapted from 40 CFR Part 122.2; Safe Drinking Water Act Section 1401; Clean Water Act Section 502.

Microbiological contaminants—Microbes used as indicators that other, potentially harmful bacteria may be present.

National Pollutant Discharge Elimination System (NPDES)—The national program for issuing, modifying, revoking and reissuing, terminating, monitoring, and enforcing permits, in addition to imposing and enforcing pretreatment requirements, under Sections 307, 402, 318, and 405 of the Clean Water Act.

Navigable waters—Waters of the United States, including territorial seas.

pCi/L—Picocuries per liter (a measure of radioactivity).

Person—An individual, corporation, partnership, association, state, municipality, commission, or political subdivision of a state, or any interstate body.

Point source—Any discernible, confined, and discrete conveyance, including but not limited to any pipe, ditch, channel, tunnel, conduit, well, discrete fissure, container, rolling stock, concentrated animal feeding operation, or vessel, or other floating craft, from which pollutants are or may be discharged. This term does not include agricultural stormwater discharges and return flows from irrigated agriculture.

Pollutant—Dredged soil, solid waste, incinerator residue, filter backwash, sewage, garbage, sewage sludge, munitions, chemical wastes, biological materials, radioactive materials (except those regulated under the Atomic Energy Act of 1954), heat, wrecked or discarded equipment, rock, sand, cellar dirt, and industrial, municipal, and agricultural waste discharged into water. It does not mean (a) sewage from vessels or (b) water, gas, or other material injected into a well to facilitate production of oil or gas, or water derived in association with oil and gas production and disposal of in a well, if the well used either to facilitate production or for disposal purposes is approved by authority of the state in which the well is located, and if the state determines that the injection or disposal will not result in the degradation of ground or surface water sources.

Public water system—A system for the provision to the public of piped water for human consumption, if such system has at least 15 service connections or regularly serves at least 25 individuals.

Publicly owned treatment works (POTW)—Any device or system used in the treatment of municipal sewage or industrial wastes of a liquid nature which is owned by a state or municipality. This definition includes sewer, pipes, or other conveyances only if they convey wastewater to a POTW providing treatment.

Recharge zone—The area through which water enters a sole or principal source aquifer.

Regulated substances—Substances that are regulated by the USEPA; they cannot be present at levels above the MCL.

Significant hazard to public health—Any level of contaminant that causes or may cause the aquifer to exceed any maximum contaminant level set forth in any promulgated National Primary Drinking Water Regulations at any point where the water may be used for drinking purposes or which may otherwise adversely affect the health of persons, or which may require a public water system to install additional treatment to prevent such adverse effect.

Sole or principal source aquifer—An aquifer that supplies 50% or more of the drinking water for an area.

Streamflow source zone—The upstream headwaters area that drains into an aquifer recharge zone.

Toxic pollutants—Pollutants that, after discharge and upon exposure, ingestion, inhalation, or assimilation into any organism, will cause death, disease, behavioral abnormalities, cancer, genetic mutations, physiological malfunctions, or physical deformations in such organisms or their offspring.

Treatment technique (TT)—A required process intended to reduce the level of a substance in drinking water.

Turbidity—A measure of the cloudiness of water; turbidity is not necessarily harmful but can interfere with the disinfection of drinking water.

Unregulated monitored substances—Substances that are not regulated by the USEPA but must be monitored so information about their presence in drinking water can be used to develop limits.

Variance—A document for water systems having technical and financial difficulty meeting National Primary Drinking Water Regulations that postpones compliance when such postponement will not result in an unreasonable risk to health.

Waters of the United States—(1) All waters that are currently used, were used in the past, or may be susceptible to use in interstate or foreign commerce, including all waters that are subject to the ebb and flow of the tide. (2) All interstate waters, including interstate wetlands. (3) All other waters such as interstate lakes, rivers, streams, mudflats, sandflats, wetlands, sloughs, prairie potholes, wet meadows, playa lakes, or natural ponds, the use, degradation, or destruction of which would affect interstate or foreign commerce.

Wetlands—Areas that are inundated or saturated by surface water or groundwater at a frequency and duration sufficient to support a prevalence of vegetation typically adapted for life in saturated soil conditions. Wetlands generally include swamps, marshes, bogs, and similar areas.

DID YOU KNOW?

State agencies are the principal organizations for enforcing water quality regulations. They have inspectors, usually located at regional offices throughout the state, who visit oil and gas well sites to ensure compliance with regulations.

9.7.2 SDWA Specific Provisions

To ensure the safety of public water supplies, the Safe Drinking Water Act requires the USEPA to set safety standards for drinking water. Standards are now in place for over 80 different contaminants. The USEPA sets a maximum level for each contaminant; however, in cases where making this distinction is not economically or technologically feasible, the USEPA specifies an appropriate treatment technology instead. Water suppliers must test their drinking water supplies and maintain records to ensure quality and safety. Most states carry the responsibility for ensuring that their public water supplies are in compliance with the national safety standards. Provisions also authorize the USEPA to conduct basic research on drinking water contamination, to provide technical assistance to states and municipalities, and to provide grants to states to help them manage their drinking water programs. To protect groundwater supplies, the law provides a framework for managing underground injection compliance. As part of that responsibility, the USEPA may disallow new underground injection wells based on concerns over possible contamination of a current or potential drinking water aquifer.

Each state is expected to administer and enforce the SDWA regulations for all public water systems. Public water systems must provide water treatment, ensure proper drinking water quality through monitoring, and provide public notification of contamination problems. As mentioned, the 1986 amendments to the SDWA significantly expanded and strengthened its protection of drinking water. Under the 1986 provisions, the SWDA required the following five basic activities:

- *Establishment and enforcement of maximum contaminant levels (MCLs)*—These are the maximum levels of certain contaminants that are allowed in drinking water from public systems. Under the 1986 amendments, the USEPA has set numerical standards or treatment techniques for an expanded number of contaminants.
- *Monitoring*—The USEPA requires monitoring of all regulated and certain unregulated contaminants, depending on the number of people served by the system, the source of the water supply, and the contaminants likely to be found.

- *Filtration*—The USEPA has criteria for determining which systems are obligated to filter water from surface water sources.
- *Use of lead materials*—The use of solder or flux containing more than 0.2% lead or pipes and pipe fittings containing more than 8% lead is prohibited in public water supply systems. Public notification is required where lead is used in construction materials of the public water supply system, or where water is sufficiently corrosive to cause leaching of lead from the distribution system or lines.
- *Well head protection*—The 1986 amendments require all states to develop Well Head Protection Programs. These programs are designed to protect public water supplies from sources of contamination.

The *National Drinking Water Standards* were developed by the USEPA to meet the requirements of the SDWA. Found in CFR 40, these regulations are subdivided into National Primary Drinking Water Regulations (40 CFR 141), which specify maximum contaminant levels (MCLs) based on health-related criteria, and National Secondary Drinking Water Regulations (40 CFR 143), which are unenforceable guidelines based on both aesthetic qualities such as taste, odor, and color of drinking water, as well as non-aesthetic qualities such as corrosivity and hardness. In setting MCLs, the USEPA is required to balance the public health benefits of the standard against what is technologically and economically feasible. In this way, MCLs are different from other set standards, such as the National Ambient Air Quality Standards (NAAQS), which must be set at levels that protect public health regardless of cost or feasibility (Masters, 2007).

The USEPA also creates unenforceable maximum contaminant level goals (MCLGs) set at levels that present no known or anticipated health effects and include a margin of safety, regardless of technological feasibility or cost. The USEPA is also required (under SDWA) to periodically review the actual MCLs to determine whether they can be brought closer to the desired MCLGs.

DID YOU KNOW?

If monitoring the contaminant level in drinking water is not economically or technically feasible, the USEPA must specify a treatment technique that will effectively remove the contaminant from the water supply or reduce its concentration. The MCLs currently cover a number of volatile organic chemicals, organic chemicals, inorganic chemicals, and radionuclides, as well as microbes and turbidity (cloudiness or muddiness). The MCLs are based on an assumed human consumption of 2 liters (roughly 2 quarts) of water per day.

DID YOU KNOW?

For non-carcinogens, MCLGs are determined by a three-step process. The first step is calculating the reference dose (RfD) for each specific contaminant. The RfD is an estimate of the amount of a chemical that a person can be exposed to on a daily basis that is not anticipated to cause adverse systemic health effects over the person's lifetime. A different assessment system is used for chemicals that are potential carcinogens. If toxicological evidence leads to the classification of the contaminant as a human or probable human carcinogen, the MCLG is set at zero (Boyce, 1997).

9.7.2.1 National Primary Drinking Water Regulations

Categories of primary contaminants include *organic chemicals, inorganic chemicals, microorganisms, turbidity,* and *radionuclides*. Except for some microorganisms and nitrate, water that exceeds the listed MCLs will pose no immediate threat to public health; however, all of these substances must be controlled, because drinking water that exceeds the standards over long periods of time may be harmful.

9.7.2.1.1 *Organic Chemicals*

Organic contaminants for which MCLs are being promulgated are classified using the following three groupings: *synthetic organic chemical* (SOCs), *volatile organic chemicals* (VOCs), and *trihalomethanes* (THMs). Table 9.3 shows a partial list of maximum allowable levels for several selected organic contaminants. As we learn more from research about the health effects of various contaminants, the number of regulated organics is likely to grow. Public drinking water supplies must be sampled and analyzed for organic chemicals at least once every three years.

Synthetic organic chemicals (SOCs) are manmade and are often toxic to living organisms. SOCs are compounds used in the manufacture of a wide variety of agricultural and industrial products, including pesticides and herbicides. This group includes PCBs, carbon tetrachloride, 2.4-D, aldicarb, chlordane, dioxin, xylene, phenols, and thousands of other synthetic chemicals. A 1995 study of 29 Midwestern cities and towns by the Washington, DC-based nonprofit Environmental Working Group found pesticide residues in the drinking water in nearly all of them. In Danville, Illinois, the level of the weed killer cyanazine (made by DuPont) was 34 times the federal standard. In Fort Wayne, Indiana, one glass of tap water contained nine kinds of pesticides. The fact is, each year approximately 2.6 billion pounds of pesticides are used in the United States (Lewis, 1996). These pesticides find their way into water supplies and thus present increased risk to public health.

TABLE 9.3

Selected Primary Standard MCLs and MCLGs for Organic Chemicals

Contaminant	Health Effects	MCL–MCLG (mg/L)	Sources
Aldicarb	Nervous system effects	0.003–0.001	Insecticide
Benzene	Possible cancer risk	0.005–0	Industrial chemicals, paints, plastics, pesticides
Carbon tetrachloride	Possible cancer risk	0.005–0	Cleaning agents, industrial wastes
Chlordane	Possible cancer	0.002–0	Insecticide
Endrin	Nervous system, liver, kidney effects	0.002–0.002	Insecticide
Heptachlor	Possible cancer	0.0004–0	Insecticide
Lindane	Nervous system, liver, kidney effects	0.0002–0.0002	Insecticide
Pentachlorophenol	Possible cancer risk, liver, kidney effects	0.001–0	Wood preservative
Styrene	Liver, nervous system effects	0.1–0.1	Plastics, rubber, drug industry
Toluene	Kidney, nervous system, liver, circulatory effects	1–1	Industrial solvent, gasoline additive chemical manufacturing
Total trihalomethanes (TTHM)	Possible cancer risk	0.1–0	Chloroform, drinking water chlorination byproduct
Trichloroethylene (TCE)	Possible cancer risk	0.005–0	Waste from disposal of dry cleaning material and manufacture of pesticides, paints, waxes; metal degreaser
Vinyl chloride	Possible cancer	0.002–0	May leach from PVC pipe
Xylene	Liver, kidney, nervous system effects	10–10	Gasoline refining byproduct, paint ink, detergent

Source: USEPA, *Is Your Drinking Water Safe?*, EPA 810-F-94-002, U.S. Environmental Protection Agency, Washington, DC, 1994; USEPA, *National Primary Drinking Water Regulations*, EPA 816-F-09-0004, U.S. Environmental Protection Agency, Washington, DC, 2009.

DID YOU KNOW?

In water, VOCs are particularly dangerous. VOCs are absorbed through the skin through contact with water—for example, every shower or bath. Hot water allows these chemicals to evaporate rapidly; they are harmful if inhaled. VOCs can be present in any tapwater, regardless of location or water source. If tapwater contains significant levels of these chemicals, they pose a health threat from skin contact, even if the water is not ingested (Ingram, 1991).

Volatile organic chemicals (VOCs) are synthetic chemicals that readily vaporize at room temperature. These include degreasing agents, paint thinners, glues, dyes, and some pesticides—more specifically, benzene, carbon tetrachloride, 1.1.1-trichloroethane (TCA), trichloroethylene (TCE), and vinyl chloride.

Trihalomethanes (THMs) are created in the water itself as byproducts of water chlorination. Chlorine (present in essentially all U.S. tapwater) combines with organic chemicals to form THMs. They include chloroform, bromodichloromethane, dibormochloromethane, and bromoform. THMs are known carcinogens—substances that increase the risk of getting cancer—and they are present at varying levels in all public tapwater.

9.7.2.1.2 Inorganic Chemicals

Several inorganic substances (particularly lead, arsenic, mercury, and cadmium) are of public health importance. These inorganic contaminants and others contaminate drinking water supplies as a result of natural processes, environmental factors, or, more commonly, human activity. Some of these chemicals are listed in Table 9.4. For most organics, MCLs are the same as MCLGs, but the MCLG for lead is zero. *Note:* In Table 9.4, the nitrate level is set at 10 mg/L, because nitrate levels above 10 mg/L pose an immediate threat to children under 1 year old. Excessive levels of nitrate can react with hemoglobin in blood to produce an anemic condition known as "blue babies." Treated water is sampled and tested for inorganics at least once per year (Nathanson, 1997).

DID YOU KNOW?

The abbreviation mg/L stands for milligrams per liter. In metric units, this is the weight of the chemical dissolved in 1 liter of water. One liter is about equal to 1 quart, and 1 ounce is equal to about 28,500 milligrams, so 1 milligram is a very small amount. About 25 grains of sugar weigh 1 milligram.

TABLE 9.4

Selected Primary Standard MCLs for Inorganic Chemicals

Contaminant	Health Effects	MCL (mg/L)	Sources
Arsenic	Nervous system effects	0.010	Geological, pesticide residues, industrial waste, smelter operations
Asbestos	Possible cancer	7 MFL[a]	Natural mineral deposits, A/C pipe
Barium	Circulatory system effects	2	Natural mineral deposits, paint
Cadmium	Kidney effects	0.005	Natural mineral deposits, metal finishing
Chromium	Liver, kidney, digestive system effects	0.1	Natural mineral deposits, metal finishing, textile and leather industries
Copper	Digestive system effects	TT[b]	Corrosion of household plumbing, natural deposits, wood preservatives
Cyanide	Nervous system effects	0.2	Electroplating, steel, plastics, fertilizer
Fluoride	Dental fluorosis, skeletal effects	4	Geological deposits, drinking water additive, aluminum industries
Lead	Nervous system and kidney effects, toxic to infants	TT	Corrosion of lead service lines and fixtures
Mercury	Kidney, nervous system effects	0.002	Industrial manufacturing, fungicide, natural mineral deposits
Nickel	Heart, liver effects	0.1	Electroplating, batteries, metal alloys
Nitrate	Blue-baby effect	10	Fertilizers, sewage, soil and mineral deposits
Selenium	Liver effects	0.05	Natural deposits, mining, smelting

Source: USEPA, *Is Your Drinking Water Safe?*, EPA 810-F-94-002, U.S. Environmental Protection Agency, Washington, DC, 1994; USEPA, *National Primary Drinking Water Regulations*, EPA 816-F-09-0004, U.S. Environmental Protection Agency, Washington, DC, 2009.

[a] Million fibers per liter.

[b] Treatment techniques have been set for lead and copper because the occurrence of these chemicals in drinking water usually results from corrosion of plumbing materials. All systems that do not meet the action level at the tap are required to improve corrosion control treatment to reduce the levels. The action level for lead is 0.015 mg/L, and for copper it is 1.3 mg/L.

9.7.2.1.3 Microorganisms (Microbiological Contaminants)

This group of contaminants includes bacteria, viruses, and protozoa, which can cause typhoid, cholera, and hepatitis, as well as other waterborne diseases. Bacteria are closely monitored in water supplies because they can be dangerous and because their presence is easily detected. Because tests designed to detect individual microorganisms in water are difficult to perform, in actual practice a given water supply is not tested by individually testing for specific pathogenic microorganisms. Instead, a simpler technique is used, based on testing water for evidence of any fecal contamination. Coliform bacteria are used as indicator organisms whose presence suggests that the water is contaminated. To test for total coliforms, the number of monthly samples required is based on the population served and the size of the distribution system. Because the number of coliform bacteria excreted in feces is on the order of 50 million per gram, and the concentration of coliforms in untreated domestic wastewater is usually several million per 100 mL, that water contaminated with human wastes would have no coliforms is highly unlikely. That conclusion is the basis for the drinking water standard for microbiological contaminants, which specifies in essence that, on the average, water should contain no more than 1 coliform per 100 mL. The SWDA standards now require that coliforms not be found in more than 5% of the samples examined during a 1-month period. Known as the *presence/absence concept*, it replaces previous MCLs based on the number of coliforms detected in the sample. Viruses are very common in water. If we removed a teaspoonful of water from an unpolluted lake, over a billion viruses would be present in the water. The two most common and troublesome protozoans found in water are called *Giardia* and *Cryptosporidium* (or *Crypto*). In water, these protozoans occur in the form of hard-shelled cysts. Their hard covering makes them resistant to chlorination and chlorine residual that kills other organisms.

9.7.2.1.4 Turbidity

Turbidity is the measure of fine suspended matter in water, which is mostly caused by clay, silt, organic particulates, plankton, and other microscopic organisms, ranging in size from colloidal to coarse dispersion. Turbidity in the water is measured in *nephelometric turbidity units* (NTUs), which measure the amount of light scattered or reflected from the water. Officially reported in standard units or equivalent to milligrams per liter of silica of diatomaceous earth that could cause the same optical effect, turbidity testing is not required for groundwater sources.

9.7.2.1.5 Radionuclides

Radioactive contamination of drinking water is a serious matter. Radionuclides (the radioactive metals and minerals that cause this contamination) come from both natural and manmade sources. Naturally occurring radioactive minerals move from underground rock strata and geologic formations into the underground streams flowing through them and primarily affect groundwater. In water, radium-226, radium-228, radon-222, and uranium are the natural radionuclides of most concern. Uranium is typically found in groundwater and, to a lesser degree, in some surface waters. Radium in water is found primarily in groundwater. Radon, a colorless, odorless gas and a known cancer-causing agent, is created by the natural decay of minerals. Radon is an unusual contaminant in water, because the danger arises not from drinking radon-contaminated water but from breathing the gas after it has been released into the air. Radon dissipates rapidly when exposed to air. When present in household water, it evaporates easily into the air, where household members may inhale it. Some experts believe that the effects of radon inhalation are more dangerous than those of any other environmental hazard. Manmade radionuclides (more than 200 are known) are believed to be potential drinking water contaminants. Manmade sources of radioactive minerals in water are nuclear power plants, nuclear weapons facilities, radioactive materials disposal sites, and docks for nuclear-powered ships.

9.7.2.2 National Secondary Drinking Water Regulations

The National Secondary Drinking Water Regulations are non-enforceable guidelines regulating contaminants that may cause cosmetic effects (such as skin or tooth discoloration) or aesthetic effects (such as taste, odor, or color) in drinking water. A range of concentrations is established for substances that affect water only aesthetically and have no direct effect on public health. Secondary regulations are provided in Table 9.5.

9.7.3 1996 Amendments to SDWA

After more than 3 years of effort, the Safe Drinking Water Act Reauthorization (one of the most significant pieces of environmental legislation passed to date) was adopted by Congress and signed into law by President Clinton on August 6, 1996. The new streamlined version of the original SDWA gives states greater flexibility in identifying and considering the likelihood for contamination in potable water supplies and in establishing monitoring criteria. It establishes increased reliance on "sound science" instead of "feel-good science," paired with more consumer information presented in readily understandable form, and calls for increased attention to assessment and protection of source waters. The significance of the 1996 SDWA amendments lies in the fact that they are a radical rewrite of the law that the USEPA, states, and water systems had been trying to implement for the past 10 years. In contrast to the 1986 amendments (which were crafted with little substantive input from the regulated community and embraced a command-and-control approach with compliance costs rooted in water rates), the 1996 amendments were developed with significant contributions from water suppliers and

TABLE 9.5

National Secondary Drinking Water Regulations

Contaminants	Suggested Levels	Contaminant Effects
Aluminum	0.05–0.2 mg/L	Discoloration of water
Chloride	250 mg/L	Salty taste; corrosion of pipes
Color	15 color units	Visible tint
Copper	1.0 mg/L	Metallic taste; blue-green staining of porcelain
Corrosivity	Noncorrosive	Metallic taste; fixture staining corroded pipes (corrosive water can leach pipe materials, such as lead, into drinking water)
Fluoride	2.0 mg/L	Dental fluorosis (brownish discoloration of the teeth)
Foaming agents	0.5 mg/L	Aesthetic: frothy, cloudy, bitter taste, odor
Iron	0.3 mg/L	Bitter metallic taste; staining of laundry, rusty color, sediment
Manganese	0.05 mg/L	Taste; staining of laundry, black to brown color, black staining
Odor	3 TON[a]	Rotten egg, musty, or chemical smell
pH	6.5–8.5	Low pH: bitter metallic taste, corrosion High pH: slippery feel, soda taste, deposits
Silver	0.1 mg/L	Argyria (discoloration of skin), graying of eyes
Sulfate	250 mg/L	Salty taste; laxative effects
Total dissolved solids	500 mg/L	Taste and possible relation between low hardness and cardiovascular disease; also an indicator of corrosivity (related to lead levels in water); can damage plumbing and limit effectiveness of soaps and detergents
Zinc	5 mg/L	Metallic taste

Source: USEPA, *Secondary Drinking Water Regulations: Guidance for Nuisance Chemicals*, EPA 816-F-10-079, U.S. Environmental Protection Agency, Washington, DC, 2012.

[a] Threshold odor number.

state and local officials and embody a partnership approach that includes major new infusions of federal funds to help water utilities—especially the thousands of smaller systems—comply with the law. Table 9.6 provides a summary of many of the major provisions of the new amendments, which are as complex as they are comprehensive.

9.7.4 Implementing SDWA

On December 3, 1998, at the oceanfront of Fort Adams State Park, Newport, Rhode Island, in remarks by President Clinton to the community of Newport, a significant part of the 1996 SDWA and amendments were announced—the expectation being that the new requirements would protect most of the nation from dangerous contaminants while adding only about $2 to many monthly water bills. The rules require approximately 13,000 municipal water suppliers to use better filtering systems to screen out *Cryptosporidium* and other microbes, ensuring that U.S. community water supplies are safe from microbial contamination. In his speech, President Clinton said:

> This past summer I announced a new rule requiring utilities across the country to provide their customers regular reports on the quality of their drinking water. When it comes to the water our children drink, Americans cannot be too vigilant.
>
> Today I want to announce three other actions I am taking. First, we're escalating our attack on the invisible microbes that sometimes creep into the water supply. ...Today, the new standards we put in place will significantly reduce the risk from *Cryptosporidium* and other microbes, to ensure that no community ever has to endure an outbreak like the one Milwaukee suffered.
>
> Second, we are taking steps to ensure that when we treat our water, we do it as safely as possible. One of the great health advances to the 20th century is the control of typhoid, cholera, and other diseases with disinfectants. Most of the children in this audience have never heard of typhoid and cholera, but their grandparents cowered in fear of it, and their great-grandparents took it as a fact of life that it would take away significant numbers of the young people of their generation. But as with so many advances, there are trade-offs. We now see that some of the disinfectants we use to protect our water can actually combine with natural substances to create harmful compounds. So today I'm announcing standards to significantly reduce our exposure to these harmful byproducts, to give our families greater peace of mind with their water.
>
> The third thing we are doing today is to help communities meet higher standards, releasing almost $800 million to help communities in all 50 states to upgrade their drinking water systems...to give 140 million Americans safer drinking water.

9.7.5 Underground Injection Control

As one aspect of the protection of drinking water supplies that impacts oil and gas drilling, production, and processing, the SDWA establishes a framework for the Underground Injection Control (UIC) program to prevent the injection of liquid wastes into underground sources of drinking water (USDWs). The USEPA and states implement the UIC program, which sets standards for safe waste injection practices and bans certain types of injection altogether. The UIC program provides these safeguards so injection wells do not endanger USDWs. The first federal UIC regulations were issued in 1980.

The USEPA currently groups underground injection wells into five classes for regulatory control purposes and has a sixth class under consideration. Each class includes wells with similar functions, construction, and operating features so technical requirements can be applied consistently to the class.

TABLE 9.6

Summary of Major Amendment Provisions of the 1996 SDWA Regulations

Definition	Constructed conveyances such as cement ditches used primarily to supply substandard drinking water to farm workers are now SDWA protected.
Contaminant Regulation	Deletes old contamination selection requirement (USEPA regulate 25 new contaminants every 3 years).
	Requires the USEPA to evaluate at least five contaminants for regulation every 5 years, addressing the most risky first, and considering vulnerable populations. The USEPA must issue a *Cryptosporidium* rule (Enhanced Surface Water Treatment Rule) and disinfection byproduct rules within agreed deadlines.
	Deletes the Senate provision giving industry veto power over the USEPA's expediting the rules. The USEPA is authorized to address "urgent threats to health" using an expedited, streamlined process.
	No earlier than 3 years after enactment, no later than the date the USEPA adopts the State II DBP rule, the USEPA must adopt a rule requiring disinfection of certain groundwater systems and provide guidance on determining which systems must disinfect. The USEPA may use cost/benefit provisions to establish this regulation.
Risk Assessment, Management, and Communication	Requires cost/benefit analysis, risk assessment, vulnerable population impact assessment, and development of public information materials for USEPA rules.
	Standard setting provision allows but does not require the USEPA to use risk assessment and cost/benefit analysis in setting standards.
Standard Setting	Reduces the Senate's process to issue standards from three to two steps, deleting the requirement for advanced notice of proposed rule making.
	Requires risks to vulnerable populations to be considered.
	Makes considering costs/benefits and risks/risks a discretionary USEPA authority. "Sound science" provision is limited to standard setting and scientific decisions.
	Requires standard to be reevaluated every 6 years instead of every 3 years.
Treatment Technologies for Small Systems	Establishes new guidelines for the USEPA to identify the best treatment technology for meeting specific regulations.
	For each new regulation, the USEPA must identify affordable treatment technologies that achieve compliance for three categories of small systems: (1) those serving 3301–10,000, (2) those serving 501–3000, and (3) those serving 500 or fewer.
	For all contaminants other than microbials and their indicators, the technologies can include package systems as well as point-of-use and point-of-entry units owned and maintained by water systems.
	The USEPA has 2 years to list such technologies for current regulations, and 1 year to list such technologies for the surface water treatment rule.
	The USEPA must identify the best treatment technologies for the same system categories for use under variances. Such technologies do not have to achieve compliance but must achieve maximum reduction, be affordable, and protect public health.
	The USEPA has 2 years to identify variance technologies for current regulations.

Limited Alternative to Filtration	Allows systems with fully controlled pristine watersheds to avoid filtration if the USEPA and state agree that health is protected through other effective inactivation of microbial contaminants. The USEPA has 4 years to regulate recycling of filter backwash.
Effective Date of Rules	Extends compliance time from 18 months (current law) to 3 years, with available extensions of up to 5 years total.
Arsenic, Sulfate, and Radon	*Arsenic*—Requires the USEPA to set new standard by 2001 using new standard setting language, after more research and consultation with the National Academy of Sciences (NAS). The law authorizes $2.5 million/year for 4 years for research. *Sulfate*—The USEPA has 30 months to complete a joint study with the Centers for Disease Control and Prevention (CDC) to establish a reliable dose–response relationship. Must consider sulfate for regulation within 5 years. If the USEPA decides to regulate sulfate, it must include public notice requirements and allow alternative supplies to be provided to at-risk populations. *Radon*—Requires the USEPA to withdraw its proposed radon standard and to set a new standard in 4 years, after NAS conducts a risk assessment and a study of risk-reduction benefits associated with various mitigation measures. Authorizes cost/benefit analysis for radon, taking into account the costs and benefits of indoor air radon control measures. States or water systems obtaining USEPA approval of a multimedia radon program in accordance with USEPA guidelines would only have to comply with a weaker "alternative maximum contaminant level" for radon that would be based on the contribution of outdoor radon to indoor air.
State Primacy	Primacy states have 2 years to adopt new or revised regulations no less stringent than federal ones and allows 2 years or more if the USEPA finds it necessary and justified. Provides states with interim enforcement authority between the time they submit their regulations to the USEPA and USEPA approval
Enforcement and Judicial Review	Streamlines USEPA administrative enforcement, increases civil penalties, clarifies enforceability of the lead ban and other previously ambiguous requirements, allows enforcement to be suspended in some cases to encourage system consolidation or restructuring, requires states to have administrative penalty authority, and clarifies provisions for judicial review of final USEPA actions.

(continued)

TABLE 9.6 (continued)

Summary of Major Amendment Provisions of the 1996 SDWA Regulations

Public Right to Know	"Consumer confidence reports" provision requires consumers be provided at least annually: (1) the levels of regulated contaminants detected in tap water; (2) what the enforceable maximum contaminant levels and the health goals are for the contaminants (and what those levels mean); (3) the levels found of unregulated contaminants required to be monitored; (4) information on the system's compliance with health standards and other requirements; (5) information on the health effects of regulated contaminants found at levels above enforceable standards; (6) information on health effects of up to three regulated contaminants found at levels below USEPA enforceable health standards where health concerns may still exist; and (7) USEPA's toll-free hotline for further information. Governors can waive the requirement to mail these reports for systems serving under 10,000 people, but systems must still publish the report in the paper. Systems serving 500 or fewer people need only prepare the report and tell their customers it is available. States can later modify the content and form of the reporting requirements. The public information provision modestly improves public notice requirements for violations (such as requiring "prominent" newspaper publication instead of buried classified ads). States and the USEPA must prepare annual reports summarizing violations.
Variances and Exemptions	Provisions for small system variances make minor changes to current provisions regarding exemption criteria and schedules. States are authorized to grant variances to systems serving 3300 or fewer people but need USEPA approval to grant variances to systems serving between 3301 and 10,000 people. Such variances are available only if the USEPA identifies an applicable variance technology and systems install it. Variances are only granted to systems that cannot afford to comply (as defined by state criteria that meet USEPA guidelines) through treatment, alternative sources, or restructuring, and when states determine that the terms of the variance ensure adequate health protection. Systems granted such variances have 3 years to comply with its terms and may be granted an extra 2 years if necessary; states must review the eligibility of such variances every 5 years thereafter. Variances are not allowed for regulations adopted prior to 1986 for microbial contaminants or their indicators. The USEPA has 2 years to adopt regulations specifying procedures for granting or denying such variances and for informing consumers of proposed variances and pertinent public hearings. They also must describe proper operation of variance technologies and eligibility criteria. The USEPA and the federal Rural Utilities Service have 18 months to provide guidance to help states define affordability criteria. The USEPA must periodically review state small-system variance programs, may object to proposed variances, and may overturn issued variances if objections are not addressed. Also, customers of a system for which a variance is proposed can petition the USEPA to object. New York may extend deadlines for certain small, unfiltered systems in nine counties to comply with federal filtration requirements.

Capacity Development	States must acquire authority to ensure that community and nontransient/noncommunity systems beginning operation after October 1, 1999, have the technical, managerial, and financial capacity to comply with SDWA regulations. States that fail to acquire authority lose 20% of their annual state revolving loan fund grants.
	States have 1 year to send the USEPA a list of systems with a history of significant noncompliance and 5 years to report on the success of enforcement mechanisms and initial capacity development efforts. State primacy agencies must also provide progress reports to governors and the public.
	States have 4 years to implement strategy to help systems acquire and maintain capacity before losing portions of their SRLF grants.
	The USEPA must review existing capacity programs and publish information within 18 months to help states and water systems implement such programs. The USEPA has 2 years to provide guidance for ensuring capacity of new systems and must describe likely effects of each new regulation on capacity.
	The law authorizes $26 million over 7 years for grants to establish small water systems technology assistance centers to provide training and technical assistance. The law also authorizes $1.5 million/year through 2003 for the USEPA to establish programs to provide technical assistance aimed at helping small systems achieve and maintain compliance.
Operator Certification	Requires all operators of community and nontransient/noncommunity systems be certified. The USEPA has 30 months to provide guidance specifying minimum standards for certifying water system operators, and states must implement a certification program within 2 years or lose 20% of Sustainability Revolving Loan Fund (SRLF) grants.
	States with such programs can continue to use them as long as the USEPA determines that they are substantially equivalent to its program guidelines.
	The USEPA must reimburse states for the cost of certification training for operators of systems serving 3300 or fewer people, and the law authorizes $30 million/year through 2003 for such assistance grants
State Supervision Program	Authorizes $100 million/year through 2003 for public water system supervision grants to states
	Allows the USEPA to reserve a state's grant should the USEPA assume primacy and, if needed, use SRLF resources to cover any shortfalls in Public Water System Supervision (PWSS) appropriations.
Drinking Water Research	Gives the USEPA authorization to conduct drinking water and groundwater research and is required to develop a strategic research plan and to review the quality of all such research.
Water Return Flows	Repeals the provision in current law that allows businesses to withdraw water from a public water system (such as for industrial cooling purposes), then to return the used water—perhaps with contamination—to the water system's pipe.
Enforcement	Expands and clarifies the USEPA's enforcement authority in primacy and nonprimacy states and provides for public hearings regarding civil penalties ranging from $5000 to $25,000.
	Provides enforcement relief to systems that submit a plan to address problems by consolidating facilities or management or transferring ownership.
	Requires states to obtain authority to issue administrative penalties, which cannot be less than $1000/day for systems serving over 10,000 people.
	The USEPA can assess civil penalties as high as $15,000/day under its emergency powers authority.

Source: Adapted from Drinan, J.E., *Water and Wastewater Treatment: A Guide for the Nonengineering Professional*, CRC Press, Boca Raton, FL, 2000, pp. 289–295.

1. Class I wells may inject hazardous and nonhazardous fluids (industrial and municipal wastes) into isolated formations beneath the lowermost USDW. Because they may inject hazardous waste, Class I wells are the most strictly regulated and are further regulated under the Resource Conservation and Recovery Act (RCRA).
2. Class II wells may inject brines and other fluids associated with oil and gas production.
3. Class III wells may inject fluids associated with solution mining minerals.
4. Class IV wells may inject hazardous or radioactive wastes into or above a USDW and are banned unless specifically authorized under other statutes for ground water remediation.
5. Class V includes all underground injection not included in Classes I to IV. Generally, most Class V wells inject nonhazardous fluids into or above a USDW and are typically onsite disposal systems, such as floor and sink drains that discharge to dry wells, septic systems, leach fields, and drainage wells. Injection practices or wells that are not covered by the UIC program include single-family septic systems and cesspools, as well as non-residential septic systems and cesspools serving fewer than 20 persons that inject *only* sanitary wastewater.
6. Class VI has been proposed specifically for the injection of CO_2 for the purpose of sequestration but has not yet been established.

Most injection wells associated with oil and gas production are Class II wells. These wells may be used to inject water and other fluids (e.g., liquid CO_2) into oil- and gas-bearing zones to enhance recovery, or they may be used to dispose of produced water. The regulation specifically prevents the disposal of waste fluids into USDWs by limiting injection only to formations that are not "underground sources of drinking water." The UIC program is designed to prevent contamination of water supplies by setting minimum requirements for state UIC programs. The basic purpose of the UIC programs is to prevent contamination of USDWs by keeping injected fluids within the intended injection zone. The injected fluids must not endanger a current or future public water supply. The UIC requirements that affect the siting, construction, operation, maintenance, monitoring, testing, and, finally, closure of injection wells have been established to address these concepts. All injection wells require authorization under general rules or specific permits.

The law was written with the understanding that states are best suited to have primary enforcement authority (primacy) for the UIC program. In the SWDA, Congress cautioned the USEPA against a "one-size-fits-all" regulatory scheme and mandated consideration of local conditions and practices. Section 1421(b)(3)(A) requires that UIC regulations permit or provide consideration of varying geological, hydrological, or historical conditions in different states and in different areas within a state. Section 1425 allows a state to obtain primacy from USEPA for oil- and gas-related injection wells, without being required to adopt the complete set of applicable federal UIC regulations. The state must be able to demonstrate that its existing regulatory program is protecting USDWs as effectively as the federal requirements (USEPA, 2003).

To date, 40 states have obtained primacy for oil and gas injection wells (Class II), although not all of these states have oil and gas production. The USEPA administers UIC programs for ten states, seven of which are oil and gas states, and all other federal jurisdictions and Indian Lands (USEPA, 1999).

9.8 Oil Pollution Act of 1990

The Oil Pollution Act (OPA) was signed into law in 1990, largely in response to rising public concern following the *Exxon Valdez* incident. The CWA and the OPA include both regulatory and liability provisions that are designed to reduce damage to natural resources from oil spills. Congress added Section 311 to the CWA, which in part authorized the President to issue regulations establishing procedures, methods, equipment, and other requirements to prevent discharges of oil from vessels and facilities (Section 311(j)(1)(C)). The OPA amended Section 311 of the CWA and contains provisions applicable to onshore facilities and operations. Section 311, as amended by the OPA, provides for spill prevention requirements, spill reporting obligations, and spill response planning. It regulates the prevention of and response to accidental release of oil and hazardous substances into navigable waters, on adjoining shorelines, or affecting natural resources belonging to or managed by the United States. This authority is primarily carried out through the creation and implementation of facility and response plans. These plans are intended to establish measures that will prevent the discharge of oil into navigable waters of the United States or adjoining shorelines as opposed to response and cleanup after a spill occurs.

DID YOU KNOW?

In addition to implementing federal statutes for the NPDES, UIC, and stormwater programs, states and tribes may impose their own requirements to protect their water resources, both surface and underground. For example, they can establish water quality standards for some or all of their surface waters. These standards are approved by the USEPA and become the baseline for CWA permits (USEPA, 2008i).

A cornerstone of the strategy to prevent oil spills from reaching the nation's waters is the Spill Prevention, Control and Countermeasure (SPCC) plan. The USEPA promulgated regulations to implement this part of the OPA:

1. SPCC plans must be prepared, certified (by a professional engineer), and implemented by facilities that store, process, transfer, distribute, use, drill for, produce, or refine oil.
2. Facilities must establish procedures and methods and install proper equipment to prevent an oil release.
3. Facilities must train personnel to properly respond to an oil spill by conducting drills and training sessions.
4. Facilities must have a plan that outlines steps to contain, clean up, and mitigate any effects of an oil spill on waterways (USEPA, 2008h).

Before a facility is subject to the SPCC rule, it must meet three criteria:

1. It must be non-transportation-related.
2. It must have an aggregate aboveground storage capacity greater than 1320 gal (31.4 bbl) and completely buried storage capacity greater than 42,000 gal (1000 bbl).
3. There must be a reasonable expectation of a discharge into or upon navigable waters of the United States or adjoining shorelines.

An SPCC plan is site specific and describes the measures the facility owner has taken to prevent oil spills and what measures are in place to contain and clean up spills. It includes information about the facility, the oil storage containment, inspections, and a site diagram showing locations of tanks (above and below ground) and drainage, and other pertinent details. Prevention measures include secondary containment around tanks and oil-containing equipment. The SPCC program is not as applicable to shale gas operations as it is to oil production sites. Shale gas operators may have to prepare plans if they store large amounts of fuel (exceeding the volumes stated above) onsite or if oil-filled equipment is present and there is a risk of that oil impacting U.S. waters.

9.9 Air Quality*

Impact on environmental air quality is regulated under the Clean Air Act (CAA). As described below, the Act sets national standards for emissions of certain pollutants and requires permits for some industrial operations. Greenhouse gases are not regulated as such; therefore, they are not

* This section is adapted from Spellman, F.R., *The Science of Air*, 2nd ed., CRC Press, Boca Raton, FL, 2008.

specifically discussed in this section. A basic explanation of air pollution, air science parameters, dispersion, dispersion modeling, and major air pollutants is provided before discussing the Clean Air Act and its interface with hydraulic fracturing shale gas operations.

9.9.1 Air Pollution

In the past, the sight of belching smokestacks was a comforting sight to many people: More smoke equaled more business, which indicated that the economy was healthy. But, many of us are now troubled by evidence that indicates that polluted air adversely affects our health. Many toxic gases and fine particles entering the air pose health hazards (e.g., cancer, genetic defects, respiratory disease). Nitrogen and sulfur oxides, ozone, and other air pollutants from fossil fuels are inflicting damage on our forests, crops, soils, lakes, rivers, coastal waters, and buildings. Chlorofluorocarbons (CFCs) and other pollutants entering the atmosphere are depleting the Earth's protective ozone layer, allowing more harmful ultraviolet radiation to reach the surface of the Earth. Fossil fuel combustion is increasing the amount of carbon dioxide in the atmosphere which can have a severe, long-term environmental impact. It is interesting to note that when ambient air is considered, the composition of "unpolluted" air is unknown to us. Humans have lived on the planet thousands of years, and they influenced the composition of the air through their many activities before it was possible to measure the constituents of the air. In theory, the air has always been polluted to some degree. Natural phenomena such as volcanoes, wind storms, the decomposition of plants and animals, and even the aerosols emitted by the ocean can be considered to pollute the air.

The pollutants we usually refer to when we talk about air pollution, though, are those generated as a result of human activity. An *air pollutant* can be considered to be a substance in the air that, in high enough concentrations, produces a detrimental environmental effect. These effects can be either health effects or welfare effects. A pollutant can affect the health of humans, as well as the health of plants and animals. Pollutants can also affect non-living materials such as paints, metals, and fabrics. An *environmental effect* is defined as a measurable or perceivable detrimental change resulting from contact with an air pollutant. Human activities have had a detrimental effect on the makeup of air. Activities such as driving cars and trucks; burning coal, oil and other fossil fuels; and manufacturing chemicals have changed the composition of air by introducing many pollutants. There are hundreds of pollutants in the ambient air. Ambient air is the air to which the general public has access (i.e., any unconfined portion of the atmosphere). The two basic physical forms of air pollutants are particulate matter and gases. Particulate matter includes small solid and liquid particles such as dust, smoke, sand, pollen, mist, and fly ash. Gases include substances such as carbon monoxide (CO), sulfur dioxide (SO_2), nitrogen oxides (NO_x), and volatile organic chemicals (VOCs).

It was once thought that air renewed itself, through interactions with vegetation and the oceans, at a sufficient rate to make up for the influx of anthropogenic pollutants. Today, however, this kind of thinking is being challenged by evidence that clearly indicates that increased use of fossil fuels, expanding industrial production, and the growing use of motor vehicles are having a detrimental effect on air and the environment. With regard to fracking and air pollution, it is important to note that, at each stage of production and delivery, tons of toxic compounds (VOCs), other hydrocarbons, and methane (fugitive natural gas) can escape and mix with nitrogen oxides (NO_x) from the exhaust of diesel-fueled, mobile, and stationary equipment to produce ground-level ozone (CDPHE, 2007; CH2M HILL, 2007; OTA, 1989; URS, 2008). In addition to this air pollution, the constant traffic of trucks hauling condensate and produced water to large waste facility evaporation pits on unpaved roads creates dust.

9.9.2 Atmospheric Dispersion, Transformation, and Deposition*

A source of air pollution is any activity that causes pollutants to be emitted into the air. There have always been natural sources of air pollution, also known as *biogenic sources*; for example, volcanoes have spewed particulate matter and gases into our atmosphere for millions of years. Lightning strikes have caused forest fires, with their resulting contribution of gases and particles, for as long as storms and forests have existed. Organic matter in swamps decays and wind storms whip up dust. Trees and other vegetation contribute large amounts of pollen and spores to our atmosphere. These natural pollutants can be problematic at times but generally are not as much of a problem as human-generated pollutants, or anthropogenic sources. The quality of daily life depends on many modern conveniences. People enjoy the freedom to drive cars and travel in airplanes for business and pleasure. They expect their homes to have electricity and their water to be heated for bathing and cooking. They use a variety of products such as clothing, pharmaceuticals, and furniture made of synthetic materials. At times, they rely on services that use chemical solvents, such as the local dry cleaner and print shop. Yet, the availability of these everyday conveniences comes at a price, because they all contribute to air pollution. Air pollutants are released from both stationary and mobile sources. Scientists have gathered much information on the sources, quantity, and toxicity levels of these pollutants. The measurement of air pollution is an important scientific skill, and such practitioners are usually well grounded in the relevant sciences, particularly with respect to modeling and analyses of air pollutants in the ambient atmosphere. To get at the very heart of air pollution, however, the practitioner must also be well versed in how to determine the origin of

* This section is adapted from USEPA, *Basic Air Pollution Meteorology*, U.S. Environmental Protection Agency, Washington, DC, 2005 (www.epa.gov/apti).

the pollutants and must understand the mechanics of pollutant dispersal, transport, and deposition. Air pollution practitioners must constantly deal with one basic fact: Air pollutants rarely stay at their release location; instead, wind flow conditions and turbulence, local topographic features, and other physical conditions work to disperse these pollutants. So, along with having a thorough knowledge and understanding of the pollutants in question, the air pollution practitioner has a definite need for detailed knowledge of the atmospheric processes that govern their subsequent dispersal and fate.

Conversion of precursor substances to secondary pollutants such as ozone is an example of chemical transformation in the atmosphere. Transformations, both physical and chemical, affect the ultimate impact of emitted air pollutants. Pollutants emitted to the atmosphere do not remain there forever. Two common deposition (depletion) mechanisms are *dry deposition*, the removal of particles and gases as they come into contact with the surface of the Earth, and *washout*, the uptake of particles and gases by water droplets and snow and their removal from the atmosphere as precipitation that falls to the ground. Acid rain is a form of pollution depletion from the atmosphere. The following sections discuss atmospheric dispersion of air pollutants in greater detail and the main factors associated with this phenomenon, including weather, turbulence, air parcels, buoyancy factors, lapse rates, mixing, topography, inversions, plume behavior, and transport.

9.9.2.1 Weather

The air contained in Earth's atmosphere is not still. Constantly in motion, air masses warmed by solar radiation rise at the equator and spread toward the colder poles, where they sink and flow downward, eventually returning to the equator. Near the surface of the Earth, as a result of the Earth's rotation, major wind patterns develop. During the day the land warms more quickly than the sea does; at night, the land cools more quickly. Local wind patterns are driven by this differential warming and cooling. Normally, breezes carry cooler, denser air from over land masses out over the water at night. Precipitation is also affected by wind patterns. Warm, moisture-laden air rising from the oceans is carried inland, where the air masses eventually cool, causing the moisture to fall as rain, hail, sleet, or snow. Even though pollutant emissions may remain relatively constant, air quality varies tremendously from day to day. The determining factors have to do with weather.

Weather conditions have a significant impact on air quality and air pollution, both favorable and unfavorable. On hot, sun-filled days, when the weather is calm with stagnating high-pressure cells, air quality suffers because of the buildup of pollutants at ground level. When local weather conditions include cool, windy, stormy weather with turbulent low-pressure cells and cold fronts, these conditions allow the upward mixing and dispersal of air pollutants.

Weather has a direct impact on pollution levels in both mechanical and chemical ways. Mechanically, precipitation works to cleanse the air of pollutants (transferring the pollutants to rivers, streams, lakes, or the soil). Winds transport pollutants from one place to another. Winds and storms often dilute pollutants with cleaner air, making pollution levels less annoying in the area of their release. In a low-pressure cell, air and its accompanying pollution are carried aloft when the air is heated by the sun. When wind accompanies this rising air mass, the pollutants are diluted with fresh air. In a high-pressure cell, the opposite occurs—air and the pollutants it carries sink toward the ground. With no wind, these pollutants are trapped and concentrated near the ground, where serious air pollution episodes may occur.

Chemically, weather can also affect pollution levels. Winds and turbulence mix pollutants together in a sort of giant chemical broth in the atmosphere. Energy from the sun, moisture in the clouds, and the proximity of highly reactive chemicals may cause chemical reactions, which lead to the formation of secondary pollutants. Many of these secondary pollutants may be more dangerous than the original pollutants.

9.9.2.2 Turbulence

In the atmosphere, the degree of turbulence (which results from wind speed and convective conditions related to the change of temperature with height above the surface of the Earth) is directly related to stability (a function of vertical distribution of atmospheric temperature). The stability of the atmosphere refers to the susceptibility of rising air parcels to vertical motion (attributed to high- and low-pressure systems, air lifting over terrain or fronts, and convection), consideration of atmospheric stability or instability is essential in establishing the dispersion rate of pollutants. When specifically discussing the stability of the atmosphere, we are referring to the lower boundary where air pollutants are emitted.

The degree of turbulence in the atmosphere is usually classified by stability class: *stable, unstable,* and *neutral.* A stable atmosphere is marked by air cooler at the ground than aloft, by low wind speeds, and consequently by a low degree of turbulence. A plume of pollutants released into a stable lower layer of the atmosphere can remain relatively intact for long distances; thus, we can say that stable air discourages the dispersion and dilution of pollutants. An unstable atmosphere is marked by a high degree of turbulence. A plume of pollutants released into an unstable atmosphere may exhibit a characteristic looping appearance produced by turbulent eddies. A neutrally stable atmosphere is an intermediate class between stable and unstable conditions. A plume of pollutants released into a neutral stability condition is often characterized by a coning appearance as the edges of the plume spread out in a V-shape.

The importance of the state of the atmosphere and the effects of stability cannot be overstated. The ease with which pollutants can disperse vertically into the atmosphere is mainly determined by the rate of change of air temperature with height (altitude); therefore, air stability is a primary factor in determining where pollutants will travel and how long they will remain aloft. Stable air discourages the dispersion and dilution of pollutants; conversely, in unstable air conditions, rapid vertical mixing takes place, encouraging pollutant dispersal, which increases air quality.

9.9.2.3 Air Parcels

Think of air inside a balloon as an analogy for the air parcel. This theoretically infinitesimal parcel is a relatively well-defined body of air (a constant number of molecules) that acts as a whole. Self-contained, it does not readily mix with the surrounding air. The exchange of heat between the parcel and its surroundings is minimal, and the temperature within the parcel is generally uniform.

9.9.2.4 Buoyancy Factors

Atmospheric temperature and pressure influence the buoyancy of air parcels. Holding other conditions constant, the temperature of air increases as atmospheric pressure increases and conversely decreases as pressure decreases. Where air pressure decreases with rising altitude, the normal temperature profile of the troposphere is one where temperature decreases with height. An air parcel that becomes warmer than the surrounding air (due to heat radiating from the surface of the Earth, for example) begins to expand and cool. As long as the temperature of the parcel is greater than the surrounding air, the parcel is less dense than the cooler surrounding air; therefore, it rises or is buoyant. As the parcel rises, it expands, thereby decreasing its pressure; therefore, its temperature decreases as well. The initial cooling of an

DID YOU KNOW?

How the atmosphere behaves when air is displaced vertically is a function of atmospheric stability. A stable atmosphere resists vertical motion; air that is displaced vertically in a stable atmosphere tends to return to its original position. This atmospheric characteristic determines the ability of the atmosphere to disperse pollutants emitted into it. To understand atmospheric stability and the role it plays in pollution dispersion, it is important to understand the mechanics of the atmosphere as they relate to vertical atmospheric motion.

air parcel has the opposite effect. In summary, warm air rises and cools, and cool air descends and warms. The extent to which an air parcel rises or falls depends on the relationship of its temperature to that of the surrounding air. As long as the temperature of the parcel is cooler, it will descend. When the temperatures of the parcel and the surrounding air are the same, the parcel will neither rise nor descend unless influenced by wind flow.

9.9.2.5 Lapse Rate

The *lapse rate* is defined as the rate of temperature change with height. With an increase in altitude in the troposphere, the temperature of the ambient air usually decreases. Temperature decreases an average of 6 to 7°C/km. This is the normal lapse rate, but it varies widely depending on location and time of day. We define a temperature decrease with height as a *negative lapse rate* and a temperature increase with height as a *positive lapse rate*.

In a dry environment, when a parcel of warm dry air is lifted in the atmosphere, it undergoes adiabatic expansion and cooling. For the most part, a parcel of air does not exchange heat across its boundaries; therefore, an air parcel that is warmer than the surrounding air does not transfer heat to the atmosphere. Any temperature changes that occur within the parcel are caused by increases or decreases in molecular activity within the parcel. Such changes occur adiabatically and are due only to the change in atmospheric pressure as a parcel moves vertically. The term *adiabatic* means "impassable from," corresponding in this instance to an absence of heat transfer. In other words, an adiabatic process is one in which there is no transfer of heat or mass across the boundaries of the air parcel. In an adiabatic process, compression results in heating and expansion results in cooling. A dry air parcel rising in the atmosphere cools at the dry adiabatic rate of 9.8°C/1000 m and has a lapse rate of –9.8°C/1000 m. Likewise, a dry air parcel sinking in the atmosphere heats up at the dry adiabatic rate of 9.8°C/1000 m and has a lapse rate of 9.8°C/1000 m. Air is considered dry, in this context, as long as any water in it remains in a gaseous state.

The *dry adiabatic lapse rate* is a fixed rate, entirely independent of ambient air temperature. A parcel of dry air moving upward in the atmosphere, then, will always cool at the rate of 9.8°C/1000 m, regardless of its initial temperature or the temperature of the surrounding air. When the ambient lapse rate exceeds the adiabatic lapse rate, the ambient rate is said to be *superadiabatic*, and the atmosphere is highly unstable. When the two lapse rates are exactly equal, the atmosphere is said to be *neutral*. When the ambient lapse rate is less than the dry adiabatic lapse rate, the ambient lapse rate is termed *subadiabatic*, and the atmosphere is stable.

The cooling process within a rising parcel of air is assumed to be adiabatic (occurring without the addition or loss of heat). A rising parcel of air (under adiabatic conditions) behaves like a rising balloon, with the air in that distinct parcel expanding as it encounters air of lesser density until its own density is equal to that of the atmosphere that surrounds it. This process is assumed to occur with no heat exchange between the rising parcel and the ambient air (Peavy et al., 1985).

A rising parcel of dry air containing water vapor will continue to cool at the dry adiabatic lapse rate until it reaches its condensations temperature, or dew point. At this point, the pressure of the water vapor equals the saturation vapor pressure of the air, and some of the water vapor begins to condense. Condensation releases latent heat in the parcel, and thus the cooling rate of the parcel slows. This new rate is called the *wet adiabatic lapse rate.* Unlike the dry adiabatic lapse rate, the wet adiabatic lapse rate is not constant but depends on temperature and pressure. In the middle troposphere, however, it is assumed to be approximately –6 to –7°C/1000 m.

DID YOU KNOW?

The actual temperature profile of the ambient air shows the environmental lapse rate. Sometimes called the *prevailing* or *atmospheric lapse rate,* it is the result of complex interactions of meteorological factors and is usually considered to be a decrease in temperature with height. It is particularly important to vertical motion because surrounding air temperature determines the extent to which a parcel of air rises or falls.

9.9.2.6 *Mixing*

Within the atmosphere, for effective pollutant dispersal to occur, turbulent mixing is important. Turbulent mixing, the result of the movement of air in the vertical dimension, is enhanced by vertical temperature differences. The steeper the temperature gradient and the larger the vertical air column in which the mixing takes place, the more vigorous the convective and turbulent mixing of the atmosphere.

9.9.2.7 *Topography*

On a local scale, topography may affect air motion. In the United States, most large urban centers are located along sea and lake coastal areas. Contained within these large urban centers is much heavy industry. Local air-flow patterns in these urban centers have a significant impact on pollution dispersion processes. Topographic features also affect local weather patterns, especially in large urban centers located near lakes, seas, and open land. Breezes from these features affect vertical mixing and pollutant dispersal. Seasonal differences in heating and cooling land and water surfaces may also precipitate the formation of inversions near the sea or lake shore.

River valley areas are also geographical locations that routinely suffer from industry-related pollution. Many early settlements began in river valleys because of the readily available water supply and the ease of transportation afforded to settlers by river systems within such valleys. Along with settlers came industry—the type of industry that invariably produces air pollutants. These air pollutants, because of the terrain and physical configuration of the valley, are not easily removed from the valley. Winds that move through a typical river valley are called *slope winds*. Slope winds, like water, flow downhill into the valley floor. At the valley floor, slope winds transform to *valley winds*, which flow down-valley with the flow of the river. These winds are lighter than slope winds, and the valley floor becomes flooded with a large volume of air which intensifies the surface inversion that is normally produced by radiative cooling. As the inversion deepens over the course of the night, it often reaches its maximum depth just before sunrise with the height of the inversion layer dependent on the depth of the valley and the intensity of the radiative cooling process. Hills and mountains can also affect local air flow. These natural topographical features tend to decrease wind speed (because of their surface roughness) and form physical barriers that prevent air movement.

9.9.2.8 Inversions

An inversion occurs when air temperature increases with altitude. Temperature inversions (extreme cases of atmospheric stability) create a virtual lid on the upward movement of atmospheric pollution. This situation occurs frequently but is generally confined to a relatively shallow layer. Plumes emitted into air layers that are experiencing an inversion (inverted layer) do not disperse very much as they are transported with the wind. Plumes that are emitted above or below an inverted layer do not penetrate that layer; rather, these plumes are trapped either above or below that inverted layer. High concentrations of air pollutants are often associated with inversions, as they inhibit plume dispersions. Two types of inversions are important from an air quality standpoint: radiation and subsidence inversions.

Radiation inversions are the most common form of surface inversion and occur when the surface of the Earth cools rapidly. They prompt the formation of fog and simultaneously trap gases and particulates, creating a concentration of pollutants. They are characteristically a nocturnal phenomenon caused by cooling of the surface of the Earth. On a cloudy night, the Earth's radiant heat tends to be absorbed by water vapor in the atmosphere. Some of this is radiated back to the surface. On clear winter nights, however, the surface more readily radiates energy to the atmosphere and beyond, allowing the ground to cool more rapidly. The air in contact with the cooler ground also cools, and the air just above the ground becomes cooler than the air above it, creating an inversion close to the ground, lasting for only a matter of hours. These radiation inversions usually begin to form at the worst time of day in large urban areas—during the late afternoon rush hour, trapping automobile exhaust at ground level and causing elevated concentrations of pollution for commuters. During evening hours, photochemical reactions cannot take place, so the biggest problem can be the accumulation of carbon monoxide. At sunrise, the sun warms the ground and the inversion begins to break up. Pollutants that have been trapped in the stable air mass are suddenly brought back to Earth in a process known as *fumigation*, which can cause a short-lived, high concentration of pollution at ground level (Masters, 2007).

The second type of inversion is the *subsidence inversion*, usually associated with anticyclones (high-pressure systems); they may significantly affect the dispersion of pollutants over large regions. A subsidence inversion is caused by the characteristic sinking motion of air in a high-pressure cell. Air in the middle of a high-pressure zone descends slowly. As the air descends, it is compressed and heated. It forms a blanket of warm air over the cooler air below, thus creating an inversion (located anywhere from several hundred meters above the surface to several thousand meters) that prevents further vertical movement of air.

9.9.2.9 Plume Behavior

One way to quickly determine the stability of the lower atmosphere is to view the shape of a smoke trail, or *plume*, from a tall stack located on flat terrain. Visible plumes usually consist of pollutants emitted from a smoke stack into the atmosphere. The formation and fate of the plume itself depend on a number of related factors: (1) the nature of the pollutants, (2) meteorological factors (combination of vertical air movement and horizontal air flow), (3) source obstructions, and (4) local topography, especially downwind. Overall, maximum ground-level concentrations will occur in a range from the vicinity of the smokestack to some distance downwind.

When the atmosphere is slightly stable or neutral, a typical plume *cones*. This is likely to occur on cloudy days or sunny days between the breakup of a radiation inversion and the development of unstable daytime conditions. When the atmosphere is highly unstable, a *looping plume* forms. In the looping plume, the stream of emitted pollutants undergoes rapid mixing, and the wind causes large eddies, which may carry the entire plume down to the ground, causing high concentrations close to the stack before dispersion is complete. In an extremely stable atmosphere, usually in the early morning during a radiation inversion, a *fanning plume* spreads horizontally, with little mixing. When an inversion layer occurs a short distance above the plume source, the plume is said to be *fumigating*. Ground-level pollutant concentrations can be very high when fumigation occurs. Sufficiently tall stacks can prevent fumigation in most cases. When inversion conditions exist below the plume source, the plume is said to be *lofting*. When conditions are neutral, the plume issuing from a smoke stack tends to rise directly into the atmosphere. When an inversion layer prevails both above and below the plume source, the plume issuing from a smokestack tends to be *trapped*.

Pollutants however, rarely come from a single point source (e.g., smokestack). In large urban areas, many plumes are generated, and they collectively combine into a large plume (*city plume*), the dispersion of which represents a huge environmental challenge. The high pollutant concentrations from the city plume frequently affect human health and welfare.

Air quality problems associated with dispersion of city plumes are compounded by the presence of an already contaminated environment. Even though conventional processes normally work to disperse emissions from point sources, they do occur within the city plume. Because of microclimates within the city and the volume of pollutants that must be handled, conventional processes often cannot disperse the pollutants effectively. Other compounding conditions present in areas where city plumes are generated—topographical barriers, surface inversions, and stagnating anticyclones—work to intensify the city plume and result in high pollutant concentrations.

Many researchers have studied plume rise over the years. The most common plume rise formulas are those developed by Gary A. Briggs, which have been extensively validated with stack plume observations (USEPA, 2005). A formula for buoyancy-dominated plumes is shown in Equation 9.1. Plume rise formulas can be used on plumes with temperatures greater than the ambient air temperature. The *Briggs' plume rise formula* is as follows:

$$h = \frac{1.6F^{1/3}x^{2/3}}{\bar{u}} \tag{9.1}$$

where:

Δh = plume rise (above stack)

F = buoyancy flux

x = downwind distance from the stack/source

$\bar{u}$ = average wind speed

$$\text{Buoyancy flux} = F = \frac{g}{\pi}V\left(\frac{T_s - T_a}{T_s}\right) \tag{9.2}$$

where:

g = acceleration due to gravity (9.8 m/s^2)

V = volumetric flow rate of stack gas

T_s = temperature of stack gas

T_a = temperature of ambient air

9.9.2.10 Transport

People living east of the Mississippi River would be surprised to find out that they are breathing air contaminated by pollutants from various sources many miles from their location. Most people view pollution as "out of sight, out of mind." As far as they are concerned, if they don't see it, it doesn't exist. Assume, for example, that a person on a farm heaps together a huge pile of assorted rubbish to be burned. The person preparing this huge bonfire is probably giving little thought to the long-range transport and consequences of any contaminants that might be generated from that bonfire. This person simply has trash, and an easy solution is to burn it. This pile of rubbish, though, is a mixture of discarded rubber tires, old compressed gas bottles, assorted plastic containers, paper, oils, grease, wood, and old paint cans. These are hazardous materials, not just household trash. When the pile of rubbish is set on fire, a huge plume of smoke forms and is carried away by a westerly wind. The firestarter looks downwind and notices that the smoke disappears just a few miles over the property line. The dilution processes and the enormity of the atmosphere work together to dissipate and move the smoke plume away, and the firestarter doesn't give it a second thought. Elevated levels of pollutants from many such fires, though, can occur hundreds to thousands of miles downwind from the combined point sources producing such plumes. The result is that people living many miles from such pollution generators end up breathing contaminated air, transported over some distance to their location. Transport or dispersion estimates are determined by using distribution equations and air quality models. These dispersion estimates are typically valid for the layer of the atmosphere closest to the ground where frequent changes occur in the temperature and distribution of the winds. These two variables have an enormous effect on how plumes are dispersed.

9.9.3 Dispersion Models

Air quality dispersion models consist of a set of mathematical equations that interpret and predict pollutant concentration due to plume dispersal and impaction. They are essentially used to predict or describe the fate of airborne gases, particulate matter, and ground-level concentrations downwind of point sources. To determine the air quality impact on a particular area, the first consideration is normal background concentrations, those pollutant concentrations from natural sources and distant, unidentified manmade sources. Each particular geographical area has a signature, or background, level of contamination considered to be the annual mean background concentration level of certain pollutants. An area, for example, might normally have a particulate matter reading of 30 to 40 µg/m^3. If particulate matter readings are significantly higher than the background level, this suggests an additional source. To establish background contaminations for a particular source, air quality data related to that site and its vicinity must be collected and analyzed.

The USEPA recognized that, in calculating the atmospheric dispersion of air pollutants, it was important to maintain consistency among air quality analyses; thus, the USEPA published two guidebooks to assist in modeling for air quality analyses: *Guidelines on Air Quality Models (Revised)* (1986) and

Industrial Source Complex (ISC) Dispersion Models User's Guide (1986). When performing dispersion calculations, particularly for health effect studies, the USEPA and other recognized experts in the field recommend following a four-step procedure (Holmes et al., 1993):

1. Estimate the rate, duration, and location of the release into the environment.
2. Select the best available model to perform the calculations.
3. Perform the calculations and generate downstream concentrations, including lines of constant concentration (isopleths) resulting from the source emissions.
4. Determine what effect, if any, the resulting discharge has on the environment, including humans, animals, vegetation, and materials of construction. These calculations often include estimates of the so-called *vulnerability zones*—that is, regions that may be adversely affected because of the emissions.

Before beginning any dispersion determination activity, the acceptable ground-level concentration of the waste pollutants must be determined. Local meteorological conduits and local topography must be considered, and having an accurate knowledge of the constituents of, for example, waste gas and its chemical and physical properties is paramount.

Air quality models provide a relatively inexpensive means of determining compliance and predicting the degree of emission reduction necessary to attain ambient air quality standards. Under the 1977 Clean Air Act amendments, the use of models is required for the evaluation of permit applications associated with permissible increments under the Prevention of Significant Deterioration (PSD) requirements, which require localities "to protect and enhance" air that is not contaminated (Godish, 1997).

Several dispersion models have been developed. These models are mathematical descriptions (equations) of the meteorological transport and dispersion of air contaminants in a particular area that allow estimates of contaminant concentrations, either at ground level or elevated (Carson and Moses, 1969). User-friendly modeling programs are available now that produce quick, accurate results from the operator's pertinent data.

The four generic types of models are Gaussian, numerical, statistical, and physical. The *Gaussian* models use the Gaussian distribution equation and are widely used to estimate the impact of nonreactive pollutants. *Numerical* models are more appropriate than Gaussian models for area sources in urban locations that involve reactive pollutants, but numerical models require extremely detailed source and pollutant information and are not widely used. *Statistical* models are used when scientific information about the chemical and physical processes of a source are incomplete or vague so the use of either Gaussian or numerical models is impractical. Finally, *physical* models require fluid modeling studies or wind tunneling. This approach involves the construction of scaled models and observing fluid flow around these models; it is very complex and requires expert technical support. For large areas with complex terrain, stack downwash, complex flow conditions, or large buildings, this type of modeling may be the best choice.

The selection of an air quality model for a particular air quality analysis is dependent on the type of pollutants being emitted, the complexity of the source, and the type of topography surrounding the facility. Some pollutants are formed by the combination of precursor pollutants; for example, ground-level ozone is formed when volatile organic chemicals (VOCs) and nitrogen oxides (NO_x) react in the presence of sunlight. Models to predict ground-level ozone concentrations would use the emission rate of VOCs and NO_x as inputs. Also, some pollutants readily react once emitted into the atmosphere. These reactions deplete the concentrations of these pollutants and may have to be accounted for in the model. Source complexity also plays a role in model selection. Some pollutants may be emitted from short stacks that are subject to aerodynamic downwash. If this is the case, a model must be used that is capable of accounting for this phenomenon. Again, topography plays a major role in the dispersal of plumes and their air pollutants and must be considered when selecting an air quality model. Elevated plumes may impact areas of high terrain. Elevated terrain heights may experience higher pollutant concentrations because they are closer to the plume centerline. A model that considers terrain heights should be used when elevated terrain exists. Probably the best atmospheric dispersion workbook for modeling published to date is that by Turner (1994); most of the air dispersion models used today are based on the Pasquill–Gifford model.

9.9.4 Major Air Pollutants

The most common and widespread anthropogenic pollutants currently emitted are sulfur dioxide (SO_2), nitrogen oxides (NO_x), carbon monoxide (CO), carbon dioxide (CO_2), volatile organic chemicals (hydrocarbons), particulates, lead, and a variety of toxic chemicals. Table 9.7 lists important air

TABLE 9.7

Pollutants and Their Sources

Pollutant	Source
Sulfur and nitrogen oxides	Fossil fuel combustion
Carbon monoxide	Primarily motor vehicles
Volatile organic chemicals	Vehicles and industry
Ozone	Atmospheric reactions between nitrogen oxides and organic compounds

Source: USEPA, *Environmental Progress and Challenges: EPA's Update*, U.S. Environmental Protection Agency, Washington, DC, 1988.

pollutants and their sources. Recall that, in the United States, the USEPA regulates air quality under the Clean Air Act and amendments that charged the federal government to develop uniform National Ambient Air Quality Standards (NAAQS). These standards include primary standards (covering criteria pollutants) designed to protect health and secondary standards to protect public welfare. Primary standards were to be achieved by 1975 and secondary standards within "a reasonable period of time." In 1971, the USEPA promulgated NAAQS for six classes of air pollutants. Later, in 1978, an air quality standard was also promulgated for lead, and the photochemical oxidant standard was revised to an ozone (O_3) standard (i.e., the ozone permissible level was increased). The particulate matter (PM) standard was revised and redesignated a PM_{10} standard in 1987. This revision reflected the need for a PM standard based on particle sizes (≤10 μm) with the potential for entering the respiratory tract and affecting human health.

Air pollutants were categorized into two groups: primary and secondary. Primary pollutants are emitted directly into the atmosphere, where they exert an adverse influence on human health or the environment. Of particular concern are primary pollutants emitted in large quantities: carbon dioxide, carbon monoxide, sulfur dioxide, nitrogen dioxide, hydrocarbons, and particulate matter. Once in the atmosphere, primary pollutants may react with other primary pollutants or atmospheric compounds such as water vapor to form secondary pollutants. Receiving a lot of press and attention is acid precipitation, which occurs when sulfur or nitrogen oxides react with water vapor in the atmosphere.

9.9.4.1 Sulfur Dioxide (SO_2)

Sulfur enters the atmosphere in the form of corrosive sulfur dioxide (SO_2) gas. Sulfur dioxide is a colorless gas possessing the sharp, pungent odor of burning rubber. On a global basis, natural sources and anthropogenic activities produce sulfur dioxide in roughly equivalent amounts. Natural sources include volcanoes, decaying organic matter, and sea spray, and anthropogenic sources include the combustion of sulfur-containing coal and petroleum products and the smelting of nonferrous ores. In industrial areas, much more sulfur dioxide comes from human activities than from natural sources (MacKenzie and El-Ashry, 1988). Sulfur-containing substances are often present in fossil fuels; SO_2 is a product of combustion that results from burning sulfur-containing materials. The largest single source (65%) of sulfur dioxide is the burning of fossil fuels to generate electricity; thus, near major industrialized areas, it is often encountered as an air pollutant.

In the air, sulfur dioxide converts to sulfur trioxide (SO_3) and sulfate particles (SO_4). Sulfate particles restrict visibility and, in the presence of water, form sulfuric acid (H_2SO_4), a highly corrosive substance that also lowers visibility. The global output of sulfur dioxide has increased sixfold since 1900 (McKenzie and El-Ashry, 1988). Most industrial nations, however, have lowered sulfur dioxide levels by 20 to 60% by shifting away from heavy industry and imposing stricter emission standards. Major sulfur dioxide reductions have come from burning coal with lower sulfur content and from using less coal to generate electricity.

Two major environmental problems have developed in highly industrialized regions of the world, where the atmospheric sulfur dioxide concentration has been relatively high: sulfurous smog and acid rain. Sulfurous smog is the haze that develops in the atmosphere when molecules of sulfuric acid accumulate, growing in size as droplets until they become sufficiently large to serve as light scatterers. The second problem, acid rain, is precipitation contaminated with dissolved acids such as sulfuric acid. Acid rain has posed a threat to the environment by causing certain lakes to become void of aquatic life.

9.9.4.2 Nitrogen Oxides (NO_x)

There are seven oxides of nitrogen that are known to occur—NO, NO_2, NO_3, N_2O, N_2O_3, N_2O_4, and N_2O_5—but only two are important in the study of air pollution: nitric oxide (NO) and nitrogen dioxide (NO_2). Nitric oxide is produced by both natural and human actions. Soil bacteria are responsible for the production of most of the nitric oxide that is produced naturally and released to the atmosphere. Within the atmosphere, nitric oxide readily combines with oxygen to form nitrogen dioxide, and together those two oxides of nitrogen are usually referred to as NO_x (nitrogen oxides). NO_x is formed naturally by lightning and by decomposing organic matter. Approximately 50% of anthropogenic NO_x is emitted by motor vehicles, and about 30% comes from power plants, with the other 20% being produced by industrial processes.

Scientists distinguish between two types of NO_x—thermal and fuel—depending on its mode of formation. Thermal NO_x is created when nitrogen and oxygen in the combustion air, such as those within internal combustion engines, are heated to a high enough temperature (above 1000 K) to cause nitrogen (N_2) and oxygen (O_2) in the air to combine. Fuel NO_x results from the oxidation of nitrogen contained within a fuel such as coal. Both types of NO_x generate nitric oxide first and then, when vented and cooled, a portion of nitric oxide is converted to nitrogen dioxide. Although both thermal NO_x and fuel NO_x can be significant contributors to the total NO_x emissions, fuel NO_x is usually the dominant source, with approximately 50% coming from power plants (stationary sources) and the other half being released by automobiles (mobile sources).

Nitrogen dioxide is more toxic than nitric oxide and is a much more serious air pollutant. Nitrogen dioxide, at high concentrations, is believed to contribute to heart, lung, liver, and kidney damage. In addition, because nitrogen dioxide occurs as a brownish haze (giving smog its reddish-brown color), it reduces visibility. When nitrogen dioxide combines with water vapor in the atmosphere, it forms nitric acid (HNO_3), a corrosive substance that, when precipitated out as acid rain, causes damage to plants and corrosion of metal surfaces. Levels of NO_x rose in several countries and then leveled off or declined during the 1970s. During this same period of time, levels of nitrogen oxide did not drop as dramatically as those of sulfur dioxide, primarily because a large part of total NO_x emissions comes from millions of motor vehicles, while most sulfur dioxide is released by a relatively small number of emission-controlled, coal-burning power plants.

9.9.4.3 Carbon Monoxide (CO)

Carbon monoxide is a colorless, odorless, tasteless gas formed when carbon in fuel is not burned completely; it is by far the most abundant of the primary pollutants, as Table 9.8 indicates. When inhaled, carbon monoxide gas restricts the blood's ability to absorb oxygen, causing angina, impaired vision, and poor coordination. Carbon monoxide has little direct effect on ecosystems but has an indirect environmental impact via contributing to the greenhouse effect and depletion of the Earth's protective ozone layer.

The most important natural source of atmospheric carbon monoxide is the combination of oxygen with methane (CH_4), which is a product of the anaerobic decay of vegetation. (Anaerobic decay takes place in the absence of oxygen.) At the same time, however, carbon monoxide is removed from the atmosphere by the activities of certain soil microorganisms, so the net result is a harmless average concentration that is less than 0.12 to 15 ppm in the Northern Hemisphere. Because stationary source combustion facilities are under much tighter environmental control than are mobile sources, the principal source of carbon monoxide that is caused by human activities is motor vehicle exhaust, which contributes to about 70% of all CO emissions in the United States.

9.9.4.4 Volatile Organic Chemicals

Volatile organic chemicals (VOCs) (also listed under the general heading of hydrocarbons) encompass a wide variety of chemicals that contain exclusively hydrogen and carbon. Emissions of volatile hydrocarbons from human resources are primarily the result of incomplete combustion of fossil fuels. Fires and the decomposition of matter are the natural sources. Of the VOCs that occur naturally in the atmosphere, methane (CH_4) is present at the highest concentrations (approximately 1.5 ppm). Even at relatively high concentrations, methane does not interact chemically with other substances, and it causes no ill health effects. In the lower atmosphere, however, sunlight causes VOCs to combine with other gases, such as NO_2, oxygen, and CO, to form secondary pollutants, such as formaldehyde, ketones, ozone, peroxyacetyl nitrate (PAN), and other types of photochemical oxidants. These active chemicals can irritate the eyes, damage the respiratory system, and damage vegetation.

TABLE 9.8

United States Emission Estimates, 1986

Source	Sulfur Oxide (teragram/yr)	Nitrogen Oxide (teragram/yr)	Volatile Organic Chemicals (teragram/yr)	Carbon Monoxide (teragram/yr)	Lead (gigagram/yr)	Particulate Matter (teragram/yr)
Transportation	0.9	8.5	6.5	42.6	3.5	1.4
Stationary source fuel	17.2	10.0	2.3	7.2	0.5	1.8
Industrial processes	3.1	0.6	7.9	4.5	1.9	2.5
Solid waste disposal	0.0	0.1	0.6	1.7	2.7	0.3
Miscellaneous	0.0	0.1	2.2	5.0	0.0	0.8
Total	21.2	19.3	19.5	60.9	8.6	6.8

Source: USEPA, *National Air Pollutant Emission Estimates, 1940–1986*, U.S. Environmental Protection Agency, Washington, DC, 1988.

9.9.4.5 Ozone and Photochemical Smog

By far the most damaging photochemical air pollutant is ozone (each ozone molecule contains three atoms of oxygen and thus is written O_3). Other photochemical oxidants, such as peroxyacetyl nitrate (PAN), hydrogen peroxide (H_2O_2), and aldehydes, play minor roles. All of these are secondary pollutants because they are not emitted but are formed in the atmosphere by photochemical reactions involving sunlight and emitted gases, especially NO_x and hydrocarbons. Ozone is a bluish gas, about 1.6 times heavier than air, and relatively reactive as an oxidant. Ozone is present in a relatively large concentration in the stratosphere and is formed naturally by ultraviolet radiation. At ground level, ozone is a serious air pollutant; it has caused serious air pollution problems throughout the industrialized world, posing threats to human health and damaging foliage and building material. Ozone concentrations in industrialized countries of North America and Europe are up to three times higher than the level at which damage to crops and vegetation begins (MacKenzie and El-Ashry, 1988). Ozone harms vegetation by damaging plant tissues, inhibiting photosynthesis, and increasing susceptibility to disease, drought, and other air pollutants.

In the upper atmosphere, where "good" (vital) ozone is produced, ozone is being depleted by the increased anthropogenic emission of ozone-depleting chemicals on the ground. With this increase, concern has been raised over a potential upset of the dynamic equilibria among stratospheric ozone reactions, with a consequent reduction in ozone concentration. This is a serious situation because stratospheric ozone absorbs much of the incoming solar ultraviolet (UV) radiation. As a UV shield, ozone helps to protect organisms on the surface of the Earth from some of the harmful effects of this high-energy radiation. If not interrupted, UV radiation could cause serious damage due to disruption of genetic material, which could lead to increased rates of skin cancers and heritable problems.

In the mid-1980s, a serious problem with ozone depletion became apparent. A springtime decrease in the concentration of stratospheric ozone (ozone holes) was observed at high latitudes, most notably over Antarctica between September and November. Scientists strongly suspected that chlorine atoms or simple chlorine compounds may be playing a key role in this ozone depletion problem.

On rare occasions, it is possible for upper stratospheric ozone (good ozone) to enter the lower atmosphere (troposphere). Generally, this phenomenon only occurs during an event of great turbulence in the upper atmosphere. On rare incursions, atmospheric ozone reaches ground level for a short period of time. Most of the tropospheric ozone is formed and consumed by endogenous photochemical reactions, which are the result of the interaction of hydrocarbons, oxides of nitrogen, and sunlight, which produces a yellowish-brown haze commonly called *smog*.

TABLE 9.9

Tropospheric Ozone Budget, Northern Hemisphere (kg/ha/yr)

Transport from stratosphere	13–20
Photochemical production	48–78
Destruction at ground	18–35
Photochemical destruction	48–55

Source: Adapted from Hov, O., *Ambio*, 13, 73–79, 1984.

Although the incursion of stratospheric ozone into the troposphere can cause smog formation, the actual formation of Los Angeles-type smog involves a complex group of photochemical interactions. These interactions are between anthropogenically emitted pollutants (NO and hydrocarbons) and secondarily produced chemicals (peroxyacetyl nitrate, aldehydes, NO_2, and ozone). The concentrations of these chemicals exhibit a pronounced diurnal pattern, depending on their rate of emission and on the intensity of solar radiation and atmospheric stability at different times of the day (Freedman, 1989). A tropospheric ozone budget for the northern hemisphere in shown in Table 9.9. The considerable range of the estimates reflects uncertainty in the calculation of the ozone fluxes. On average, stratospheric incursions account for about 18% of the total ozone influx to the troposphere, while endogenous photochemical production accounts for the remaining 82%. About 31% of the tropospheric ozone is consumed by oxidative reactions in vegetated landscapes at ground level, while the other 69% is consumed by photochemical reactions in the atmosphere (Freedman, 1989).

9.9.4.6 Carbon Dioxide

Carbon-laden fuels, when burned, release carbon dioxide (CO_2) into the atmosphere. Much of this carbon dioxide is dissipated and then absorbed by ocean water, some is taken up by vegetation through photosynthesis, and some remains in the atmosphere. Today, the concentration of carbon dioxide in the atmosphere is approximately 350 ppm and is rising at a rate of approximately 20 ppm every decade. The increasing rate of combustion of coal and oil has been primarily responsible for this occurrence, which may eventually have an impact on global climate.

9.9.4.7 Particulate Matter

Atmospheric particulate matter is defined as any dispersed matter, solid or liquid, in which the individual aggregates are larger than single small molecules but smaller than about 500 μm. Particulate matter is extremely diverse

TABLE 9.10

Atmospheric Particulates

Term	Description
Aerosol	General term for particles suspended in air
Mist	Aerosol consisting of liquid droplets
Dust	Aerosol consisting of solid particles that are blown into the air or are produced from larger particles by grinding them down
Smoke	Aerosol consisting of solid particles or a mixture of solid and liquid particles produced by chemical reactions such as fires
Fume	Generally means the same as smoke but often applies specifically to aerosols produced by condensation of hot vapors, especially of metals
Plume	The geometrical shape or form of the smoke coming out of a stack or chimney
Fog	Aerosol consisting of water droplets
Haze	Any aerosol, other than fog, that obscures the view through the atmosphere
Smog	Popular term originating in England to describe a mixture of smoke and fog; implies photochemical pollution

and complex; thus, size and chemical composition, as well as atmospheric concentrations, are important characteristics (Masters, 2007). A number of terms are used to categorize particulates, depending on their size and phase (liquid or solid). These terms are listed and described in Table 9.10. Dust, spray, forest fires, and the burning of certain types of fuels are among the sources of particulates in the atmosphere. Even with the implementation of stringent emission controls, which have worked to reduce particulates in the atmosphere, the U.S. Office of Technology Assessment (Postel, 1987) estimated that particulates and sulfates in ambient air may cause the premature death of 50,000 Americans every year.

9.9.4.8 Lead

Lead is emitted to the atmosphere primarily from human sources, such as burning leaded gasoline, in the form of inorganic particulates. In high concentrations, lead can damage human health and the environment. Once lead enters an ecosystem, it remains there permanently. In humans and animals, lead can affect the neurological system and cause kidney disease. In plants, lead can inhibit respiration and photosynthesis as well as block the decomposition of microorganisms. Since the 1970s, stricter emission standards have caused a dramatic reduction in lead output.

9.10 Clean Air Act

When you look at a historical overview of air quality regulations, you might be surprised to discover that most air quality regulations are recent; for example, in the United States, the first attempt at regulating air quality came about through passage of the Air Pollution Control Act of 1955 (Public Law 84-159). This act was a step forward but that was about all; it did little more than move us toward effective legislation. Revised in 1960 and again in 1962, the act was supplanted by the Clean Air Act (CAA) of 1963 (Public Law 88-206), which has been amended several times, most recently in 1990. The CAA is the primary means by which the USEPA regulates potential emissions that could affect air quality. The CAA encouraged state, local, and regional programs for air pollution control but reserved the right of federal intervention should pollution from one state endanger the health and welfare of citizens residing in another state. In addition, the CAA initiated the development of air quality criteria upon which the air quality and emissions standards of the 1970s were based.

The move toward air pollution control gained momentum in 1970, first by creation of the Environmental Protection Agency and second by passage of amendments to the Clean Air Act (Public Law 91-604), which the USEPA was responsible for implementing. The Act was important because it set primary and secondary ambient air quality standards. Primary standards (based on air quality criteria) allowed for an extra margin of safety to protect public health, whereas the secondary standards (also based on air quality criteria) were established to protect public welfare—animals, property, plants, and materials. The 1977 amendments to the Clean Air Act (Public Law 95-95) further strengthened the existing laws and set the nation's course toward cleaning up our atmosphere. In 1990, further amendments to the Clean Air Act were passed to

- Encourage the use of market-based principles and other innovative approaches, such as performance-based standards and emissions banking and trading.
- Promote the use of clean, low-sulfur coal and natural gas, as well as the use of innovative technologies to clean high-sulfur coal through the acid rain program.
- Reduce enough energy waste and create enough of a market for clean fuels derived from grain and natural gas to cut dependency on oil imports by 1 million bbl/day.
- Promote energy conservation through an acid rain program that gives utilities flexibility to obtain needed emission reductions through programs that encourage customers to conserve energy.

Components of the 1990 amendments to the CAA include the following:

- Title 1, which specifies provisions for attainment and maintenance of National Ambient Air Quality Standards (NAAQS)
- Title 2, which specifies provisions for mobile sources of pollutants
- Title 3, which covers air toxics
- Title 4, which covers specifications for acid rain control
- Title 5, which addresses permits
- Title 6, which specifies stratospheric ozone and global protection measures
- Title 7, which discusses provisions relating to enforcement

9.10.1 Clean Air Act Titles

9.10.1.1 Title 1—Attainment and Maintenance of NAAQS

The 1977 amendments to the Clean Air Act brought about significant improvements in U.S. air quality, but the urban air pollution problems of smog (ozone), carbon monoxide (CO), and particulate matter (PM_{10}) persisted. In 1990, over 100 million Americans were living in cities that had not attained the public health standards for ozone, and a new strategy for attacking the urban smog problem was needed. The 1990 amendments to the Clean Air Act created such a strategy. Under these new amendments, states were given more time to meet the air quality standards (e.g., up to 20 years for ozone in Los Angeles) but they had to make steady, impressive progress in reducing emissions. Specifically, the 1990 amendments required the federal government to reduce emissions from (1) cars, buses, and trucks; (2) consumer products such as window-washing compounds and hair spray; and (3) ships and barges during loading and unloading of petroleum products. In addition, the federal government was directed to develop the technical guidance required by states to control stationary sources. With regard to urban air pollution problems involving smog (ozone), carbon monoxide (CO), and particulate matter (PM_{10}), the new amendments clarified how areas are designated and redesignated as achieving attainment. The USEPA is also allowed to define the boundaries of nonattainment areas (geographical areas whose air quality does not meet federal air quality standards designed to protect public health). The 1990 amendments also established provisions defining when and how the federal government can impose sanctions on areas of the country that have not met certain conditions.

For ozone specifically, the amendments established nonattainment area classifications ranked according to the severity of the area's air pollution problem:

- Marginal
- Moderate
- Serious
- Severe
- Extreme

The USEPA assigns each nonattainment area one of these categories, thus prompting varying requirements the areas must comply with in order to meet the ozone standard. Again, nonattainment areas have to implement different control measures, depending on their classifications. Those closest to meeting the standard, for example, are the marginal areas, which are required to conduct an inventory of their ozone-causing emissions and institute a permit program. Various control measures must be implemented by nonattainment areas with more serious air quality problems; that is, the worse the air quality, the more controls areas will have to implement.

For carbon monoxide and particulate matter, the 1990 CAA amendments established similar programs for areas that do not meet the federal health standard. Areas exceeding the standards for these pollutants are divided into the classifications of moderate and serious. Areas that exceed the carbon monoxide standard are required primarily to implement programs introducing oxygenated fuels or enhanced emission inspection programs. Likewise, areas exceeding the particulate matter standard have to (among other requirements) implement either reasonably available control measures (RACMs) or best available control measures (BACMs).

Title 1 attainment and maintenance of NAAQS requirements have gone a long way toward improving air quality in most locations throughout the United States; however, in 1996, in an effort to upgrade NAAQS for ozone and particulate matter, the USEPA put into effect two new NAAQS for ozone (62 FR 38855) and $PM_{2.5}$, particulate matter smaller than 2.5 μm in diameter (62 FR 38651). They were the first update in 20 years for ozone (smog) and the first in 10 years for particulate matter (soot).

Table 9.11 lists the National Ambient Air Quality Standards, including the new requirements. Note that the NAAQS are important but are not enforceable by themselves. The standards set ambient concentration limits for the protection of human health and environment-related values; however, it is important to remember that it is a very rare case where any one source of air pollutants is responsible for the concentrations in an entire area.

9.10.1.2 Title 2—Mobile Sources

Even though great strides have been made since the 1960s in reducing vehicle emissions, cars, trucks, and buses account for almost half the emissions of the ozone precursors volatile organic chemicals (VOCs) and nitrogen oxides, and up to 90% of CO emissions in urban areas. A large portion of the emission reductions gained from motor vehicle emission controls has been offset by the rapid growth in the number of vehicles on the highways and the total miles driven. Because of the unforeseen growth in automobile emissions in urban areas, compounded with serious air pollution problems in many urban areas, Congress made significant changes to the motor vehicle provisions found in the 1977 amendments to the Clean Air Act. The 1990 amendments established even tighter pollution standards for emissions from motor vehicles. These

TABLE 9.11

National Ambient Air Quality Standards (NAAQs)

Pollutant	Standard Value	
Carbon monoxide (CO)		
8-hour average	9 ppm	10 mg/m^3
1-hour average	35 ppm	40 mg/m^3
Nitrogen dioxide (NO_2)		
Annual arithmetic mean	0.053 ppm	100 µg/m^3
Ozone (O_3)		
1-hour average	0.12 ppm	235 µg/m^3
8-hour average	0.08 ppm	157 µg/m^3
Lead (Pb)		
Quarterly average	1.5 µg/m^3	
Particulate matter (PM_{10})[a]		
Annual arithmetic mean	50 µg/m^3	
24-hour average	150 µg/m^3	
Particulate matter ($PM_{2.5}$)[b]		
Annual arithmetic mean	15 µg/m^3	
24-hour average	65 µg/m^3	
Sulfur dioxide (SO_2)		
Annual arithmetic mean	0.03 ppm	80 µg/m^3
24-hour average	0.14 ppm	365 µg/m^3
3-hour average	0.50 ppm	1300 µg/m^3

Source: USEPA, *National Ambient Air Quality Standards (NAAQS)*, U.S. Environmental Protection Agency, Washington, DC, 2007.

[a] Particles with diameters of 10 µm or less.
[b] Particles with diameters of 2.5 µm or less.

standards were designed to reduce tailpipe emissions of hydrocarbons, nitrogen oxides, and carbon monoxide on a phased-in basis beginning with model year 1994. Automobile manufacturers are also required to reduce vehicle emissions resulting from the evaporation of gasoline during refueling operations. The latest version of the Clean Air Act (1990, with 1997 amendments for ozone and particulate matter) also requires fuel quality to be controlled. New programs requiring cleaner or reformulated gasoline were initiated in 1995 for the nine cities with the worst ozone problems. Other cities were given the option to "opt in" to the reformulated gasoline program. In addition, a clean fuel car pilot program was established in California, which required the phasing-in of tighter emission limits for several thousand vehicles in model year 1996 and up to 300,000 by model year 1999. The law allows these standards to be met with any combination of vehicle technology and cleaner fuels. The standards became even stricter in 2001.

9.10.1.3 Title 3—Air Toxics

Toxic air pollutants, which are hazardous to human health or the environment (typically carcinogens, mutagens, and reproductive toxins) were not specifically covered under the 1977 amendments to the Clean Air Act. This situation is quite surprising when we consider that information generated as a result of Title III of the Superfund Amendments and Reauthorization Act (SARA) (Section 313) indicates that in the United States more than 2 billion pounds of toxic air pollutants are emitted annually. The 1990 amendments to the Clean Air Act offered a comprehensive plan for achieving significant reductions in emissions of hazardous air pollutants from major sources. The new law improved the USEPA's ability to address this problem effectively and dramatically accelerated progress in controlling major toxic air pollutants. The 1990 amendments include a list of 189 toxic air pollutants whose emissions must be reduced. The USEPA was required to publish a list of source categories that emit certain levels of these pollutants and was also required to issue maximum achievable control technology (MACT) standards for each listed source category. The law also established a Chemical Safety Board to investigate accidental releases of extremely hazardous chemicals.

9.10.1.4 Title 4—Acid Deposition

Let's talk about the acid rain problem for a moment. Consider the following: In the evening, when you stand on your porch and look out over your terraced lawn and that flourishing garden of perennials during a light rainfall, you probably feel a sense of calm and relaxation that's difficult to describe—but not hard to accept. Maybe it's the sound of raindrops falling on the roof of the porch, the lawn, the sidewalk, and the street and that light wind blowing through the boughs of the evergreens that are soothing you. Whatever it is that makes you feel this way, rainfall is a major ingredient. But those who are knowledgeable or trained in environmental science might take another view of such a seemingly peaceful event. They might wonder to themselves whether the rainfall is as clean and pure as it should be. Is this actually just rainfall—or is it rain carrying acids as strong as lemon juice or vinegar and capable of harming both living and nonliving things such as trees, lakes, and buildings? This may seem strange to some folks who might wonder why anyone would be concerned about such off-the-wall matters.

Such a concern was unheard of before the Industrial Revolution, but today the purity of rainfall is a major concern for many people, especially with regard to its acidity. Most rainfall is slightly acidic because of decomposing organic matter, the movement of the sea, and volcanic eruptions, but the principal factor is atmospheric carbon dioxide, which causes carbonic acid to form. *Acid rain* (pH <5.6) is produced by the conversion of the primary pollutants sulfur dioxide and nitrogen oxides to sulfuric acid and nitric acid, respectively. These processes are complex, depending on the physical dispersion processes and the rates of the chemical conversions.

Contrary to popular belief, acid rain is not a new phenomenon nor does it result solely from industrial pollution. Natural processes—volcanic eruptions and forest fires, for example—produce and release acid particles into the air, and the burning of forest areas to clear land in Brazil, Africa, and other countries also contributes to acid rain; however, the rise in manufacturing that began with the Industrial Revolution literally dwarfs all other contributions to the problem. The main culprits are emissions of sulfur dioxide from the burning of fossil fuels, such as oil and coal, and nitrogen oxide, formed mostly from internal combustion engine emissions, which is readily transformed into nitrogen dioxide. These mix in the atmosphere to form sulfuric acid and nitric acid.

In dealing with atmospheric acid deposition, the Earth's ecosystems are not completely defenseless; they can deal with a certain amount of acid through natural alkaline substances in soil or rocks that buffer and neutralize acid. Highly alkaline soil (limestone and sandstone) in the American Midwest and southern England provides some natural neutralization; however, areas with thin soil and those laid on granite bedrock have little ability to neutralize acid rain.

Scientists continue to study how living beings are injured or even killed by acid rain. This complex subject has many variables. We know from various episodes of acid rain that pollution can travel over very long distances. Lakes in Canada and New York are feeling the effects of coal burning in the Ohio Valley. For this and other reasons, the lakes of the world are where most of the scientific studies have taken place. In lakes, the smaller organisms often die off first, leaving the larger animals to starve to death. Sometimes the larger animals (e.g., fish) are killed directly; as lake water becomes more acidic, it dissolves heavy metals, leading to toxic and often lethal concentrations. Have you ever wandered up to the local lake shore and observed thousands of fish belly-up? Not a pleasant sight or smell, is it? Loss of life in lakes also disrupts the system of life on the land and the air around them. In some parts of the United States, the acidity of rainfall has fallen well below 5.6. In the northeastern United States, for example, the average pH of rainfall is 4.6, and rainfall with a pH of 4.0, a level 1000 times more acidic than distilled water, has occurred.

Despite intensive research into most aspects of acid rain, there are still many areas of uncertainty and disagreement. That is why progressive, forward-thinking countries emphasize the importance of further research into acid rain, and that is why the 1990 amendments to the Clean Air Act initiated a permanent reduction in SO_2 levels. One of the interesting features of the Clean Air Act is that it allowed utilities to trade allowances within their systems or buy and sell allowances to and from other affected sources. Each source must have sufficient allowances to cover its annual emissions. If not, the source is subject to excess emissions fees and a requirement to offset the excess emissions in the following year. The 1990 amendments also include specific requirements for reducing emissions of nitrogen oxides for certain boilers.

9.10.1.5 Title 5—Permits

The 1990 CAA amendments also introduced an operating permit system similar to the National Pollutant Discharge Elimination System (NPDES). The permit system has a twofold purpose: (1) to ensure compliance with all applicable requirements of the CAA, and (2) to enhance the USEPA's ability to enforce the Act. Under the Act, air pollution sources must develop and implement the program, and the USEPA must issue permit program regulations, review each state's proposed program, and oversee the state's effort to implement any approved program. The USEPA must also develop and implement a federal permit program when a state fails to adopt and implement its own program.

9.10.1.6 Title 6—Ozone and Global Climate Protection

Ozone is formed in the stratosphere by radiation from the sun and helps to shield life on Earth from some of the sun's potentially destructive ultraviolet radiation. In the early 1970s, scientists suspected that the ozone layer was being depleted. By the 1980s, it became clear that the ozone shield is indeed thinning in some places and at times even has a seasonal hole in it, notably over Antarctica. The exact causes and actual extent of the depletion are not yet fully known, but most scientists believe that various chemicals in the air are responsible.

Most scientists identify the family of chlorine-based compounds, most notably chlorofluorocarbons (CFCs) and chlorinated solvents (carbon tetrachloride and methyl chloroform), as the primary culprits involved in ozone depletion. Molina and Rowland (1974) hypothesized that the chlorine-containing CFCs were responsible for ozone depletion. They pointed out that chlorine molecules are highly active and readily and continually break apart three-atom ozone into the two-atom form of oxygen generally found close to Earth, in the lower atmosphere. The Interdepartmental Committee for Atmospheric Sciences (ICAS, 1975) estimated that a 5% reduction in ozone could result in nearly a 10% increase in cancer. This already frightening scenario was made even more frightening by 1987 when evidence showed that CFCs destroy ozone in the stratosphere above Antarctica every spring. The ozone hole had become larger, with more than half of the total ozone column wiped out, and essentially all ozone disappeared from some regions of the stratosphere (Davis and Cornwell, 1991).

In 1988, it was reported that, on a worldwide basis, the ozone layer had shrunk approximately 2.5% in the preceding decade (Zurer, 1988). This obvious thinning of the ozone layer, with its increased chances of skin cancer and cataracts, is also implicated in suppression of the human immune system and damage to other animals and plants, especially aquatic life and soybean crops. The urgency of the problem spurred the 1987 signing of the Montreal Protocol by 24 countries, which required signatory countries to reduce their consumption of CFCs by 20% by 1993 and by 50% by 1998, marking a significant achievement in solving a global environmental problem.

The 1990 amendments to the Clean Air Act borrowed from USEPA requirements already on the books in other regulations and mandated phase-out of the production of substances that deplete the ozone layer. Under these provisions, the USEPA was required to list all regulated substances along with their ozone-depletion potential, atmospheric lifetime, and global warming potentials.

9.10.1.7 Title 7—Enforcement

A broad array of authorities is contained within the Clean Air Act to make the law more readily enforceable. The 1990 amendments gave the USEPA new authority to issue administrative penalties with fines, and field citations (with fines) for smaller infractions. In addition, sources must certify their compliance, and the USEPA has authority to issue administrative subpoenas for compliance data.

9.10.2 Use of Meteorology in Air Quality Regulatory Programs*

The Clean Air Act amendments require that State Implementation Plans (SIPs) be developed, the impact upon the atmosphere be evaluated for new sources, and air quality modeling analyses be performed. These regulatory programs require knowledge of the air quality in the region around a source, air quality modeling procedures, and the fate and transport of pollutants in the atmosphere. Implicit in air pollution programs is knowledge of the climatology of the area in question.

9.10.2.1 State Implementation Plans

State Implementation Plans (SIPs) are federally approved plans developed by state (or local) air quality management authorities to attain and maintain the National Ambient Air Quality Standards. Generally, these SIPs are a state's (local) air quality rules and regulations that are considered an acceptable control strategy once approved by the USEPA. The purpose of SIPs is to control the amount and types of pollution for any given area or region of the United States.

In these types of control strategies, emission limits should be based on ambient pollutant concentration estimates for the averaging time that results in the most stringent control requirements. In all cases, these concentration estimates are assumed to be the sum of the pollutant concentrations contributed by the source and an appropriate background concentration. An air quality model is used to determine which averaging time (e.g., annual, 24-hour, 8-hour, 3-hour, 1-hour) results in the highest ambient impact. For

* This section is adapted from USEPA, *Basic Air Pollution Meteorology*, U.S. Environmental Protection Agency, Washington, DC, 2005 (www.epa.gov/apti).

example, if the annual average air quality standard is approached by a greater degree (percentage) than standards for other averaging times, the annual average is considered the restrictive standard. In this case, the sum of the highest estimated annual average concentration and the annual average background concentration provides the concentration that should be used to specify emission limits; however, if a short-term standard is approached by a greater degree and is thus identified as the restrictive standard, other considerations are required because the frequency of occurrence must also be taken into account.

9.10.2.2 New Source Review

New major stationary sources or major modifications to existing sources of air pollution are required by the Clean Air Act to obtain an air quality permit before construction is started. This process is called a New Source Review (NSR), and it is required for any new major stationary source or major modification to an existing source regardless of whether or not the NAAQS are exceeded. Sources located in areas that exceed the NAAQS (nonattainment areas) would undergo a nonattainment NSR. New Source Reviews for major sources in areas where the NAAQS are not violated (attainment areas) would involve the preparation of a Prevention of Significant Deterioration (PSD) permit. Some sources will have the potential to emit pollutants for which their area is in attainment (or unclassifiable) as well as the potential to emit pollutants for which their area is in nonattainment. When this is the case, the source's permit will contain terms and conditions to meet both the PSD and nonattainment area major NSR requirements because these requirements are pollutant specific. In most cases, any new source must obtain a nonattainment NSR permit if it will emit, or has the potential to emit, 100 tons per year or more of any regulated NSR pollutant for which that area is in nonattainment, from marginal to extreme. In areas where air quality problems are more severe, the USEPA has established lower thresholds for three criteria pollutants: ozone (VOCs), particulate matter (PM_{10}), and carbon monoxide. The significance levels are lower for modifications to existing sources.

In general, a new source located in an attainment or unclassifiable area must get a PSD permit if it will emit, or has the potential to emit, 250 tons per year (tpy) or more of any criteria or NSR regulated pollutant. If the source is on the USEPA's list of 28 PSD source categories, a PSD permit is required if it will or may emit 100 tpy or more of any NSR regulated pollutant. The significance levels are lower for modifications to existing sources. In addition, PSD review would be triggered, with respect to a particular pollutant, if a new source or major modification is constructed within 10 kilometers of a Class 1 area (see below) and would have an impact on such area equal to or greater than 1 mg/m^3 (24-hour average) for the pollutant, even though the emissions of such pollutants would not otherwise be considered significant.

DID YOU KNOW?

Ozone is not emitted by industrial sources; it is formed by volatile organic chemicals (VOCs) and nitrogen oxides (known as ozone precursors) in the presence of heat and sunlight. VOCs emissions are regulated as a surrogate for ozone.

Some new sources or modifications to sources that are in attainment areas may be required to perform an air quality modeling analysis. This *air quality impact analysis* should determine if the source will cause a violation of the NAAQS or cause air quality deterioration that is greater than the available PSD increments. PSD requirements provide an area classification system based on land use for areas within the United States. These three areas are Class I, Class II, and Class III, and each class has an established set of increments that cannot be exceeded. Class I areas consist of national parks and wilderness areas that are only allowed a small amount of air quality deterioration. Due to the pristine nature of these areas, the most stringent limits on air pollution are enforced in the Class I areas. Class II areas consist of normal, well-managed industrial development. Moderate levels of air quality deterioration are permitted in these regions. Class III areas allow the largest amount of air quality deterioration to occur. When a PSD analysis is performed, the PSD increments set forth a maximum allowable increase in pollutant concentrations, which limit the allowable amount of air quality deterioration in an area. This in turn limits the amount of pollution that enters the atmosphere for a given region. In order to determine if a source of sulfur dioxide, for example, will cause an air quality violation, the air quality analysis uses the highest estimated concentration for annual averaging periods, and the second highest estimated concentration for averaging periods of 24 hours or less. The new NAAQS for PM and ozone contain specific procedures for determining modeled air quality violations. For reviews of new or modified sources, the air quality impact analysis should generally be limited to the area where the source's impact is significant, as defined by regulations. In addition, due to the uncertainties in making concentration estimates of large downwind distances, the air quality impact analysis should generally be limited to a downwind distance of 50 km, unless adverse impacts in a Class I area may occur at greater distances.

9.10.2.3 Air Quality Monitoring

Air quality modeling is necessary to ensure that a source is in compliance with the SIP and New Source Review requirements. When air quality modeling is required, the selection of a model is dependent on the source characteristics, pollutants emitted, terrain, and meteorological parameters. The USEPA's *Guideline on Air Quality Modeling* (40 CFR 1, Appendix W) summarizes the available models, techniques, and guidance in conducting air quality modeling analyses used in regulatory programs. This document was written to promote consistency among modelers so that all air quality modeling activities would be based on the same procedures and recommendations.

When air quality modeling is required, the specific model used (from a simple screening tool to a refined analysis) will require meteorological data. The data can vary from a few factors such as average wind speed and Pasquill–Gifford stability categories to a mathematical representation of turbulence. Whatever model is chosen to estimate air quality, the meteorological data must match the quality of the model used; for example, average wind speed used in a simple screening model will not be sufficient for a complex refined model. An air quality modeling analysis incorporates the evaluation of terrain, building dimensions, ambient monitoring data, relevant emissions from nearby sources, and the aforementioned meteorological data. For a dispersion model to provide useful and valid results, the meteorological data used in the model must be representative of the transport and dispersions characteristics in the vicinity of the source that the model is trying to simulate. The representativeness of the meteorological data is dependent on the following:

- The proximity of the meteorological monitoring site to the area under consideration
- The complexity of the terrain in the area
- The exposure of the meteorological monitoring site
- The period of time during which the data are collected

In addition, the representativeness of the data can be adversely affected by large distances between the source and the receptor of interest. Similarly, valley/mountain, land/water, and urban/rural characteristics affect the accuracy of the meteorological data for the source under consideration. For control strategy evaluations and New Source Reviews, the minimum meteorological data required to describe transport and dispersion of air pollutants in the atmosphere are wind direction, wind speed, mixing height, and atmospheric stability (or related indicators of atmospheric turbulence and mixing). Because of the question of representativeness of meteorological data, site-specific data are preferable to data collected offsite. Typically, 1 year of onsite data is required. If an offsite database is used (from a nearby airport for example), 5 years of data are normally required. With 5 years of data, the model can incorporate most of the possible variations in the meteorological conditions at the site.

9.10.2.4 Visibility

Visibility is the distance an observer can see along the horizon. The scattering and absorption of light by air pollutants in the atmosphere impair visibility. There are generally two types of air pollution that impair visibility.

The first type consists of smoke, dust, or gaseous plumes which obscure the sky or horizon and are emitted from a single source or small group of sources. The second type is a widespread area of haze that impairs visibility in every direction over a large area and originates from a multitude of sources. Regardless of the type of air pollution that impairs the visibility at a particular location, any change in the meteorology or source emissions that would increase the pollutant concentration in the atmosphere will result in increased visibility impairment. PSD Class I areas have the most stringent PSD increments and therefore must be protected from not only high pollutant concentrations but also the additional problems pollutants in the atmosphere can cause. Under the Clean Air Act, PSD Class I areas must be evaluated for visibility impairment. This may involve a visibility impairment analysis. According to USEPA regulations, visibility impairment is defined as any humanly perceptible change in visibility (visual range, contrast, or coloration) from natural conditions; therefore, any location is susceptible to visibility impairment due to air pollution sources. Because PSD Class I areas (national parks and wilderness areas) are known for their aesthetic quality any change or alteration in the visibility of the area must be analyzed.

9.10.2.5 Pollutant Dispersion

Pollutant dispersion is the process of pollutants being removed from the atmosphere and deposited onto the surface of the Earth. Stack plumes contain gases and a small amount of particles that are not removed from the gas stream. When the plume emerges from the stack, these particles are carried with it. Once airborne, the particles begin to settle out and become deposited on the ground and on surface objects. There are basically two ways the particles can be deposited: dry deposition (gravitational settling) or wet deposition (precipitation scavenging). Depending on the meteorological conditions during the time of pollutant emission, these may

1. Settle out quickly due to their weight and the effect of gravity
2. Be transported further downwind of the source due to buoyancy and wind conditions
3. Be washed out of the atmosphere by precipitation or clouds (wet deposition)

In any case, the deposition of these pollution particles is important to understand and quantify because pollutants deposited upon the ground can impact human health, vegetation, and wildlife. Pollutant deposition concentrations must be predicted to minimize the risk to human health. In order to quantify the amount of pollutant deposition that occurs from stack emissions, air quality models can be used. These models determine pollution deposition based on the chemical reactivity and solubility of various gases and by using detailed data on precipitation for the areas in question.

9.10.2.6 Vapor-Plume-Induced Icing

Vapor plumes are emitted from cooling towers and stacks and consist mainly of water vapor. Although pollutant concentrations are not a major concern with vapor plumes, other problems arise when vapor plume sources are located close to frequently traveled roads and populated areas. Vapor emitted from a stack is warm and moist. When meteorological conditions are favorable, the moisture in the vapor plume condenses out and settles on cooler objects (e.g., road surfaces). This phenomenon is similar to the moisture that collects on the sides of a glass of water on a warm day. If temperatures are at or below freezing when the moisture condenses, road surfaces can freeze rapidly, creating hazardous driving conditions. In addition, light winds can cause the plume to remain stagnant, creating a form of ground fog that can cause low visibility as well. Water vapor plumes that lower visibility can create hazards for aircraft, especially during critical phases of flight, including landings and takeoffs.

9.11 Clean Air Act and Hydraulic Fracturing Operations

The Clean Air Act requires the USEPA to set national standards to limit levels of certain pollutants. USEPA regulates those pollutants by developing human health-based and environmentally and scientifically based criteria for setting permissible levels. Air regulations do not normally include exceptions for a company's size, the age of a field, or the type of operations. Typically, the air rules are silent on issues such as conventional versus unconventional shale gas plays, old versus new fields, and the depth of a well. For the most part, the air emissions, applicable regulations, and associated emissions controls for a shale play are not different than those for any other natural operations. There may be differences due to location (some areas of the country have better air quality than others), equipment needs (some shale plays may produce a wetter gas than others), and sulfur content level of the gas. Geographic areas that do not meet USEPA standards for a given pollutant are designated as nonattainment areas (USEPA, 2008e). This is the case for the Barnett Shale play, much of which is located in or near the Dallas–Fort Worth ozone nonattainment area. As a result, Barnett Shale production activities must often comply with much more stringent regulations than similar operations proposed outside of a nonattainment area. As a result of implementation of the CAA, air quality has improved dramatically across the United States during the last few decades, and existing regulations should continue to reduce air pollution emissions during the next 20 years or longer (USEPA, 2008f). As with other U.S. industries, shale gas producers must comply with existing and new air regulations including those resulting

DID YOU KNOW?

The CAA is a program of "cooperative federalism" between the USEPA and the states, designed to achieve attainment and maintenance of the nation's air quality goals. The CAA achieves this form of "cooperative federalism" by allocating regulatory responsibility between the USEPA and the states. The USEPA is responsible for establishing the National Ambient Air Quality Standards (NAAQS) for certain common air pollutants that are commonly referred to as *criteria pollutants*, and states are responsible for ensuring that state air quality meets the NAAQS (Jacus, 2011).

from the 1990 CAA amendments. These rules pose an ongoing challenge to company resources as producers strive to understand and comply with enforcement, fines, public reaction, and possibly even project cancellations in light of new standards.

The USEPA has established National Emission Standards for Hazardous Air Pollutants (NESHAPs), which are nationally uniform standards to control specific air emissions. In 2007, USEPA implemented a new standard regarding maximum achievable control technology (MACT) for hazardous air pollutants (HAPs); the standard targeted small area sources such as shale gas operations located in areas near larger populations and limits HAP emissions (primarily benzene) from process vents on glycol dehydration units, storage vessels with flash emissions, and equipment leaks. Another example of new or amended federal regulations that will have a direct impact on controlling emissions from shale gas operations are the Standards of Performance for Stationary Spark Ignition Internal Combustion Engines and National Emission Standards for Hazardous Air Pollutants for Reciprocating Internal Combustion Engines, which regulate new and refurbished engines (USEPA, 2007b). These rules (described below), passed in 2007, target all internal combustion engines regardless of horsepower rating, location, or fuel (electric engines are not included) and include extensive maintenance, testing, monitoring, recordkeeping, and reporting requirements (USEPA, 2007a).

9.11.1 Stationary Internal Combustion Engines

In 2007, the USEPA issued a rule to reduce emissions of criteria and air toxic pollutants from stationary internal combustion engines. These engines are used at facilities such as power plants and chemical and manufacturing plants to generate electricity and power pumps and compressors. They are also used in emergencies to produce electricity and to pump water for flood and fire control. In shale gas fracking, stationary internal combustion engines are used for prime movers to run the main power source of the drilling rig—and the frack pumps, mud pumps, generators (to supply electricity to the doghouse and other onsite buildings), air packages, cement pumps, and stimulation pumps. The power required to run the drawworks, which functions to reel out and reel in the drill line, a large-diameter wire rope, in a controlled fashion, is a function of total borehole depth, with 10 hp required for every 30.5 m. Industry sources report that the ideal and most commonly used rig in Marcellus Shale drilling is a 1000-hp unit (NYECE, 2009; Rigzone, 2010). For stationary drilling and completion, engine emissions are calculated from Equation 9.3:

$$E_i = (ER \times LF \times OT \times ETT) \times EF \tag{9.3}$$

where:

E_i = carbon emissions of species i

ER = equipment rating (hp)

LF = fractional load factor

OT = operating time

ETT = equipment thermal efficiency

EF = efficiency factor

Note that combustion emissions associated with other streams (i.e., production, processing, and transmission/distribution) in the life cycle are difficult to estimate without details of site equipment and fuel usage data (Santoro et al., 2011).

9.11.2 Reciprocating Internal Combustion Engines

In 2010, the USEPA issued a final rule that will reduce emissions of toxic air pollutants from existing gas-fired stationary reciprocating internal combustion engines (RICEs). These engines are known as spark ignition (SI) engines. Industrial facilities such as power plants and chemical and manufacturing plants use these engines to generate electricity for compressors and pumps. These engines are used in the oil and gas industry, both for production and transport by pipeline. They also are used in emergencies to produce electricity to pump water for flood and fire control. Operators of existing stationary SI engines will be required to install emissions control equipment that would limit air toxic emissions for the following engines:

- Stationary non-emergency four-stroke lean-burn (4SLB) engines with a site rating between 100 hp and 500 hp and located at a major source of HAP emissions
- Stationary non-emergency four-stroke rich-burn (4SRB) and 4SLB engines with a side rating greater than 500 hp and located at an area source of HAP emissions

DID YOU KNOW?

The USEPA is not large enough to regulate every air emissions source nationwide, let alone consider the local and regional differences; therefore, the agency typically delegates that role to local, state, and tribal agencies. This delegation of authority can include rule implementation, permitting, reporting, and compliance. Any state given such delegation of authority can pass more restrictive rules, but they are prohibited from passing a rule that is less stringent than its federal counterpart.

9.11.3 Air Quality Permits

Air quality permits are legal documents that facility owners and operators must abide by. A permit specifies what construction is allowed, what emission limits must be met, how the emissions sources must be operated, and what conditions (e.g., monitoring, recordkeeping, reporting requirements) must be maintained to ensure ongoing compliance. Shale gas producers may need air quality permits for a number of emissions sources, including gas compressor engines, glycol dehydrators, and flares. A company's permitting responsibility does not end with the issuance of its initial air permit. They must be constantly vigilant that a new regulation, modification, replacement, or process change does not impact their existing permit and require a permit amendment or a more stringent permit. Although these permits may differ across the country, they all contain specific conditions designed to ensure that state and federal standards are met and to prevent any significant degradation in air quality as a result of a proposed activity.

9.12 Land (Soil) Quality

Shale gas operations can impact land because of the accompanying need for solid waste disposal and surface disturbances that, in turn, may disturb the visual landscape or may affect wildlife habitat. Operations on federal lands are a special case with unique requirements that are discussed below.

9.12.1 Resource Conservation and Recovery Act

The 1976 *Resource Conservation and Recovery Act* (RCRA) is the United State's single most important law dealing with the management of hazardous waste. RCRA and its 1984 amendments, *Hazardous and Solid Waste Amendments,* deal with the ongoing management of solid wastes throughout the country—with an emphasis on hazardous waste. Keyed to the waste side of hazardous materials, rather than broader issues dealt with in other acts, RCRA is primarily concerned with land disposal of hazardous wastes. The goal is to protect groundwater supplies by creating a "cradle-to-grave" management system with three key elements (Masters, 1991):

1. *Tracking system*—Requires a manifest document to accompany any waste transported from one location to another.
2. *Permitting system*—Helps ensure safe operation of facilities that treat, store, or dispose of hazardous wastes.
3. *Disposal control system*—Controls and restrictions governing the disposal of hazardous wastes onto, or into, the land.

The RCRA regulates five specific areas for the management of hazardous waste, with a focus on treatment, storage, and disposal (Griffin, 1989):

1. Identifying what constitutes a hazardous waste and providing classification of each
2. Publishing requirements for generators to identify themselves, which includes notification of hazardous waste activities and standards of operation for generators
3. Adopting standards for transporters of hazardous wastes
4. Adopting standards for treatment, storage, and disposal facilities
5. Providing for enforcement of standards through a permitting program and legal penalties for noncompliance

In 1978, the USEPA proposed hazardous waste management standards that included reduced requirements for some industries, including oil and gas, with large volumes of wastes. The USEPA determined that these large-volume "special wastes" were lower in toxicity than other wastes being regulated as hazardous waste under the RCRA (USEPA, 2008g). That same year, the USEPA proposed regulations for managing hazardous waste under Subtitle C of RCRA (43 FR 58946). Included in these proposed regulations was a deferral of hazardous waste requirements for six categories of waste (i.e., special wastes) until further study and assessment could be completed to determine their risk to human health and the environment:

- Cement kiln dust
- Mining waste
- Oil and gas drilling muds and oil production brines
- Phosphate rock mining, beneficiation (i.e., processes whereby extracted ore from mining is separated into mineral and gangue), and processing waste
- Uranium waste
- Utility waste (i.e., fossil fuel combustion waste)

These wastes typically are generated in large volumes and, at the time, were believed to pose less risk to human health and the environment than the wastes being identified for regulation as hazardous waste.

In 1980, Congress enacted the Solid Waste Disposal Act Amendments (Public Law 96-482), which amended the RCRA in several ways. Pertinent to special wastes was the addition of Sections 3001(b)(2)(A) and 3001(b)(3)(A). These new sections—frequently referred to as the Bensten and Bevill Amendments—exempted "special wastes" from regulation under Subtitle C of RCRA until further study and assessment of risk could be performed. Specifically, the Bentsen Amendment (Section 3001(b)(2)(A)) exempted drilling fluids, produced waters, and other wastes associated with the exploration, development, and production of crude oil or natural gas or geothermal energy. The Bevill Amendment (Section 3001(b)(3)(A)(i–iii)) exempted the special wastes listed above.

In 1987, the USEPA issued a Report to Congress that outlined the results of a study on the management, volume, and toxicity of wastes generated by the oil, natural gas, and geothermal industries. In 1988, the USEPA issued a final regulatory determination stating that control of oil and gas exploration and production wastes under RCRA Subtitle C was not warranted. The USEPA made this determination because it found that other state and federal programs could protect human health and the environment more effectively. In lieu of regulation under subtitle C, the USEPA implemented a three-pronged strategy to ensure that the environmental and programmatic issues were addressed:

1. Improve other federal programs under existing authorities.
2. Work with states to improve some programs.
3. Work with congress to develop any additional statutory authorities that may be required.

These wastes have remained exempt from Subtitle C regulations, but this does not preclude these wastes from control under state regulations or other federal regulations (USEPA, 2002). The exemption applies only to the federal requirements of RCRA Subtitle C. A waste that is exempt from Subtitle C regulation might be subject to more stringent or broader state hazardous and nonhazardous waste regulations and other state and federal program regulations. For example, oil and gas exploration and production wastes may be subject to regulation under RCRA Subtitle D, the Clean Air Act, the Clean Water Act, the Safe Drinking Water Act, and the Oil Pollution Act of 1990 (USEPA, 1988, 2008j).

In 1989, the USEPA worked with the Interstate Oil and Gas Compact Commission (IOGCC), state regulatory officials, industry representatives, and nationally recognized environmental groups to establish a Council on Regulatory Needs. The purpose of the council was to review existing state oil and gas exploration and production waste management programs and to develop guidelines to describe the elements necessary for an effective state program. This effort was begun by the USEPA as part of the second prong of the agency's approach. These groups then worked together with state regulatory agencies to review state programs, on a voluntary basis, against these guidelines and to make recommendations for improvement. This state review program continues today under the guidance of a nonprofit organization called STRONGER. The state programs reviewed to date represent over 90% of the onshore domestic production (STRONGER, 2008b).

Working with the IOGCC, STRONGER has continued to update the guidelines consistent with developing environmental and oilfield technologies and practices. Under the state review process, state programs have continued to improve, and the follow-up reviews have shown significant improvement where states have successfully implemented the recommendations of the review committees. Arguably, the RCRA is our single most important law dealing with the management of hazardous waste—it certainly is the most comprehensive piece of legislation that the USEPA has promulgated to date.

9.12.2 Endangered Species Act

The Endangered Species Act (ESA) passed in 1973 (Public Law 93-205) protects plants and animals that are listed by the federal government as "endangered." Sections 7 and 9 are central to regulating oil and gas activities. Section 9 makes it unlawful for anyone to "take" a listed animal, and this includes significantly modifying its habitat (*Babbitt v. Sweet Home Chapter of Communities for a Great Oregon*, 515 U.S. 687, 1995). This applies to private parties and private land; a landowner is not allowed to harm an endangered animal or its habitat on his or her property. Section 7 applies not to private parties, but to federal agencies. This section covers not only federal activities but also the issuance of federal permits for private activities, such as Section 404 permits issued by the Army Corps of Engineers, to people who want to do construction work in waters or wetlands. Section 7 imposes an affirmative duty on federal agencies to ensure that their actions (including permitting) are not likely to jeopardize the continued existence of a listed species (plant or animal) or result in the destruction or modification of critical habitat. Both Sections 7 and 9 allow "incidental" takes of threatened or endangered species, but only with a permit.

To "take" is to harass, harm, pursue, hunt, shoot, wound, kill, trap, capture, or collect a plant or animal of any threatened or endangered species. Harm includes significant habitat modification when it kills or injures a member of a listed species through impairment of essential behavior (e.g., nesting or reproduction). For any non-federal industrial activity, the burden is on the owner or operator to determine if an incidental take permit is needed. This is typically accomplished by contacting the U.S. Fish and Wildlife Service (FWS) to determine whether any listed species are present or will potentially inhabit the project site. A biological survey may be required to determine whether protected species are present on the site and whether

DID YOU KNOW?

On April 9, 2011, *The Cypress Times* published an article entitled "Natural Gas Company Pleads Guilty in Arkansas in Connection with Fayetteville Shale Pipeline Construction Activities." The article discussed the activities of a wholly owned subsidiary of a Houston-based energy company. Specifically, the subsidiary company pleaded guilty to three counts of violating the Endangered Species Act. The subsidiary was engaged in gathering, conditioning, and treating activities related to the development of natural gas properties in the Fayetteville Shale in Arkansas. In a plea agreement, the subsidiary admitted that it did not adequately control erosion during construction of the pipelines in the Little Red River Watershed. This lack of erosion control allowed silt to run downhill to the streams, causing sediment to build up at the stream crossing and downstream. This erosion and sedimentation occurred in waters containing the endangered speckled pocketbook mussel (*Lampsilis streckeri*) and caused a take of at least one mussel by harassment in the South Fork, Little Fork, and Archey Fork of the Little Red River.

a Section 9 permit may be required (ALL Consulting and MBOGC, 2004; CICA, 2008). The FWS as well as state fish and game agencies offer services to help operators determine whether a given project is likely to result in a take and whether a permit is required. The FWS can also provide technical assistance to help design a project so as to avoid impacts; for example, the project could be designed to minimize disturbances during nesting or mating seasons (CICA, 2008).

A Section 9 permit must include a habitat conservation plan (HCP) consisting of an assessment of impacts; measures that will be undertaken to monitor, minimize, and mitigate any impacts; alternative actions considered and an explanation of why they were not taken; and any additional measures that the FWS may require (USFS, 2011). Mitigation measures, which are actions that reduce or address potential adverse effects of a proposed

DID YOU KNOW?

All 50 states have fish and game or wildlife agencies that work in cooperation with U.S. Fish and Wildlife Service district offices with regard to the incidental take permitting process. Many states also have their own endangered and threatened species lists that may include species not on the federal lists, and they may have their own requirements for protecting endangered species (CICA, 2008).

activity upon species, must be designed to address the specific needs of the species involved and be manageable and enforceable. Mitigation measures may take many forms, such as preservation (via acquisition or conservation easement) of existing habitat, enhancement or restoration of degraded or former habitat, creation of new habitats, establishment of buffer areas around existing habitats, modifications of land use practices, and restriction on access (USFS, 2011).

9.13 Oil and Gas Operations on Public Lands

Oil and Gas Operations Regulations (43 CFR 3160) govern operations associated with the exploration, development, and production of oil and gas on federal and Indian lands. The U.S. Department of Interior's Bureau of Land Management (BLM) is responsible for permitting and managing most onshore oil and gas activities on federal lands. The BLM carries out its responsibility to protect the environment throughout the process of oil and gas resource exploration and development on public lands. Resource protection is considered throughout the land use planning process when Resource Management Plans (RMPs) are prepared and when an Application for Permit to Drill (APD) is processed (ALL Consulting and MBOGC, 2002). The BLM's inspection and enforcement and monitoring program is designed to ensure that operators comply with relevant laws and regulations as well as specific stipulations set forth during the permitting process.

Because most shale gas activity in the near future is expected to occur in the eastern U.S. basins, it is not likely that much of this development will occur on federal lands. While there are some federal lands, such as national parks, national forests, and military installations, these are much less extensive in the east than in the west. Where shale gas operations do occur on federal lands, the BLM has a well-established program for managing these activities to protect human health and the environment.

DID YOU KNOW?

The amount of state-owned land varies considerably from state to state, and each state manages these lands differently. In most states, leasing of state-owned minerals occurs through lease auctions. Because states are already set up to manage oil and gas operations within their borders, no special permitting or enforcement systems are required. Some states do have environmental policy acts that require a review of environmental impacts that may result from leasing or operations on state lands or of any state action that may affect the environment.

9.14 Superfund (CERCLA)

The Comprehensive Environmental Response, Compensation, and Liability Act (CERCLA) of 1980 was enacted by Congress to clean up hazardous waste disposal mistakes of the past and to cope with emergencies of the present. It is more often referred to as the Superfund Law because of its key provision, a large trust fund (about $1.6 billion). Later, in 1986, when the law was revised, this fund was increased to almost $9 billion. The revised law is referred to as the Superfund Amendments and Reauthorization Act (SARA). The key requirements under CERCLA include the following:

1. CERCLA authorizes the USEPA to deal with both short-term problems (emergency situations triggered by a spill or release of hazardous substances) and long-term problems involving abandoned or uncontrolled hazardous waste sites for which more permanent solutions are required.
2. CERCLA has set up a remedial scheme for analyzing the impact of contamination on sites under a hazard ranking system, from which a list of prioritized disposal and contaminated sites is compiled. This list is known as the National Priorities List (NPL). The NPL identifies the worst sites in the nation, based on such factors as the quantities and toxicity of wastes involved, the exposure pathways, the number of people potentially exposed, and the importance and vulnerability of the underlying groundwater.
3. CERCLA also forces those parties who are responsible for hazardous waste problems to pay the entire cost of cleanup.
4. Title III of SARA requires federal, state, and local governments and industry to work together in developing emergency response plans and reporting on hazardous chemicals. This requirement is commonly referred to as the Community Right-to-Know Act, which allows the public to obtain information about the presence of hazardous chemicals in their communities and releases of these chemicals into the environment.

CERCLA Section 101(14) excludes certain substances from the definition of hazardous substance, thus exempting them from CERCLA regulation. These substances include petroleum, meaning crude oil or any fraction thereof that is not specifically listed as a hazardous substance, natural gas, natural gas liquids, liquefied natural gas, and synthetic gas usable for fuel. If a release of one of these substances occurs, CERCLA notification is not required. Thus, CERCLA reporting will only apply to shale gas production and processing sites if hazardous substances other than crude oil or natural gas are spilled in reportable quantities; such are not usually present at these sites.

DID YOU KNOW?

Many states have separate requirements regarding hazardous substances. Reporting of releases of the materials exempted under CERCLA may be required under state law.

This particular exclusion applies only to CERCLA Section 103(a) reporting requirements; it does not exempt a facility from the Emergency Planning and Community Right-to-Know Act (EPCRA) Section 304 reporting requirements. A release of a petroleum product containing certain substances is potentially reportable under EPRCA Section 304 if more than an RQ (reportable quantity) of that substance is released (USEPA, 1998).

9.15 Emergency Planning and Community Right-to-Know Act

The Emergency Planning and Community Right-to-Know Act (EPCRA) of 1986 was created to help communities plan for emergencies involving hazardous substances. The community right-to-know provisions of EPCRA are the most relevant part of the law for shale gas producers. They help increase the public's knowledge and access to information on chemicals at individual facilities, along with the uses and potential releases into the environment. Under Sections 311 and 312 of EPCRA, facilities manufacturing, processing, or storing designated hazardous chemicals must make Material Safety Data Sheets (MSDSs) describing the properties and health effects of these chemicals available to state and local officials and local fire departments. Facilities must also provide state and local officials and local fire departments with inventories of all onsite chemicals for which MSDSs exist. Information about chemical inventories at facilities and MSDSs must be available to the public. Facilities that store over 10,000 pounds of hazardous chemicals are subject to this requirement. Any hazardous chemicals above the threshold stored at shale gas production and processing sites must be reported in this manner.

Section 313 of EPCRA authorizes the USEPA's Toxic Release Inventory (TRI), a publicly available database that contains information on toxic chemical release and waste management activities reported annually by certain industries as well as federal facilities. The USEPA issues a list of industries that must report releases for the database. To date, the USEPA has not included oil and gas extraction as an industry that must report under the TRI. This is not an exemption in the law; rather, it is a decision by the USEPA that this industry is not a high priority for reporting under the TRI. Part of the rationale for this decision is based on the fact that most of the information required under the TRI is already reported by producers to state agencies that make it publicly available. Also, TRI reporting from the hundreds of thousands of oil and gas sites would overwhelm the existing USEPA reporting system and make it difficult to extract meaningful data from the massive amount of information submitted (IOGCC, 1996; IPAA, 2000).

Section 304 of EPCRA requires reporting of releases to the environment of certain materials that are subject to this law. As noted in the section above, this requirement would apply to any releases of petroleum products that exceed reporting thresholds, even if those products are exempt from CERCLA reporting. Although shale gas production facilities do not normally store the materials subject to EPCRA reporting (known as EPCRA *extremely hazardous substances* and as CERCLA *hazardous substances*), a limited number of chemicals used in the hydraulic fracturing process, such as hydrochloric acid, are classified as hazardous under CERCLA. These chemicals may be brought onsite for a few days at most during fracturing or work-over operations. Businesses must report non-permitted releases—into the atmosphere, surface water, or groundwater—of any listed chemicals above threshold amounts (the reportable quantity, or RQ) to federal, state, and local authorities. Although every precaution is taken to prevent chemical spills, in the event of an accidental release above the reportable quantity a report must be made to these authorities by the operator.

9.16 OSHA Oil and Gas Drilling, Servicing, and Storage Standards

9.16.1 Occupational Safety and Health Act of 1970

Although some federal safety legislation was passed prior to 1970, this legislation affected only a small fraction of the American workforce. At the end of the 1960s, two shortcomings became blatantly obvious: (1) a new national policy needed to be established that would encompass the majority of industries, and (2) states had generally failed to meet their voluntary obligations for health and safety in the workplace. To solve these shortcomings, the OSH Act (designed to "ensure so far as possible every working man and woman in the Nation safe and healthful working conditions and to preserve our human resources") was signed by President Nixon on December 29, 1970. Since its effective date on April 28, 1971, this single act has had enormous impact on the safety and health movement within the United States, more than any other legislation. The law affects approximately 60 million employees in over 4 million establishments but excludes employees of state and federal government, who are protected under regulations similar to those within the Occupational Safety and Health Administration (OSHA). Under the Secretary of Labor, the Assistant Secretary of Labor for Occupational

Safety and Health has the responsibility to guide and administer OSHA. Under the provisions of the Act, each employer covered by the Act has the following duties:

1. The general duty to furnish each of his/her employees employment and places of employment that are free from recognized hazards that are causing or likely to cause death or serious physical harm (which means that even if a hazard in the workplace is not specifically covered by a regulation, the employer must protect the employee anyway). This is commonly referred to as the Act's *General Duty Clause*; safety professionals and in particular OSHA professionals view it as a "safety net."
2. The specific duty of complying with safety and health standards promulgated under the act.

Each employee has the duty to comply with the safety and health standards, as well as all rules, regulations, and orders applicable to his own actions and conduct on the job. Experience has shown that when employees are informed of this requirement under OSHA, they are surprised. They often view the Act as applying only to the employer.

The OSHA regulations (compliance with these regulations being the major concern of this text) take two basic forms: They are either *specific standards* or *performance standards*. Specific standards explain exactly how to comply; for example, the OSHA regulation covering means of egress from buildings very specifically lists requirements for means of egress, exit access, exit discharge, and so forth. A performance standard lists the ultimate goal of compliance but does not explain exactly how to accomplish it. A good example of a performance standard is the General Duty Clause, which states that the employer must protect the health and safety of the employee even if no OSHA regulation currently covers the work activity in question. These standards do not explain how to accomplish this—that is left up to the employer.

In general, safety and health regulations governing labor practices are listed under Title 29 of the Code of Federal Regulations (CFR). The occupational safety and health regulations are found in Parts 1900 to 1999. The actual workplace regulations we are concerned with in this text are 29 CFR Part 1910 (General Industry Standards) and Part 1926 (Construction Standards). To gain an understanding of what is contained in one of these parts, let's take a look at Part 1910, which is divided into subparts A to Z, as shown in Figure 9.1. OSHA determines how well a program is working by reviewing the company's injury data and insurance costs. If the program is effective, the injury and insurance costs will reflect that.

Obviously, to get accurate data upon which to base its judgment, OSHA requires extensive recordkeeping for written programs, injuries, illnesses, safety audits, inspections, corrections, and training. Training is a major part of the OSH Act. Almost every regulation requires some sort of transmission

29 CFR 1910 Subparts A–Z	
Subpart A. *General*	Provides provisions for OSHA's initial implementation of regulations
Subpart B. *Adoption and Extension of Established Federal Standards*	Explains which businesses are covered by OSHA regulations
Subpart C. *General Safety and Health Provisions*	Provides the right for an employee to gain access to exposure and medical records
Subpart D. *Walking and Working Surfaces*	Establishes requirements for fixed and portable ladders, scaffolding, manually propelled ladder stands, and general walking surfaces
Subpart E. *Means of Egress*	Establishes general requirements for employee emergency plans and fire prevention plans
Subpart F. *Powered Platforms, Man Lifts, and Vehicle-Mounted Work Platforms*	Mandates the minimum requirements for an elevated safe work platform
Subpart G. *Occupational Health and Environmental Control*	Mandates engineering controls of physical hazards such as ventilation for dusts, control of noise, and control of ionizing and non-ionizing radiation
Subpart H. *Hazardous Materials*	Provides requirements for the use, handling, and storage of hazardous materials
Subpart I. *Personal Protective Equipment*	Provides general requirements for personal protective equipment
Subpart J. *General Environmental Controls*	Mandates the requirements for sanitation, accident prevention signs and tags, confined space entry, and hazardous energy lockout/tagout
Subpart K. *Medical and First Aid*	Requires that an employer provide first-aid facilities or personnel trained in first aid to be at the facility

Subpart L. *Fire Protection*	Mandates portable or fixed fire suppression systems for work places
Subpart M. *Compressed Gas and Compressed Air Equipment*	Provides requirements for air receivers
Subpart N. *Materials Handling and Storage*	Covers the use of mechanical lifting devices, changing a flat tire, and forklift and helicopter operations
Subpart O. *Machinery and Machine Guarding*	Provides requirements for guarding rotating machinery
Subpart P. *Hand and Portable Powered Tools and Other Hand-Held Equipment*	Provides requirements for hand-held equipment
Subpart Q. *Welding, Cutting, and Brazing*	Requires the use of eye protection, face shields with lenses, proper handling of oxygen and acetylene tanks
Subpart R. *Special Industries*	Provides special requirements for textiles, bakery equipment, laundry machinery, sawmills, pulpwood logging, grain handling, and telecommunications
Subpart S. *Electrical*	Requires the use of protection mechanisms for electrical installations
Subpart T. *Commercial Diving Operation*	Mandates requirements for dive teams
Subparts U–Y. Not currently assigned	—
Subpart Z. *Toxic and Hazardous Substances*	Requires monitoring and protective methods for controlling hazardous airborne contaminants

FIGURE 9.1
Occupational Safety and Health Administration Standards (29 CFR 1910 Subparts A–Z).

of information and training. Why? Simply because injury statistics show that newer employees without adequate training are far more likely to be injured on the job than those with more experience and training.

Enforcement of the OSH Act is carried out through inspections (audits), citations, and levying civil penalties. These three increasingly punitive steps are designed to achieve a safe workplace by requiring the removal of hazards. If hazardous situations are discovered, follow-up inspections ensure that the appropriate corrections are made. OSHA investigates and writes citations based on inspections of the work site. An OSHA inspector may visit a site based on the following:

- An employee complaint (don't you just love them?)
- A report that an injury or fatality has occurred
- A random visit to a high-risk business

If the inspection uncovers one or more violations, the OSHA compliance officer provides an explanation on a written inspection report. The types of violations include the following:

- *de minimis*—A condition that has no direct or immediate relationships to job safety and health (e.g., an error in interpretation of a regulation).
- *General*—Inadequate or nonexistent written programs, lack of training, training records, etc.
- *Repeated*—Violations where, upon reinspection, another violation is found of a previously cited section of a standard, rule, order, or condition violating the general duty clause.
- *Serious*—A violation that could cause serious harm or permanent injury to the employee and where the employer did not know or could not have known of the violation.
- *Willful*—A violation where evidence shows that the employer knew that a hazardous condition existed that violated an OSHA regulation but made no reasonable effort to eliminate it.
- *Imminent danger*—A condition where there is reasonable certainty that an existent hazard can be expected to cause death or serious physical harm immediately or before the hazard can be eliminated through regular procedures.

When a compliance officer believes an employer has violated a safety or health requirement of the act, or any standard, rule, or order promulgated under it, he or she will issue a citation. Any citation issued for noncompliance must be posted in clear view near the place where the violation occurred for 3 working days or until corrected, whichever is longer. Does the employer have any recourse when cited by OSHA? Actually, the employer can take either of the following courses of action regarding citations:

1. The employer can agree with the citation and correct the problem by the date given on the citation and pay any fines.
2. The employer can contest the citation, proposed penalty, or correction date, as long as it is done within 15 days of the date the citation was issued.

Specific standards have been developed by OSHA to reduce potential safety and health hazards in the oil and gas drilling, servicing, and storage industry. States also have requirements that provide further work and public safety protections. Before discussing each of the pertinent OSHA standards related to safe shale gas fracking operations, it is important to discuss and describe actual process practices, industry profile data, nature of the work, types of recorded on-the-job injuries specific to the industry, and the number and type of citations issued by OHSA during scheduled or unscheduled audits.

9.16.2 SIC Industry Group 138—Oil and Gas Field Services: Process Description*

Oil and gas well drilling and servicing are part of Major Group 13 in the Standard Industrial Classifications (SIC). The classification is further defined by three subdivisions within Industry Group 138 (Oil and Gas Field Services): SIC 1381 (Drilling and Gas Wells), SIC 1382 (Oil and Gas Field Exploration Services), and SIC 1389 (Oil and Gas Services, Not Elsewhere Classified). SIC 1381 includes establishments primarily engaged in drilling wells for oil or gas field operations for others on a contract or fee basis. This industry includes contractors that specialize in the following:

- Directional drilling of oil and gas wells on a contract basis
- Redrilling oil and gas wells on a contract basis
- Reworking oil and gas wells on a contract basis
- Spudding (starting to drill) oil and gas wells on a contract basis
- Well drilling of gas, oil, and water intake on a contract basis

SIC 1382 includes companies primarily engaged in performing oil and gas geophysical, geological, or other exploration services on a contract or fee basis:

- Aerial geophysical exploration, oil and gas field on a contract basis
- Exploration, oil and gas field on a contract basis
- Geological exploration, oil and gas field on a contract basis
- Geophysical exploration, oil and gas field on a contract basis
- Seismograph surveys, oil and gas field on a contract basis

* This section is adapted from USDOL, *Oil and Gas Well Drilling, Servicing and Storage Standards*, U.S. Department of Labor, Washington, DC, 2012 (http://www.osha.gov/SLTC/oilgaswelldrilling/index.html).

SIC 1389 includes establishments primarily engaged in performing oil and gas field services, not elsewhere classified, for others on a contract or fee basis. Services included are excavating slush pits and cellars; grading and building foundations at well locations; well surveying; running, cutting, and pulling casings, tubes, and rods; cementing wells; shooting wells; perforating well casings; acidizing and chemically treating wells; and cleaning out, bailing, and swabbing wells. Establishments that have complete responsibility for operating oil and gas wells for others on a contract or fee basis are classified according to the product extracted rather than as oil and gas field services:

- Acidizing wells on a contract basis
- Bailing wells (removing water, sand, mud, drilling cuttings, or oil from cable-tool drilling) on a contract basis
- Building oil and gas well foundations on a contract basis
- Cementing oil and gas well casings on a contract basis
- Chemically treating wells on a contract basis
- Cleaning lease tanks, oil and gas field on a contract basis
- Cleaning wells on a contract basis
- Derrick building, repairing, and dismantling oil and gas on a contract basis
- Dismantling of oil well rigs (oil field service) on a contract basis
- Erecting lease tanks, oil and gas field on a contract basis
- Excavating slush pits and cellars on a contract basis
- Fishing for tools, oil and gas field on a contract basis
- Gas compressing natural gas at the field on a contract basis
- Gas well rig building, repairing, and dismantling on a contract basis
- Grading oil and gas well foundations on a contract basis
- Hard banding service on a contract basis
- Hot oil treating of oil field tanks on a contract basis
- Hot shot service on a contract basis
- Hydraulic fracturing wells on a contract basis
- Impounding and storing saltwater in connection with petroleum
- Lease tanks, oil and gas field: erecting, cleaning, and repairing on a contract basis
- Logging wells on a contract basis
- Mud service, oil field drilling on a contract basis
- Oil sampling service for oil companies on a contract basis
- Oil well logging on a contract basis
- Perforating well casings on a contract basis

TABLE 9.12
Top 10 OSHA Violations Cited (FY 2005)

OSHA Standard	Number of Citations	Description
29 CFR 1901.1200	62	Hazard Communication
29 CFR 1910.146	54	Permit-Required Confined Spaces
OSH Act, Section 5(a)(1)	52	General Duty Clause
29 CFR 1910.132	42	Personal Protective Equipment, General
29 CFR 1910.305	42	Wiring Methods, Components, and Equipment
29 CFR 1910.23	40	Guarding Floor and Well Openings and Holes
29 CFR 1910.134	39	Respiratory Protection
29 CFR 1910.151	37	Medical Services and First Aid
29 CFR 1910.141	33	Sanitation
29 CFR 1910.157	30	Portable Fire Extinguishers

Source: OSHA, *Profile: Oil and Gas Well Drilling and Servicing*, U.S. Occupational Safety and Health Administration, Washington, DC, 2006.

- Pipe testing service, oil and gas field on a contract basis
- Plugging and abandoning wells on a contract basis
- Pumping of oil and gas wells on a contract basis
- Removal of condensate gasoline from field gathering lines on a contract basis
- Roustabout service on a contract basis
- Running, cutting, and pulling casings, tubes, and rods on a contract basis
- Servicing oil and gas wells on a contract basis
- Shooting wells on a contract basis
- Shot-hole drilling service, oil and gas field on a contract basis
- Surveying wells on a contract basis, except seismographic
- Swabbing wells on a contract basis

Table 9.12 lists the top ten OSHA citations issued during Fiscal Year 2005 for Industry Group 138. Table 9.13 lists some of the potential hazards and their sources related to Industry Group 138.

Case Study 9.2—DANGER: Confined Space Entry and Hydrogen Sulfide Exposure

A few years ago, a worker for a large metropolitan utility company in the Midwest was killed in a confined space. This incident, though not related to oil and gas operations, points to the dangers of confined space entry and hydrogen sulfide exposure with its characteristic deadly

TABLE 9.13

Some Potential Hazards and Their Sources

Hazard	Source
Struck by	Falling/moving pipe; tongs and/or spinning chain, Kelly, rotary table, etc.; high-pressure hose connection failure causing employees to be struck by whipping hose; tools or debris dropped from elevated location in rig; vehicles
Caught in/between	Collars and tongs, spinning chain, and pipe; clothing getting caught in rotary table or oil string
Fire/explosion/high-pressure release	Well blowout, drilling/tripping out/swabbing, etc., resulting in release of gas which might be ignited if not controlled at the surface; welding or cutting near combustible materials; uncontrolled ignition sources near the well head (e.g., heater in the doghouse); unapproved or poorly maintained electrical equipment; aboveground detonation of perforating gun
Rig collapse	Overloading beyond the rated capacity of the rig; improper anchoring or guying; improper raising and lowering of the rig; existing maintenance issues with the rig structure that impact its integrity
Falls	Falls from elevated areas of the rig (e.g., stabbing board, monkey board, ladder); falls from rig floor to grade
Hydrogen sulfide (H_2S) exposure	H_2S release during drilling, swabbing, perforating operations, etc., resulting in employee exposures; in production gauging operations, gauges sometimes exposed to H_2S

rotten egg odor. The case highlighted here evolved around an organizational practice whereby utility workers, during their rounds of sewage control stations, had literally been taught (via on-the-job training) to avoid using a confined space permit, the two-person rule, and other safety practices whenever certain vaults had to be entered to adjust a certain valve in each vault. The valves were wall-mounted just inside the vaults and were designed to be manipulated (from inside the vaults) by 8-inch-diameter valve wheel handles. These particular valves (in 22 similar vaults) had to be adjusted each day by the assigned station checkers. One thing a safety professional quickly learns about workers is that if there is an easy way to accomplish a certain task it's a safe bet that workers will find it. After the valves were initially installed, in only a short time the workers discovered that they could manipulate the valve wheel (and thus the valve itself) by remaining outside the vault door and reaching inside, stretching, grasping the valve wheel, and turning it. This was a difficult procedure and required a certain amount of dexterity on the worker's part; however, the workers felt that opening or closing the valve from the outside was easier than having to take the time to fill out a confined space permit, get a team together to sample the air within the vault, put on a self-contained breathing apparatus, wear a safety harness attached to a tending line, and then finally enter the space to perform the work at hand. They did so despite the fact that the air inside the vault routinely registered more than 300 ppm of hydrogen sulfide, above the lower exposure limit (LEL) for methane; an engineering study dealing with the inherent ventilation problems of the vaults had been in process for several years. In fact, management encouraged it. The job was accomplished more quickly and inexpensively by having a single worker adjust the valves rather than requiring a confined space team.

You can probably figure out the eventual outcome of this practice. On the day that a worker died in one of these confined spaces, he was in a hurry. He fully understood that the sewage and other waste in the bottom of the well within this vault not only was foul smelling but also generated hydrogen sulfide and methane gases. He understood this. He also understood that his health, safety, and wellbeing could be adversely affected if he was not careful. He understood that he could lose his life, as hydrogen sulfide in high enough concentrations is a killer. Also, not only is methane gas explosive but it can also asphyxiate. The worker understood this, but he was in a hurry. Besides, for years he had been working these valves the way he had been taught. Why worry, right? This day was no different than any of the other times he had done the same task. Why would he want to change a practice that had served him well in the past? Besides, at that moment he wasn't thinking of any reason why he should change; he was just going through his daily work routine as quickly as possible.

He opened the rusty metal door, swinging it wide on its stiff hinges, which was quite a task in itself as the door was the original door, installed about 50 years before. The horrible stench almost overwhelmed him. Outside the vault, in clean, sweet-smelling air, he took a deep breath, closed his mouth, and held his breath. He didn't inhale as he (carefully, with his back almost facing inward) reached in with his left hand to the right side of the door toward the wall with the valve wheel. He grasped the valve wheel and tried to turn it clockwise (in the open direction). As usual, this particular valve was difficult to turn. The worker strained hard to break it loose so that it would turn. It wouldn't budge.

He tried harder.

He stopped, stepped completely outside the vault, exhaled, and breathed in a huge amount of fresh air. The stench that drifted through the open doorway was almost more than he could handle, but handle it he did, at least for the moment. After a short rest, he returned to the task at hand and tried the valve wheel again.

The valve still would not budge.

He leaned inward a bit more to get a better grip on the valve wheel and to put more force into his motion.

The valve still would not move.

He leaned inside a few inches more, his lungs bursting, and tried to turn the valve again. It still would not budge, so he tried it again—but his hand slipped and he fell hard to the floor. When he hit the floor, the remaining air in his lungs was forced out by the impact. The impact and reflex forced him to wildly exhale, and reflex made him inhale just as wildly, taking in a deep breath of hydrogen sulfide at over 600 parts per million (ppm) and methane at 20% above the LEL, a potent and deadly brew. The worker's body was completely inside the vault, where it remained for 2 more hours before his remains were discovered and removed by another station checking crew (properly equipped with self-contained breathing apparatus, harnesses, and tending lines) that had been dispatched to find the worker.

This example clearly illustrates the reasoning and the logic behind OSHA's ensuring that the definition of confined "entry" (discussed in detail later) includes clear language defining what constitutes "entry"—that is, when any part of the entrant's body breaks the plane of an opening into the space, an entry has been made and is subject to the confined space entry regulations.

9.17 OSHA Standards Applicable to Shale Gas Operations

With regard to ensuring employee safety and health, all of OSHA's subparts listed in Figure 9.1 could be employed by employers not only to protect workers but also to ensure OSHA compliance. In this section, we specifically list and describe those standards that are most often cited in Industry Group 138 for lack of compliance or willful violation of Occupational Safety and Health Standards (29 CFR 1910) found during OSHA audits and determined to be causal factors during post-accident/fatality investigations.

9.17.1 Subpart D, Walking–Working Surfaces

- 1910.22 General Requirements
- 1910.23 Guarding floor and wall openings and holes
- 1910.24 Fixed industrial stairs
- 1910.27 Fixed ladders

Note that noncompliance with OSHA's requirements for guarding hole and well openings is listed as one of the top ten violations in Table 9.12.

General requirements

Housekeeping

1910.22(a)(1)—Housekeeping in all places of employment, passageways, storerooms, and service rooms shall be kept clean and orderly in a sanitary condition. Good housekeeping includes cleaning up grindings, shavings, and general debris from work areas on a daily basis.

1910.22(a)(2)—Housekeeping includes maintaining the floor of every work space in a clean, dry condition. Where wet processes are used, dry standing places should be provided where practicable.

Aisles and passageways

1910.22(b)(1)—In aisles and passageways, where mechanical handling equipment is used safe clearance shall be provided in aisles, at loading docks, through doorways, and wherever turns or passage must be made.

Covers and guardrails

1910.22(d)(1)—Floor loading protection requires that every structure have floors or mezzanines approved for load bearing when using for storage. These areas shall be marked on plans of approved design.

Guarding floor and wall openings and holes

Protection for floor openings

1910.23(a)(9)—Floor holes into which a person can accidentally walk shall be protected by a cover that leaves no opening more than 1 inch wide. The cover must be securely held in place.

Protection of open-sided floors, platforms, and runways

1910.23(c)(1)—Open-sided floors and platforms 4 feet or more above an adjacent floor or ground level shall be guarded by a standard railing. The railing shall be provided with a toeboard where there is a fall hazard.

1910.23(c)(2)—Every runway shall be guarded by a standard railing 4 feet or more above floor or ground level. Wherever tools, machine parts, or materials are likely to be used on the runway, a toeboard shall be provided on each exposed side.

Stairway railings and guards

1910.23(d)(1)—Every flight of stairs having four or more risers shall be equipped with standard railings and handrails.

Fixed industrial stairs

Stair treads

1910.24(f)—Treads on all stairs shall be reasonably slip resistant.

Fixed ladders

Rungs and cleats

1910.27(b)(1)(ii)—The distance between rungs, cleats, and steps shall not exceed 12 inches and shall be uniform throughout the length of the ladder.

1910.27(b)(1)(iii)—The minimum clear length of rungs shall be 16 inches from left to right.

DID YOU KNOW?

Choosing the right ladder for the job is an important part of working safely. Whether for maintenance or for operational reasons, ladders are devices that are used extensively in most industrial settings. In order to be used safely, ladders must be sturdy and in good repair. In addition to noting inoperable eye washes and showers in treatment plant safety audits, the audit will usually detect several ladders that are unsafe for use. Ladders are generally ignored until they are needed. A worker who needs a ladder generally grabs the first one available and uses it. Unfortunately, there is no guarantee that the ladder chosen is safe. If the plant or industrial site does not have an effective ladder inspection program, it is likely that unsafe ladders will be available for workers to use.

Protection from deterioration

1910.27(b)(6)—Metal ladders shall be painted or treated to resist corrosion and rusting, particularly ladders formed by individual metal rungs embedded in concrete. Rungs shall have a minimum diameter of 1 inch.

Clearance

1910.27(c)(2)—A clear width of at least 15 inches shall be provided on either side from the centerline of the ladder in the climbing space, except when cages or wells are necessary.

9.17.2 Subpart E, Means of Egress

1910	Subpart E Appendix—Exit Routes, Emergency Action Plans, and Fire Prevention Plans
1910.36	Design and Construction Requirements for Exit Routes
1910.37	Maintenance, Safeguards, and Operational Features for Exit Routes

9.17.2.1 Subpart E Appendix: Emergency Response

Even though no OSHA standards dedicated specifically to the issue of planning for emergencies exist at present, all OSHA standards are written for the purpose of promoting a safe, healthy, accident-free, and hence emergency-free workplace. For this reason, OSHA standards play a significant role in emergency prevention. A first step when developing emergency response plans is to review these OSHA standards. This can help organizations identify and then correct conditions that might exacerbate emergency situations before they occur.

Typically, when we think of emergency response plans for the workplace, we often conjure up thoughts about the obvious. For example, the first workplace emergency that might come to mind is *fire*—a major concern because fire in the workplace is something that can happen, that happens more often than we might want to think, and because fire can be particularly devastating—in ways we know all too well. Most employees do not need to be informed about the dangers of fire, but employers still have the responsibility to do just that—to provide training for employees on fire, fire prevention, and fire protection. Many local codes go beyond this information requirement, insisting that employers develop and implement fire emergency response and evacuation plans.

The primary emphasis has been on the latter, but developing an emergency response plan is critical. Employers that equip their workplaces with fire extinguishers and other firefighting equipment and expect their employees to respond aggressively to extinguish workplace fires must have emergency response plans in place. Also, the employer must ensure that all company personnel called upon to fight a fire are completely trained on how to do so safely (29 CFR 1910.156(c); 29 CFR 1910.157(g)).

Medical emergencies are another commonly considered workplace emergency that must be addressed in emergency response plans. Many facilities satisfy this requirement simply by directing employees to call 911 or some other emergency number whenever a medical emergency occurs in the workplace. Other facilities, though, may require employees to provide emergency first aid. When the employer chooses the employee-supplied first aid option, certain requirements must be met before any employee can legally administer first aid. The first aid responder must be trained and certified to administer first aid. This training must also include training on OSHA's bloodborne pathogen standard, which requires that employees be trained on the dangers inherent in handling and being exposed to human body fluids. Employees must be trained on how to protect themselves from contamination. If the first aid responder or anyone else is exposed to and contaminated by body fluids, the employer must make available the hepatitis B vaccine and vaccination series to all employees who have occupational exposure, as well as post-exposure evaluation and follow-up to all employees who have had an exposure incident (29 CFR 1910.1030).

Under 29 CFR 1910.120 (Hazardous Waste Operations and Emergency Response, or HAZWOPER), another type of emergency that must be covered by an emergency response plan is the release of hazardous materials. Unless the facility operator can demonstrate that the operation does not involve employee exposure or the reasonable possibility for employee exposure to safety or health hazards, the following operations are covered:

1. Cleanup operations required by a governmental body involving hazardous substances conducted at uncontrolled hazardous waste sites, state priority site lists, sites recommended by the USEPA, National Priorities List, and initial investigations of government identified sites that are conducted before the presence or absence of hazardous substance has been ascertained
2. Corrective actions involving cleanup operations at sites covered by the Resource Conservation and Recovery Act of 1976 (RCRA)
3. Voluntary cleanup operations at sites recognized by federal, state, local, or other governmental bodies as uncontrolled hazardous waste sites
4. Operations involving hazardous waste conducted at treatment, storage, and disposal (TSD) facilities regulated by the RCRA
5. Emergency response operations for releases of, or substantial threats of releases of, hazardous substances without regard to the location of the hazard

The final requirement impacts the largest number of facilities that meet the criteria requiring full compliance with 29 CFR 1910.120 (HAZWOPER), because many such facilities do not normally handle, store, treat, or dispose of hazardous waste but do use or produce hazardous materials in their processes. Because the use of hazardous materials could lead to an emergency from the release or spill of such materials, facilities using these materials must develop and employ an effective site emergency response plan.

Before discussing the basic goals of an effective emergency response plan, we should define *emergency response*. Considering that individual facilities are different, with different dangers and different needs, defining emergency response is not always easy. For our purposes, however, we use the following definition: "Emergency response is defined as a limited response to abnormal conditions expected to result in unacceptable risk requiring rapid corrective action to prevent harm to personnel, property, or system function" (CoVan, 1995). Another important point about emergency response, one critical for the safety engineer, is that "although emergency response and engineering tend toward prevention, emergency response is a skill area that safety engineers must be familiar with because of regulations and good engineering practice" (CoVan, 1995).

Now that we have defined emergency response, let's move on to the basic goals of an effective emergency response plan. Much of the currently available literature on this topic generally lists the goals as twofold:

1. Minimize injury to facility personnel.
2. Minimize damage to the facility and return to normal operation as soon as possible.

Obviously, these goals make a great deal of good sense, but you may be wondering about the language used, particularly "facility personnel" and "damage to the facility." Remember that we are talking about OSHA requirements here. Under OSHA, the primary emphasis is on protecting the worker; protecting the worker's health and safety is OSHA's only focus. What about people who live offsite—the site's neighbors? What about the environment?

Such questions emphasize the fact that OSHA is not normally concerned with the environment, unless contamination of the environment (at the worksite) might adversely impact worker safety and health. What about the neighbors? Again, OSHA's focus is on the worker. One OSHA compliance office explained that, if employers take every step necessary to protect their employees from harm resulting from the use or production of hazardous materials, then the surrounding community should have little to fear.

This statement was puzzling, so the same OSHA compliance officer was queried about incidents beyond the control of the employer—accidents that could put employees in harm's way and endanger the surrounding community. The answer? "Well, that's the EPA's bag—we only worry about the worksite and the worker."

Fortunately, OSHA, in combination with the USEPA, has taken steps to overcome this blatant shortcoming (we like to think of it as an oversight). Under OSHA's Process Safety Management (PSM) and USEPA's Risk Management Planning (RMP) directive, chemical spills and other chemical accidents that could impact both the environment and neighbors have now been properly addressed. What PSM and RMP really accomplish is changing the typical twofold goal of an effective emergency response plan to a threefold goal.

Let us point out that accomplishment of these two- or threefold goals or objectives is essential in any emergency response. Accomplishing these goals or objectives requires an extensive planning effort prior to the emergency (*prior* being the keyword here, because the attempt to develop an emergency response plan when a disaster is occurring or after one has occurred is both futile and stupid). The safety official must never forget that hazards in any facility can be reduced, but risk is an element of everyday existence and therefore cannot be totally eliminated. The safety engineer's goal must be to keep risk to an absolute minimum. To accomplish this advance planning is critical—and essential. We pointed out earlier that most plans address fire, medical emergencies, and the accidental release or spills of hazardous materials; however, the development of emergency response plans should also factor in other possible emergencies, such as natural disasters, floods, or explosions. Site emergency response plans should include the following:

- Assessment of risk
- Chain of command for dealing with emergencies
- Assessment of resources
- Training
- Incident command procedures
- Site security
- Public relations

The Federal Emergency Management Agency (FEMA), the U.S. Army Corps of Engineers, and several other agencies, as well as numerous publications, provide guidance on how to develop a site emergency response plan. Local agencies, such as fire departments, emergency planning commissions and agencies, HazMat teams, and Local Emergency Planning Committees (LEPCs), also provide information on how to design a site plan. All of these agencies typically recommend that a site's plan contain the elements listed in Table 9.14.

In the safety official's effort to incorporate and manage a facility emergency response plan, and in the response itself, two elements—security and public relations—must be given special attention. If not handled correctly, a lack of effective security measures and improper public relations can turn an already disastrous incident into a mega-disaster. Provisions should be made to have a well-trained security team limit site access to only the people and equipment that can assist in coping with and resolving the emergency. Public relations can be a tricky enterprise. The person identified to interface with the media must have thorough knowledge of the site, process, and personnel involved. The public relations representative must also have access to the highest levels of site management; otherwise, that person will not be able to deal with the public and media in an effective manner.

9.17.2.2 Egress Requirements

Design and construction requirements for exit routes

An exit door must be unlocked

1910.36(d)(1)—Employees must be able to open an exit route door from the inside at all times without keys, tools, or special knowledge even in the dark. A device such as a panic bar that locks only from the outside is permitted on exit discharge doors.

TABLE 9.14

Site Emergency Response Plan

Component	Description
Emergency Response Notification	List of who to call and information to pass on when an emergency occurs
Record of Changes	Table of changes and dates for them
Table of Contents/Introduction	Purpose, objective, scope, applications, policies, and assumptions for the plan
Emergency Response Operations	Details regarding what actions must take place
Emergency Assistance Telephone Numbers	Current list of people and agencies that may be needed in an emergency
Legal Authority and Responsibility	Laws and regulations that provide authority for the plan
Chain of Command	Response organization structure and responsibilities
Disaster Assistance and Coordination	Where additional assistance may be obtained when the regular response organizations are over-burdened
Procedures for Changing or Updating the Plan	Who makes changes and how they are made and implemented
Plan Distribution	List of organizations and individuals who have been given a copy of the plan
Spill Cleanup Techniques	Detailed information about how response teams should handle cleanups
Cleanup/Disposal Resources	List of what is available, where it is obtained, and how much is available
Consultant Resources	List of special facilities and personnel who may be valuable in a response
Technical Library/References	List of libraries and other information sources that may be valuable for those preparing, updating, or implementing the plan
Hazard Analysis	Details regarding the kinds of emergencies that may be encountered, where they are likely to occur, what areas of the community may be affected, and the probability of occurrence
Documentation of Spill Events	Various incident and investigative reports on spills that have occurred
Hazardous Materials Information	Listing of hazardous materials, their properties, response data, and related information
Dry Runs	Training exercises for testing the adequacy of the plan, training personnel, and introducing changes

Source: FEMA, *Planning Guide and Checklist for Hazardous Materials Contingency Plans*, FEMA-10, Federal Emergency Management Agency, Washington, DC, 1981.

A slide-hinged exit door must be used

1910.36(e)(2)—The door that connects any room to an exit route must swing out in the direction of exit travel if the room is designed to be occupied by more than 50 people or if the room is a high hazard area.

An exit route must meet minimum height and width requirements

1910.36(g)(2)—An exit access must be at least 28 inches (71.1 cm) wide at all points. Where there is only one exit access leading to an exit or exit discharge, the width of the exit and exit discharge must be at least equal to the width of the exit access.

1910.36(g)(4)—Objects that project into the exit route must not reduce the width of the exit route to less than the minimum width requirements for exit routes.

An outdoor exit route is permitted

1910.36(h)(1)—The outdoor exit route must have guardrails to protect unenclosed sides if a fall hazard exists (three or more rise treads).

Maintenance, safeguards, and operational features for exit routes

The dangers to employees must be minimized

1910.37(a)(3)—Exit routes must be free and unobstructed. No materials or equipment may be placed, either permanently or temporarily, within the exit route.

1910.37(a)(4)—Safeguards designed to protect employees during an emergency must be in proper working order at all times (e.g., emergency lighting, alarm systems, sprinkler systems, fire doors, exhaust systems).

Lighting and marking must be adequate and appropriate

1910.37(b)(1)—Each exit route must be adequately lighted so that a person with normal vision can see along the exit route (including exterior lights to a safe location).

1910.37(b)(2)—Each exit must be clearly visible and marked by a sign reading "EXIT" (except a main entrance/exit door that is readily obvious).

1910.37(b)(4)—If the direction of travel to the exit or exit discharge is not immediately apparent, signs must be posted along the exit access indicating the direction of travel to the nearest exit and exit discharge. Additionally, the line-of-sight to an exit sign must clearly be visible at all times. (Sample citation might read, "The direction of travel to the exit or exit discharge was not immediately apparent at the south end.")

1910.37(b)(5)—Each doorway or passage along an exit route access that could be mistaken for an exit must be marked "Not an Exit" or similar designation, or be designated by a sign indicating its actual use (e.g., "Closet").

1910.37(b)(6)—If emergency lighting is available in the building, then exit signs must be illuminated by emergency lighting or internally illuminated.

9.17.3 Subpart F, Powered Platforms, Manlifts, and Vehicle-Mounted Work Platforms*

1910.66 Powered Platforms for Building Maintenance

1910.66 Appendix C, Personal Fall Arrest/Protection

Falls associated with Industry Group 138 include falls from elevated areas of a drilling rig (e.g., stabbing board, monkey board, ladder); thus, fall arrest and protection systems are important in maintaining the safety of shale gas drilling and operations personnel.

9.17.3.1 What Is Fall Protection?

Fall protection is the series of steps taken to cause reasonable elimination or control of the injurious effects of an unintentional fall while accessing or working (Ellis, 1988, p. xvi):

> Fall hazard distance begins and is measured from the level of a workstation on which a worker must initially step and where a fall hazard exists. It ends with the greatest distance of possible continuous fall, including steps, openings, projections, roofs, and direction of the fall (interior or exterior). Protection is required to keep workers from striking objects and to avoid pendulum swings, crushing and impact with any part of the body to which injury could occur. The object of elevated fall protection is to convert the hazard to a slip or minor fall at the very worst—a fall from which hopefully no injury occurs.

Because injuries received from falls in the workplace are such a common occurrence—in a typical year more than 10,000 workers will lose their lives in falls—safety officials need to be aware of not only fall hazards but also the need to institute a fall protection safety program (Kohr, 1989). Just how frequent and serious are accidents related to falls? Let's look at a few telling facts about falls in the workplace. The National Safety Council's annual report typically predicts 1400 or more deaths and more than 400,000 disabling injuries to occur each year due to falls. Falls are the leading cause of disabling injuries in the United States, accounting for close to 18% of all workers' compensation claims. A Bureau of Labor Statistics 24-state survey reported that 60% of elevated falls were under 10 feet, and 50% of those were under 5 feet (Pater, 1985). The primary causes of falls have been identified as the following (Kohr, 1989):

* This section is adapted from Spellman, F.R. and Whiting, N., *Safety Engineering: Principles and Practices*, 2nd ed., Government Institutes Press, Latham, MD, 2005.

1. A foreign object on the walking surface
2. A design flaw in the walking surface
3. Slippery surfaces
4. An individual's impaired physical condition

Which industries have the most injuries as a result of falls? The construction industry experiences the largest percentage of injuries, accounting for 42% of all injuries resulting from falls. The National Safety Council reported that 70% of reported falls were from scaffolds, 14% from roofs, and another 14% from barrels, boxes, equipment, or furniture (NSC, 1985, 1986). Eisma (1990) reported that 85% of falls from elevation resulted in lost workdays, and 20% resulted in death. In a more recent report (USDHHS, 2000), falls from elevations were the fourth leading cause of occupational fatalities from 1980 through 1994. The 8102 such deaths that occurred during this period accounted for 10% of all fatalities, and an average of 540 deaths per year. In 1994, slips and trips caused more than 70,000 injuries, or 18% of all occupational injuries.

When attempting to initiate a fall protection safety program at any organization, safety officials must first define the needs of the organization. The actual needs of any type of fall protection program are going to be driven mainly by the type of work the organization does. Obviously, if the company is involved in construction, the needs are rather straightforward, because much of the work conducted will include the necessity of doing elevated work; however, this might also be the case for various trades as well, such as carpentry. Public utility and transportation work might also require elevated work.

To define the problem associated with all types of falls, let's examine what falls are all about. None of us has a problem understanding what a fall from a high-rise construction project involves—it is simply a fall from elevation. In many workplaces, though, worker injuries result from types of falls other than those from elevations. Falls in the workplace also include slips, trips, and stair falls, as well as elevated falls. *Slips* and *trips* are falls on the same level. *Stair falls* are falls on one or more levels. *Elevated falls* are from one level to another. In the following sections, each of these types of falls is discussed in greater detail, but first we discuss the physical factors at work in causing a fall. Remember that safety officials must address and work to reduce or eliminate *all* types of falls.

9.17.3.2 Physical Factors at Work in a Fall

We've all heard someone say, "The bigger they are, the harder they fall," or "It's not the fall that's so bad; it's the sudden stop at the end." Many would-be practitioners in the safety field are often surprised, however, to find out that science plays a role in falls and that slips, trips, and falls actually involve three well-known laws of science:

- *Friction* is the resistance between things, such as between work shoes and a walking surface. Without friction, workers are likely to slip and fall. Probably the best example of this phenomenon is slipping on ice. On icy surfaces, shoes can't grip the surface normally, causing a loss of traction and a fall.
- *Momentum* is the product of the mass of a body and its linear velocity. Simply put, momentum is affected by the speed and size of the moving object. Momentum is best understood if we translate the saying above to: "The more you weigh and the faster you move, the harder you'll fall if you slip or trip."
- *Gravity* is the force of attraction between any object in the Earth's gravitational field and the Earth itself. Simply put, gravity is the force that pulls you to the ground once a fall is in progress. If someone loses balance and begins to fall, they are going to hit the ground. The human body is equipped with mechanisms that work to prevent falls, including the eyes, ears, and muscles, all of which work to keep the human body close to its natural center of balance. When this center of balance shifts too far, a fall will occur if balance is not restored to normal. Because gravity obviously has the same effect on all of us here on Earth, it is always surprising to discover how such a well-known basic law of science is so often and conveniently ignored by various industries. It is not unusual to encounter company owners or workplace foremen who ignore the laws of gravity and require their workers to perform daring (and extremely dangerous) feats in the workplace. Workers (who need the job and the security it provides) are led to believe that somehow gravity is something that is not important to them. Obviously, this is a dangerous mindset and practice that company safety officials must not tolerate.

9.17.3.3 Slips

In its simplest form, a slip is a loss of balance caused by too little friction between the feet and the surface being walked or worked on. The more technical explanation refers to a slip as resulting in a sliding motion when the friction between the feet (shoe sole surface) and the surface is too little. This slip (loss of traction), in turn, often leads to a loss of balance, resulting in a fall. Slips can be caused by a number of design factors and work practices, individually or in combination. Design factors include footwear, floor surfaces, personal characteristics, and the work task. Footwear is an important consideration in the prevention of a slip or fall. Not only is the condition of the footwear important in fall prevention, but also the composition, shape, and style. For industrial applications, the organizational safety professional should ensure that only approved safety shoes are worn. Safety shoes should have toe protection and slip-resistant soles.

Floor surfaces, design, installation, composition, condition, gradient, modifications by protective coatings and cleaning/waxing agents, and illumination are all important elements that must be taken into consideration in providing safe floor surfaces in the workplace. Common ways to make floor surfaces slip resistant include grooving, gritting, matting, and grating.

Personal characteristics such as physical condition, age, health, emotional state, agility, and attentiveness are also factors to consider when making walking and working surfaces slip resistant for workers. Work-task design also plays an important role in causing and preventing slip-falls. Some work practices can cause walking surfaces to be constantly wet (such as from frequent spills), and weather hazards such as snow and ice can also make walking surfaces slippery. Workplace supervisors and workers (and safety officials) must follow safe work practices and exercise vigilance to reduce the occurrence of such conditions and to remediate them as quickly as possible when they do occur. This type of problem is much more common than we might realize.

Unfortunately, it is not unusual for workers to spill oil or some other slippery substance on the workplace floor and then walk away from the spill, leaving behind a slip hazard for another worker. A common workplace safe work practice and housekeeping rule should be to clean up spills right away. Another unsafe work practice that commonly leads to slips and falls is when a worker is in a hurry, rushes to finish a particular task, and overlooks safe work practices.

9.17.3.4 Trips

Trips normally occur whenever a worker's foot contacts an object that causes him or her to lose balance; however, you do not always have to come into contact with an object to trip. Trips may also be caused by too much friction between footwear and the walking surface. Like slips, trips commonly occur when a worker is in a hurry. The problem with hurrying, of course, is that the potential victim's attention is usually focused on anything but possible trip hazards. Another common factor that leads to a trip is the practice of carrying objects that are so large that the worker cannot see the walking surface. Lighting also plays a critical role in preventing trips. Inadequate lighting fixtures, burned-out bulbs, and lights that are turned off all increase the opportunity for trips to occur. Again, as in the prevention of slips, housekeeping plays an important role in prevention. Good workplace housekeeping practices include keeping passageways clean and uncluttered; arranging equipment so that it does not interfere with walkways or pedestrian traffic; keeping working areas clear of extension or power tool cords; eliminating loose footing on stairs, steps, and floors; and properly storing gangplanks and ramps.

9.17.3.5 Stair Falls

For information about falls from stairs, probably the best reference is the Bureau of Labor Statistics' *Injuries Resulting from Falls on Stairs* (Bulletin 2214). This particular booklet is excellent because not only does it provide statistical data but it is also an eye-opener for how these injuries occur. It is widely known and accepted, for example, that stairs are a high-risk area. It is also accepted that a loss of balance can occur from a slip or trip while a worker (or any person) is traveling up or down a stairway. Safety officials must consider why stairs are so hazardous. What are the causal factors?

Bulletin 2214 comes in handy when trying to answer questions like these; for example, it points out that the vast majority of falls on stairs occur when people are traveling down the stairs and are not holding onto the handrail. This is an important point about handrails for two reasons: (1) The safety person will know to focus training on this important topic, and (2) the safety person can ensure that handrails not only are in place in all stairways but are also in good repair. Loss of traction is the common cause of the highest number of stairway slipping and falling accidents. Again, this is where good housekeeping practices come into play. Many of the stairway slipping and falling accidents happen because of water or other liquid on steps. Along with improper housekeeping practices, stairs can also become hazardous whenever they are improperly designed or installed or are neglected. Safe work practices should also be considered. A work practice that allows the worker to carry or reach for objects while climbing stairs is not a good one.

9.17.3.6 Elevated Falls from One Level to Another

When workers are working from elevated scaffolds, ladders, platforms, and other surfaces, the risk of serious injury from an elevated fall is increased exponentially whenever a worker loses his balance as a result of a slip or trip. Unfortunately, it is the practice of too many companies that require workers to perform work from elevated areas to only provide some type of device (handrail or hand-line) that workers are supposed to grab onto to break their fall. In the judgment of most experienced safety professionals, this is *not* fall protection. These types of jerry-rigged devices are not acceptable substitutes for guardrails, appropriate midrails, and toeboards. OSHA requires guardrails to be 42 inches nominal, midrails 21 inches, and toe boards 4 inches. Ellis (1988) made a good point in observing that, "unlike many workplace hazards, few, if any, 'near-miss' incidents help people learn to appreciate the seriousness of elevated falls" (p. 28). When you consider that losing one's balance from an elevation of 10 to 200 feet or more usually leaves little chance to avoid serious or fatal injury, Ellis' statement makes a lot of sense.

9.17.3.7 Fall Protection Measures

Under 29 CFR 1926.501 (Duty to Have Fall Protection), employers must assess the workplace to determine if the walking or working surfaces on which employees are to work have the strength and structural integrity to safely support workers. Accordingly, the real goal should be to prevent slips, trips, and falls from elevation from occurring in the first place. To accomplish this, the following steps are recommended: (1) preplan before beginning any elevated work (e.g., on scaffolds); (2) establish a written policy and develop rules; and (3) implement safe work practices to prevent falls. Preplanning is all about thinking through the job at hand; for example, for exterior refurbishing work on a chemical storage tank that is 80 feet in height, scaffolding will almost certainly be required. Preplanning and a great deal of skill are both necessary to properly erect scaffolding. If scaffolding is to be used, the organization responsible for erecting the scaffolding should have a written scaffold safety program. A sample scaffold safety program (one that has been used for years with great success) is provided in the next section.

9.17.3.8 Fall Protection Policies

As with any organizational safety program, the safety official should ensure that a fall protection policy is established and that applicable rules are developed. Keep in mind that written safety policies are intended to protect both employees and contractors. Establishing and developing any organizational safety policy is only part of the job; making sure that every employee has been trained on the policy is the other, more difficult task. Many organizations that the author has audited have excellent written policies but when employees are asked about them, they shake their heads. "A fall protection policy? Gee, I don't think we have one of those," they might say. This kind of misinformation or lack of information is the safety official's worst nightmare. Below is a sample written policy for elevated work operations. This sample (adapted from a policy used by Kaiser Aluminum) should give readers an appreciation for the type of written policy companies should have.

Elevated Work Policy

I. Purpose

To establish a standard policy to ensure that elevated work at heights of eight (8) feet or greater conducted by Company and Contractor personnel are done so in the safest manner possible.

II. Objectives

A. To prevent employee death and/or minimize employee injury resulting from falls from elevated work locations.

B. To identify and label elevated work sites requiring fall protection equipment.
C. To establish minimum standards for all protection equipment and systems and their applications.

III. Scope

All Free Fall Hazards will be guarded against by the use of permanent or semipermanent guardrail assemblies. When a free fall hazard cannot be prevented through such measures, personal fall protection equipment must be used. This policy applies to working at elevated work locations that are eight (8) feet or more above floor or grade level. It covers activities such as (but not limited to) work in pipe racks, on sloped roofs, on unguarded scaffolding, or in ship cargo holds; working from suspended scaffolds, floats, or boatswains chairs; working on tank tops; and when working inside or outside any process structure not equipped with appropriate guarded work platforms.

IV. Procedure

A. Supervisor Planning
1. The supervisor will be responsible for evaluating the need for a personal fall protection system as an integral part of preplanning a job.
2. When work must be performed at recognized unguarded elevated heights, the supervisor may select either option below:
   a. *Option 1—Fall Prevention*
   Eliminate the free fall hazard during all phases of the job (traveling to and from elevated work areas as well as during the performance of the task at the elevated work area) by means of temporary scaffolding, platforms, railings, man lifts, and ladders, etc.

Note: Every effort must be taken to minimize the potential for free fall hazards to individuals installing temporary or permanent fall protection systems.

   b. *Option 2—Fall Protection System*
   By selecting and installing a personal fall protection system, eliminating the free fall hazard (greater than 8 feet) when traveling to and from the elevated work area, as well as during performance of the task at the elevated work areas.

Note: Every effort must be taken to minimize the potential of the free fall hazards to individuals installing temporary or permanent fall protection system.

B. User Responsibility
1. Each employee assigned to work at elevated heights has the responsibility of thoroughly inspecting the personal fall protection system's anchor points, connecting means (i.e., lanyard or device), and body holding devices (i.e., harnesses) prior to using the system. Any problems noted with any of the above must be brought to the attention of the supervisor.
2. Any questions concerning the type of personal fall protection systems best suited for a particular job, as well as systems installation (e.g., anchor point type/strength), should be directed to the safety official.

Blaming falls on worker carelessness or accident is commonly used to sidestep important safety precautions. Although falls are sometimes caused by inattention or carelessness, other factors are also at work. Materials that affect footing, misplaced equipment, and improperly managed and assembled scaffolding are all contributors to the possibility of injury from falls. Safe work practices and worker awareness of workplace hazards can go far in alleviating these hazards.

9.17.4 Subpart G, Occupational Health and Environmental Control

1910.95 Occupational Noise Exposure

It is important to note that 1910.95(o) states: "Paragraphs (c) through (n) of this section shall not apply to employers engaged in oil and gas well drilling and servicing operations."

9.17.4.1 OSHA Noise Hazard Requirements

In 1983, OSHA adopted a hearing conservation amendment to 29 CFR 1910.95 that requires employers to implement *hearing conservation programs* in any work setting where employees are exposed to an 8-hour time-weighted average of 85 dBA and above. Employers are required to implement *hearing conservation procedures* in settings where the noise level exceeds a time-weighted average of 90 dBA. They are also required to provide personal protective equipment for employees who show evidence of hearing loss, regardless of the noise level at their worksites. In addition to concerns over noise levels, the OSHA standard also addresses the issue of the duration of exposure (LaBar, 1989):

> Duration is another key factor in determining the safety of workplace noise. The regulation has a 50 percent 5 dBA logarithmic tradeoff. That is, for every 5-decibel increase in the noise level, the length of exposure must be reduced by 50 percent. For example, at 90 decibels (the sound

level of a lawnmower or shop tools), the limit on "safe" exposure is 8 hours. At 95 dBA, the limit on exposure is 4 hours, and so on. For any sound that is 106 dBA and above—this would include such things as a sandblaster, rock concert, or jet engine—exposure without protection should be less than 1 hour, according to OSHA's rule.

Although not all of the standard's requirements are pertinent to oil and gas drilling operations, the basic requirements of OSHA's hearing conservation standard are explained here (LaBar, 1989):

- *Monitoring noise levels.* Noise levels should be monitored on a regular basis. Whenever a new process is added, an existing process is altered, or new equipment is purchased, special monitoring should be undertaken immediately.
- *Medical surveillance.* The medical surveillance component of the regulation specifies that employees who will be exposed to high noise levels should be tested upon being hired and again at least annually.
- *Noise controls.* The regulation requires that steps be taken to control noise at the source. Noise controls are required in situations where the noise level exceeds 90 dBA. Administrative controls are sufficient until noise levels exceed 100 dBA. Beyond 100 dBA, engineering controls must be used.
- *Personal protection.* Personal protective devices are specified as the next level of protection when administrative and engineering controls do not reduce noise hazards to acceptable levels. They are to be used in addition to rather than instead of administrative and engineering controls.
- *Education and training.* The regulation requires the provision of education and training to ensure that employees understand (1) how the ear works, (2) how to interpret the results of audiometric tests, (3) how to select personal protective devices that will protect them against the types of noise hazards to which they will be exposed, and (4) how to properly use personal protective devices (LaBar, 1989).

9.17.4.2 Occupational Noise Exposure

Noise is commonly defined as any unwanted sound. Noise literally surrounds us every day and is with us just about everywhere we go; however, the noise we are concerned with here is that produced by industrial processes. Excessive amounts of noise in the work environment (and outside of it) cause many problems for workers, including increased stress levels, interference with communication, disrupted concentration, and, most importantly, varying degrees of hearing loss. Exposure to high noise levels also adversely affects job performance and increases accident rates.

TABLE 9.15

Permissible Noise Exposures (29 CFR 1910.95)

Duration per Day (hr)	Sound Level (dBA)
8	90
6	92
4	95
3	97
2	100
1-1/2	102
1	105
1/2	110
1/4 or less	115

Note: When the daily noise exposure is composed of two or more periods of noise exposure of different levels, their combined effect should be considered, rather than the individual effect of each. If the sum of the following fractions C(1)/T(1) + C(2)/T(2) + C(n)/T(n) exceeds unity, then the mixed exposure should be considered to exceed the limit value. C(n) indicates the total time of exposure at a specified noise level, and T(n) indicates the total time of exposure permitted at that level. Exposure to impulsive or impact noise should not exceed 140-dB peak sound pressure level.

One of the major problems with attempting to protect workers' hearing acuity is the tendency of many workers to ignore the dangers of noise. Because hearing loss, like cancer, is insidious, it is easy to ignore. It sort of sneaks up slowly and often is not apparent until after the damage is done. Alarmingly, hearing loss from occupational noise exposure has been well documented since the 18th century, and since the advent of the industrial revolution the number of exposed workers has greatly increased (Mansdorf, 1993). Today, though, the picture of hearing loss is not as bleak as it has been in the past, as a direct result of OSHA's requirements. Now that noise exposure must be controlled in all industrial environments, well-written and well-managed hearing conservation programs must be put in place, and employee awareness regarding the dangers of exposure to excessive levels of noise has been raised, job-related hearing loss is coming under control (see Table 9.15).

9.17.4.3 Hearing Protection

The *hearing protection* element of a hearing conservation program provides hearing protection devices for employees and training in how to wear them effectively, as long as hazardous noise levels exist in the workplace. Hearing protection comes in various sizes, shapes, and materials, and the cost of this equipment can vary dramatically. Two general types of hearing protection are used widely in industry: the cup muff (commonly referred to as *Mickey Mouse ears*) and the plug insert type. Because feasible engineering noise controls have not been developed for many types of industrial equipment, hearing protection devices are the best option for preventing noise-induced hearing loss in these situations. As with the other elements of a hearing conservation program, the hearing protective device element must be in writing and included in the program.

9.17.4.4 Safe Work Practices for Hearing Protection

Safe work practices are an important element in any hearing conservation program. Written safe work practices for hearing conservation should focus on relaying noise hazard information to the employee. If an employee is required to perform some kind of maintenance function in a high noise hazard area, the written procedure for that maintenance should include a statement warning the employee about the noise hazard and should list the personal protective devices that employees should use to protect themselves from the noise. Experience has shown that when such warnings (safe work practices) are placed in preventive maintenance procedures (e.g., noise hazard area, confined space, lockout/tagout required), the program is much more efficient and the repeated reminder helps workers to maintain compliance with regulatory standards.

9.17.5 Subpart H, Hazardous Materials

1910.106 Flammable and Combustible Liquids
1910.110 Storage and Handling of Liquefied Petroleum Gases
1910.120 Hazardous Waste Operations and Emergency Response

9.17.5.1 Hazardous Materials

A hazardous material is a substance (gas, liquid, or solid) capable of causing harm to people, property, and the environment. The U.S. Department of Transportation (DOT) uses the term *hazardous materials* to cover nine categories identified by the United Nations Hazard Class Number System, including:

- Explosives
- Gases (compressed, liquefied, dissolved)
- Flammable liquids
- Flammable solids
- Oxidizers
- Poisonous materials

- Radioactive materials
- Corrosive materials
- Miscellaneous materials

9.17.5.2 Flammable and Combustible Liquids

In addition to basic fire prevention, emergency response training, and fire extinguisher training, employees must be trained on the hazards involved with flammable and combustible liquids; 29 CFR 1910.106 addresses this area. Industrial facilities typically use all types of flammable and combustible liquids. These dangerous materials must be clearly labeled and stored safely when not in use. The safe handling of flammable and combustible liquids is a topic that needs to be fully addressed by the facility safety engineer and workplace supervisor. Worker awareness of the potential hazards that flammable and combustible liquids pose must be stressed. Employees need to know that flammable and combustible liquid fires burn extremely hot and can produce copious amounts of dense, black smoke. Explosion hazards exist under certain conditions in enclosed, poorly ventilated spaces where vapors can accumulate. A flame or spark can cause vapors to ignite, creating a flash fire with the terrible force of an explosion. One of the keys to reducing the potential spread of flammable and combustible fires is to provide adequate containment. All storage tanks should be surrounded by storage dikes or containment systems, for example. Correctly designed and built dikes will contain spilled liquid. Spilled flammable and combustible liquids that are contained are easier to manage than those that have free run of the workplace. Properly installed containment dikes can prevent environmental contamination of soil and groundwater. Flammable liquids have a flash point below 100°F. Both flammable and combustible liquids are divided into the three classifications shown below:

Flammable Liquids

Class IA—Flash point below 73°F, boiling point below 100°F

Class IB—Flash point below 73°F, boiling point at or above 100°F

Class IC—Flash point at or above 73°F, but below 100°F

Combustible Liquids

Class II—Flash point at or above 100°F, but below 140°F

Class IIIA—Flash point at or above 140°F, but below 200°F

Class IIIB—Flash point at or above 200°F

9.17.5.3 What Is a Hazardous Waste?

A general rule of thumb states that any hazardous substance that is spilled or released into the environment is no longer classified as a hazardous substance but as a hazardous waste. The USEPA uses the same definition for hazardous wastes as it does for hazardous substances. The four characteristics of reactivity, ignitability, corrosivity, and toxicity can be used to identify hazardous substances as well as hazardous wastes. The USEPA lists substances that it considers to be hazardous waste; these lists take precedence over any other method used to identify and classify a substance as hazardous. If a substance is included on one of the USEPA's lists described below, it is a hazardous substance, no matter what.

Hazardous wastes are organized by the USEPA into three categories: nonspecific source wastes, specific source wastes, and commercial chemical products. All listed wastes are presumed to be hazardous, regardless of their concentrations. USEPA developed these lists by examining different types of wastes and chemical products to determine whether they met any of the following criteria:

- Exhibit one or more of the four characterizations of a hazardous waste
- Meet the statutory definition of hazardous waste
- Are acutely toxic or acutely hazardous
- Are otherwise toxic

These lists can be described briefly as follows:

- *Nonspecific source wastes* are generic wastes commonly produced by manufacturing and industrial processes. Examples from this list include spent halogenated solvents used in degreasing and wastewater treatment sludge from electroplating processes, as well as dioxin wastes, most of which are "acutely hazardous" wastes because of the danger they present to human health and the environment.
- *Specific source wastes* are from specially identified industries such as wood preserving, petroleum refining, and organic chemical manufacturing. These wastes typically include sludges, still bottoms, wastewaters, spent catalysts, and residues, such as wastewater treatment sludge from pigment production.
- *Commercial chemical products* (also called "P" or "U" list wastes because their code numbers begin with these letters) include specific commercial chemical products or manufacturing chemical intermediates. This list includes chemicals such as chloroform and creosote, acids such as sulfuric and hydrochloric, and pesticides such as DDT and kepone (40 CFR 261.31, 261.32, 261.33).

The USEPA ruled that any waste mixture containing a listed hazardous waste is also considered a hazardous waste and must be managed accordingly. This applies regardless of what percentage of the waste mixture is composed of listed hazardous wastes. Wastes derived from hazardous wastes (residues from the treatment, storage, and disposal of a listed hazardous waste) are considered hazardous waste as well. Hazardous wastes are derived from several waste generators. Most of these waste generators are in the manufacturing and industrial sectors and include chemical manufacturers, the printing industry, vehicle maintenance shops, leather products manufacturers, the construction industry, and metal manufacturing, among others. These industrial waste generators produce a wide variety of wastes, including strong acids and bases, spent solvents, heavy metal solutions, ignitable wastes, cyanide wastes, and many more.

From the responsible safety official's perspective, any hazardous waste release that could alter the environment or impact the health and safety of employees in any way is a major concern. The specifics of the safety engineer's concern lie in the acute and chronic toxicity to organisms, bioconcentration, biomagnification, genetic change potential, etiology, pathways, change in climate or habitat, extinction, persistence, esthetics such as visual impact, and, most importantly, the impact on the health and safety of employees.

Remember, we have stated consistently that when a hazardous substance or hazardous material is spilled or released into the environment, it becomes a hazardous waste. This is important because specific regulatory legislation has been put in place regarding hazardous wastes, responding to hazardous waste leak and spill contingencies, and the proper handling, storage, transportation, and treatment of hazardous wastes. The goal, of course, is protecting the environment and ultimately the health and safety of our employees and the surrounding community. Why are we so concerned about hazardous substances and hazardous wastes? This question is relatively easy to answer based on experience, publicity, and actual hazardous materials incidents, which have resulted in tragic consequences to the environment and to human life.

Humans are strange in many ways. We may know that a disaster is possible, is likely, could happen, and is predictable, but do we act before someone dies? Not often enough. We often ignore the human element—we forget a victim's demise. We simply do not want to think about it, because if we think about it, we must come face to face with our own mortality. The safety engineer, though, must think about constantly potential disasters to prevent them from ever occurring. Because of the Bhopal incident and other similar but less catastrophic chemical spill events, the U.S. Congress (pushed by public concern) developed and passed certain environmental laws and regulations to regulate hazardous substances and wastes in the United States. This section focuses on the two regulatory acts most crucial to the current management programs for hazardous wastes. The first, mentioned already throughout this text, is the Resource Conservation and Recovery Act (RCRA). Specifically, the RCRA provides guidelines for prudent management of new and future hazardous substances and wastes. The second act (more briefly mentioned) is the Comprehensive Environmental Response, Compensation, and Liability Act (CERCLA), otherwise known as Superfund, which deals primarily with mistakes of the past (i.e., inactive and abandoned hazardous waste sites).

9.17.6 Subpart I, Personal Protective Equipment

1910.132	General Requirements
1910,133	Eye and Face Protection
1910.134	Respiratory Protection
1910.135	Head Protection
1910.136	Occupational Foot Protection

9.17.6.1 Personal Protective Equipment

Noncompliance with the requirements of OSHA's respiratory protection standard is one of the top ten citations listed in Table 9.12. In the following statement, Mansdorf (1993) makes a number of important statements concerning personal protective equipment (PPE) worth taking some time to consider carefully:

> The primary objective of any health and safety program is worker protection. It is the responsibility of management to carry out this objective. Part of this responsibility includes protecting workers from exposure to hazardous materials and hazardous situations that arise in the workplace. It is best for management to try to eliminate these hazardous exposures through changes in workplace design or engineering controls. When hazardous workplace exposures cannot be controlled by these measures, personal protective equipment (PPE) becomes necessary. When looking at hazardous workplace exposures, keep in mind that government regulations consider PPE the last alternative in worker protection because it does not eliminate the hazards. PPE only provides a barrier between the worker and the hazard. If PPE must be used as a control alternative, a positive attitude and strong commitment by management is required.

"It is best for management to try to eliminate these hazardous exposures through changes in workplace design or engineering controls." Sound familiar? We consistently make this same point throughout this text. A hazard, any hazard, if possible, should be engineered out of the system or process. Determining when and how to engineer out a hazard is one of the safety official's primary functions; however, the safety official can much more effectively accomplish this if he or she is included in the earliest stages of design. Remember, it does little good (and is often very expensive) to attempt to engineer out any hazard once the hazard is in place.

"When hazardous workplace exposures cannot be controlled by these measures, personal protective equipment (PPE) becomes necessary." While the goal of safety officials is certainly to engineer out all workplace hazards, we realize that this goal is virtually impossible to achieve. Even in this day of robotics, computers, and other automated equipment and processes, the man–machine–process interface still exists. When people are included in the work equation, the opportunity for their exposure to hazards is very real—as injury statistics make clear.

"[C]onsider PPE the last alternative in worker protection because it does not eliminate the hazards." This is extremely important for two reasons: First, the safety official's primary goal is (as we have said before) to engineer out the problem. If this is not possible, the second alternative is to implement administrative controls. When neither is possible, PPE becomes the final choice. The key words here are "the final choice." Second, PPE is sometimes incorrectly perceived—by both the supervisor and the worker—as their first line of defense against all hazards. This, of course, is incorrect and dangerous. The worker must be made to understand (by means of enforced company rules, policies, and training) that PPE affords only minimal protection against most hazards—*it does not eliminate the hazard.*

"PPE only provides a barrier between the worker and the hazard." Experience shows that when some workers put on their PPE, they also don a "Superperson" mentality. What does this mean? Often, when workers use eye, hand, foot, head, hearing, or respiratory protection, they take on an "I can't be touched" attitude. They feel safe, as if the PPE somehow magically protects them from the hazard, so they act as if they are protected, are invincible, are beyond injury. They feel, however illogically, that they are well out of harm's way. Nothing could be further from the truth.

9.17.6.2 OSHA's PPE Standard

In the past, many OSHA standards have included PPE requirements, ranging from very general to very specific. It may surprise the reader to know, however, that not until recently (1993–1994) did OSHA incorporate a stand-alone PPE standard into its 29 CFR 1910/1926 guidelines. This relatively new personal protective equipment standard is covered under 1910.132–138, but you can find PPE requirements elsewhere in the General Industry Standards. For example, 29 CFR 1910.156, OSHA's Fire Brigade Standard, has requirements for firefighting gear. In addition, 29 CFR 1926.95–106 cover the construction industry. The PPE standard focuses on head, feet, eye, hand, respiratory, and hearing protection. Common PPE classifications and examples include the following:

1. Head protection (hard hats, welding helmets)
2. Eye protection (safety glasses, goggles)
3. Face protection (face shields)
4. Respiratory protection (respirators)
5. Arm protection (protective sleeves)
6. Hearing protection (ear plugs, muffs)
7. Hand protection (gloves)

8. Finger protection (cots)
9. Torso protection (aprons)
10. Leg protection (chaps)
11. Knee protection (kneeling pads)
12. Ankle protection (boots)
13. Foot protection (boots, metatarsal shields)
14. Toe protection (safety shoes)
15. Body protection (coveralls, chemical suits)

Respiratory and hearing protection have their own standards. Respiratory protection is covered under 29 CFR 1910.134 and hearing protection under 1910.95. Using PPE is often essential, but it is generally the last line of defense after engineering controls, work practices, and administrative controls. Engineering controls involve physically changing a machine or work environment. Administrative controls involve changing how or when employees do their jobs, such as scheduling work and rotating employees to reduce exposures. Work practices involve training workers how to perform tasks in ways that reduce their exposure to workplace hazards.

9.17.6.3 OSHA's PPE Requirements

Several requirements for both the employer and the employee are mandated under OSHA's personal protective equipment standard. OSHA's requirements include the following:

1. Employers are required to provide employees with PPE that is sanitary and in good working condition.
2. The employer is responsible for examining all PPE used on the job to ensure that it is of a safe (and approved) design and in proper condition.
3. The employer must ensure that employees use PPE.
4. The employer must provide a means for obtaining additional and replacement equipment; defective and damaged PPE is not to be used.
5. The employer must ensure that PPE is inspected on a regular basis.
6. The employee must ensure that he or she dons PPE when required.
7. Where employees provide their own PPE, the employer must ensure that it is adequate and that it is properly maintained and sanitized.

Although the employer must ensure that employees wear PPE when required, both employers and employees should factor in three things: (1) the PPE used must not degrade performance unduly, (2) it must be reliable, and (3) it must be suitable for the hazard involved.

9.17.6.4 PPE Hazard Assessment

How does a safety official determine when and where an employer should provide PPE and when the employee should use it? This can be determined in three ways:

1. From the manufacturer's guidance. When it comes to equipment and processes produced by a manufacturer, the manufacturer is considered the expert on that equipment or process and is normally best suited to determine the hazards associated with them.
2. If the process or equipment involves chemicals, the Material Safety Data Sheets (MSDSs) for the chemicals involved list the required PPE to be used.
3. OSHA mandates that the employer perform a hazard assessment of the workplace. The purpose of the hazard assessment is to determine if hazards are present or likely to be present that necessitate the use of PPE. If a facility presents such hazards, the employer is required to (a) select and have each affected employee use the types of PPE that will protect the affected employee from the hazards identified in the hazard assessment; (b) communicate selection decisions to each affected employee; and (c) select PPE that properly fits each affected employee. The employer is required to verify that the workplace hazard assessment has been conducted through a written certification of hazard assessment that identifies the workplace evaluated, the person certifying that the evaluation has been performed, and the date of the hazard assessment.

9.17.6.5 PPE Training Requirements

The employer must provide training to each employee required to use PPE. This training must instruct the employee on when the PPE is necessary; what PPE is necessary; how to properly don, doff, adjust, and wear PPE; the limitations of the PPE; and the proper care, maintenance, useful life, and disposal of the PPE.

During an OSHA audit of a facility, the auditor may want to look at a copy of the facility's PPE training program. Almost certainly, the auditor will want to review the company's training records for PPE training. Remember: You can conduct all the training in the world and have it performed by well-known experts in the field, but if you did not document the training then, in OSHA's eyes, it never occurred. You *must* have proof of training conducted.

After workers complete PPE training, OSHA requires each employee to demonstrate his or her understanding of the training. This is usually best accomplished through a written examination (records of which should also be maintained). If the employer has reason to believe that any affected employee who has already been trained does not have the understanding and skill required, the employer must retrain each such employee. In this retraining requirement, remember that everything in life is dynamic, including the workplace and work assignments. OSHA understands this and thus requires employers to retrain employees when new processes or equipment have been installed—any new element in a job task that might render previous training obsolete. Changes also occur in PPE itself. When a new type or model of PPE is introduced and used in the workplace, the employer must ensure that employees using such PPE are fully trained on it.

9.17.6.6 Respiratory Protection

> The basic purpose of any respirator is, simply, to protect the respiratory system from inhalation of hazardous atmospheres. Respirators provide protection either by removing contaminants from the air before it is inhaled or by supplying an independent source of respirable air. The principal classifications of respirator types are based on these categories.
>
> **—NIOSH (1987)**

> Written procedures shall be prepared covering safe use of respirators in dangerous atmospheres that might be encountered in normal operations or in emergencies. Personnel shall be familiar with these procedures and the available respirators.
>
> **—OSHA 29 CFR 1910.134(c)**

Respirators allow workers to breathe safely without inhaling particles or toxic gases. Two basic types are (1) *air-purifying*, which filter dangerous substances from the air, and (2) *air-supplying*, which deliver a supply of safe breathing air from a tank (SCBA), from a group of tanks (cascade system), or from an uncontaminated area nearby via a hose or airline to the mask. Respiratory protection might be a requirement in ensuring safe confined space entry. Often the organization's safety official holds the responsibility for making this determination. If the safety official determines that respiratory protection is required, then it is incumbent upon him or her to implement a written respiratory protection program that is in compliance with OSHA's respiratory protection standard (29 CFR 1910.134). Remember, though, that respiratory protection is often necessary to protect workers who may not ever be called upon to enter a confined space with an atmosphere containing airborne contaminants. Workers may need protection from airborne contaminants at any worksite where airborne contaminants are health hazards. This text has continuously stressed the vital need to attempt first to engineer out such hazards; however, when engineering and other methods of control or proper selection and use of respiratory protection cannot eliminate airborne hazards, it becomes part of the safety official's responsibility. Unlike past practices, where respiratory protection entailed nothing more than providing respirators to workers who could be exposed to airborne hazards and expecting workers to

DID YOU KNOW?

For permit-required confined space entry operations, respiratory protection is a key piece of safety equipment, one always required for entry into an immediately dangerous to life or health (IDLH) space and one that must be readily available for emergency use and rescue if conditions change in a non-IDLH space. Remember, however, that only air-supplying respirators should be used in confined spaces where there is not enough oxygen.

use the respirator to protect themselves, supplying respirators today without the proper training, paperwork, and testing is illegal. Employers are sometimes unaware that by supplying respirators to their employees without having a comprehensive respiratory protection program they are making a serious mistake. By issuing respirators, they have implied that a hazard actually exists. In a lawsuit, they then become fodder for the lawyers.

The respiratory protection program mandated by OSHA must not only follow OSHA's guidelines but must also be well planned and properly managed. A well-planned, well-written respiratory protection program must include the eleven elements shown in the sample program provided below. Selecting the proper respirator for the job, the hazard, and the worker is very important, as is thorough training in the use and limitations of respirators. Compliance with OSHA's respiratory protection standard begins with developing written procedures covering all applicable aspects of respiratory protection.

9.17.6.6.1 Sample Written Respiratory Protection Program

Respiratory Protection Program

I. Introduction

The Occupational Safety and Health Act (OSH Act) requires that every employer provide a safe and healthful work environment. This includes ensuring that workers are protected from unacceptable levels of airborne hazards. Although most air is safe to breathe, certain work operations and locations have characteristic problems of air contamination. Control measures are required to reduce airborne hazard concentrations to safe levels. When controls are not feasible or while they are being implemented, workers must wear approved respiratory protection.

The Company has adopted this Respiratory Protection Program to comply with OSHA regulations (as set forth in 29 CFR 1910.134) and to do all that is possible to protect those employees who are filling a job classification that requires respirator use in the performance of their duties. All departments and work centers are included and must adhere to the requirements set forth in this program. The company's Respiratory Protection Program is an organized approach to ensuring that employees have a safe workplace by providing specific requirements in these areas:

1. Designation of individual departmental responsibilities.
2. Definition of various terms used in the Respiratory Protection Program
3. Designation of types of respirators and their applications
4. Designation of procedures for respirator selection and distribution
5. Designation of procedures to be used for inspection and maintenance of respirators
6. Designation of procedures for employee respirator fit testing
7. Designation of a procedure for medical surveillance
8. Designation of a training program for personnel participating in the company's Respiratory Protection Program
9. Documentation procedure for personnel participating in the company's Respiratory Protection Program

II. **Responsibilities**

A. Department directors are responsible for the following:
   1. Implement and ensure compliance of departmental personnel with the company's Respiratory Protection Program.
   2. Specify the job classifications that use respirators and ensure that this job requirement is included in job descriptions for these classifications.

B. The company's Safety Division has the following responsibilities under the company's Respiratory Protection Program:
   1. Develop and modify as necessary the company's written Respiratory Protection Program.
   2. Check and review quarterly all workcenter programs, including the workcenter respirator inspection record.
   3. Compile and maintain a master respirator inventory list for the company.
   4. Implement an ongoing respirator training program.
   5. Conduct *initial* and *annual* employee fit testing.
   6. Provide initial and annual spirometric evaluation to ensure that employees are capable of wearing a respirator under their given work conditions.
   7. Provide technical assistance in determining the need for respirators and in the selection of appropriate types of respirators.
   8. Forward an employee's training, fit test, initial and annual spirometric evaluation, and medical doctor's evaluation for suitability to wear a respirator to the Human Resources Manager for inclusion in the employee's personnel record.

9. Inspect quarterly the accuracy and proper maintenance of records specified in this program.
10. Conduct air quality tests annually on internal combustion engine-driven airline respirator compressors to ensure proper air quality.

C. Company supervisory personnel are responsible for the following:
1. Ensure that respirators are available to employees as needed.
2. Ensure that employees wear appropriate respirators as required.
3. Ensure inspection of cartridge-type respirators on a monthly basis and self-contained breathing apparatus (SCBA) and airline hose mask systems on a weekly and monthly basis. Ensure that records of respirator inspections are maintained.
4. Ensure that employees are fit tested and receive initial and annual spirometric evaluation prior to using a respirator.

D. The employee is responsible for the following:
1. Use supplied respirators in accordance with instructions and training.
2. Clean, disinfect, inspect, and store assigned respirators properly.
3. Perform self-fit test prior to each use and ensure that manageable physical obstructions such as facial hair do not interfere with respirator fit.
4. Report respirator malfunctions to supervision and conduct after-use inspection of SCBA respirators.
5. Report any poor health conditions that may preclude safe respirator usage.

E. The company Human Resources Manager is responsible for the following:
1. Schedule required initial medical examination and spirometric evaluation for all new employees who fill job classifications requiring the use of respirators.
2. Maintain records of employee medical, spirometric, and fit test results.

III. Definition of Terms

The company's Respiratory Protection Program defines various terms as follows:

Aerosol—A suspension of solid particles or liquid droplets in a gaseous medium.

Asbestos—A broad mineralogical term applied to numerous fibrous silicates composed of silicon, oxygen, hydrogen, and metallic ions such as sodium, magnesium, calcium, and iron. At least six forms of asbestos occur naturally and are currently regulated—actinolite, amosite, anthophylite, chrysotile, crocidolite, and tremolite.

Banana oil—A liquid that has a strong smell of bananas; it is used to check for general sealing of a respirator during fit-testing.

Blasting abrasive—A chemical contaminant composed of silica, silicates, carbonates, lead, cadmium, or zinc and classified as a dust.

Breathing resistance—The resistance that can build up in a chemical respirator cartridge that has become clogged by particulates.

Chemical hazard—Any chemical that has the capacity to produce injury or illness when taken into the body.

Cleaning respirators—Involves washing with mild detergent and rinsing with potable water.

Dust—A dispersion of tiny solid airborne particles produced by grinding or crushing operations.

Fit-testing—An evaluation of the ability of a respiratory device to interface with the wearer in such a manner as to prevent the workplace atmosphere from entering the worker's respiratory system.

Forced expiratory volume–1 second (FEV_1)—The volume of air that can be forcibly expelled during the first second of expiration.

Forced vital capacity (FVC)—The maximum volume of air that can be exhaled forcefully after maximal inhalation.

Fume—Solid particles generated by condensation from the gaseous state.

Gas—A substance that is in the gaseous state at ordinary temperature and pressure.

Immediately dangerous to life or health (IDLH)—Any condition that poses an immediate threat to life or which is likely to result in acute or immediately severe health effects.

Irritant smoke (stannic oxychloride)—A chemical used to check for general sealing of a respirator during a fit test.

Mist—A dispersion of liquid particulates.

Oxygen deficiency—Any level below the PEL of 19.5%.

Particulates—Dusts, mists, and fumes.

Permissible exposure limit (PEL)—The maximum time-weighted average concentration of a substance in air that a person can be exposed to during an 8-hour shift.

Respirator—A face mask that filters out harmful gases and particles from the air, enabling a person to breathe and work safely.

Respiratory hazard—Any hazard that enters the human body by inhalation.

Saccharin—A chemical sometimes used to check for general sealing of a respirator during fit testing.

Smoke—Particles that result from incomplete combustion.

Spirometric evaluation—A test used to measure pulmonary function. A measurement of forced vital capacity (FVC) and forced expiratory volume–1 second (FEV_1) of 70% or greater is satisfactory. A measurement of less than 70% may require further pulmonary function evaluation by a medical doctor.

Vapor—The gaseous state of a substance that is liquid or solid at ordinary temperature and pressure.

IV. Types of Respirators

A. Chemical cartridge respirators

1. *Description*—Chemical cartridge respirators may be considered low-capacity gas masks. They consist of a facepiece that fits over the nose and mouth of the wearer. Attached directly to the facepiece is a small, replaceable filter-chemical cartridge.
2. *Application*—Usually this type of respiratory protection equipment is used where there is exposure to solvent vapors or dust and particulate matter, as with sandblasting, spray coating, or degreasing. They may not be worn in IDLH atmospheres.

B. Airline respirators (helmet, hoods, and masks) cascade-fed or compressor-fed

1. *Description*—These devices provide air to the wearer through a small-diameter, high-pressure hose line from a source of uncontaminated air. The source is usually derived from a compressed airline with a valve in the hose to reduce the pressure. A filter must be included in the hose line (between the compressed airline and the respirator) to remove oil and water mists, oil vapors, and any particulate matter that may be present in the compressed air. Internally lubricated compressors require that precautions be taken against overheating, as the heated oil will break down and form carbon monoxide. Where the air supply for airline respirators is taken from the compressed airline, a carbon monoxide alarm must be installed in the air supply system. Completion of a prior-to-operation preventive maintenance check on the carbon monoxide alarm system is critical.
2. *Application*—Airline respirators used in industrial applications for confined space entry (IDLH atmosphere) must be equipped with an emergency escape bottle.

C. Self-contained breathing apparatus (SCBA)

1. *Description*—This type of respirator provides Grade D breathing air (not pure oxygen), either from compressed air or breathing air cylinders or by chemical action in the canister

attached to the apparatus. It enables the wearer to be independent of any outside source of air. This equipment may be operable for periods from 1/2 to 2 hours. The operation of the self-contained breathing apparatus is fairly complex; therefore, it is necessary that the wearer have special training before being permitted to use it in an emergency situation.

2. *Application*—Because the oxygen-producing mechanism is self-contained in the apparatus, it is the only type of equipment that provides complete protection and at the same time permits the wearer to travel for considerable distances from a source of respirable air. SCBA (with the exception of hot work activities) can be used in many industrial applications.

V. Respirator Selection and Distribution Procedures

Workcenter supervisors select respirators. Selection is based on matching the proper color-coded cartridge with the type of protection desired. Selection is also dependent upon the quality of fit and the nature of the work being done. Cartridge type respirators are issued to the individuals who are required to use them. Each individually assigned respirator is identified in a way that does not interfere with its performance. Questions about the selection process are to be referred to the Safety Division.

VI. Respirator Inspection, Maintenance, Cleaning, and Storage

To retain their original effectiveness, respirators should be periodically inspected, maintained, cleaned, and properly stored.

Note: In the following sections, several references are made to various inspection records. Site-specific record forms and inspection records should be designed for use with your respiratory protection program.

A. *Inspection*

1. Respirators should be inspected before and after each use, after cleaning, and whenever cartridges or cylinders are changed. Appropriate entries should be made in a respirator "Inspection After Each Use" record.
2. If a half-face air-purifying respirator is taken out of use, indicate it on the inspection records. The respirator must be inspected thoroughly before it is put back in use.
3. Workcenter supervisors shall ensure that all cartridge-type respirators are inspected once per month and make appropriate entries in a "Supervisor's Monthly Respirator Inspection Checklist." The work center supervisor or designated person shall inspect all SCBA and airline respirators weekly and monthly, and make appropriate entries in a "SCBA/Airline

Respirator Weekly and Monthly Inspection and Maintenance Checklist" record. These records are to be kept by each workcenter for a period of 3 years.

4. Safety Division personnel will inspect these records quarterly.

B. *Maintenance*—Respirators that do not pass inspection must be replaced or repaired prior to use. Respirator repairs are limited to the changing of canisters, cartridges, cylinders, filters, head straps, and those items as recommended by the manufacturer. No attempt should be made to replace components or make adjustments, modifications, or repairs beyond the manufacturer's recommendations.

C. *Cleaning*—Individually assigned cartridge respirators are cleaned as frequently as necessary by the assignee to ensure that proper protection is provided. SCBA respirators are cleaned after each use. The following procedure is used for cleaning respirators:
   1. Filters, cartridges, or canisters are removed before washing the respirator and are discarded and replaced as necessary.
   2. Cartridge-type and SCBA respirator facepieces are washed in a detergent solution, rinsed in clean potable water, and allowed to dry in a clean area. A clean brush is used to scrub the respirator to remove adhering dirt.

D. *Storage*—After inspection, cleaning, and necessary repairs, respirators are stored to protect against dust, sunlight, heat, extreme heat, extreme cold, excessive moisture, or damaging chemicals. Respirators are to be stored in plastic bags or the original case. Individuals assigned respirators are to store their respirator in assigned personal locker. General use SCBA respirators are to be stored in designated cabinets, racks, or lockers with other protective equipment. Respirators are not to be stored in toolboxes or in the open. Individual cartridges or masks with cartridges are to be sealed in plastic bags to preserve their effectiveness.

VII. Respirator Fit-Testing

The Respiratory Protection Program provides standards for respirator fit-testing. The goals of respirator fit-testing are to (1) provide the employee with a face seal on a respirator that exhibits the greatest protective and comfortable fit, and (2) to instruct the employee on the proper use of respirators and their limitations. The three levels of fit testing are Initial, Annual, and Pre-Use Self-Testing.

A. The Initial and Annual fit tests are rigorous procedures used to determine whether the employee can safely wear a respirator. The Initial and Annual fit tests are conducted by the Safety Division. Both tests utilize cartridge-type and SCBA respirators to check each employee's suitability for wearing either type. Fit-testing requires special equipment and test chemicals such as banana oil, irritant smoke, or saccharin. In general, any change to the face or mouth may alter respirator fit and may require the use of a specially fitted respirator; the company's Safety Division will make this determination. Upon completion of Initial fit-testing, the Safety Division forwards the original of the employee's "Fit Test Record" to the Human Resources Manager for inclusion in the employee's file. A copy will be forwarded to the affected work center supervisor.

B. Pre-Use Self-Testing is a routine requirement for all employees who wear respirators. Each time the respirator is used, it must be checked for positive and negative seals. The Safety Division will train supervisors on this procedure. Supervisors are responsible for training employees in their individual workcenters.
   1. *Positive Pressure Check Procedure* (cartridge-type respirator): After the respirator has been put in place and straps adjusted for firm but comfortable tension, the exhalation valve is blocked by the wearer's palm. The wearer takes a deep breath, gently exhales a *little* air, and holds his or her breath for ten (10) seconds. If the mask fits properly, it will feel as if it wants to pop away from the face, but no leakage will occur.
   2. *Negative Pressure Check Procedure* (cartridge-type respirator): While still wearing the respirator, the wearer covers both filter cartridges with his or her palms, inhales slightly to partially collapse the mask, and holds this negative pressure for 10 seconds. If no air leaks into the mask, it can be assumed that the mask is fitting properly.

Note: Self-test fit testing can be conducted for both positive and negative pressure checks on SCBA respirators by crimping the hoses with one's fingers and blocking airways with the palms of one's hands. If either test shows leakage, the following procedure should be followed:

1. Ensure that the mask is clean. A dirty or deteriorated mask will not seal properly, nor will one that has been stored in a distorted position. Proper cleaning and storage procedures must be used.
2. Adjust the head straps to achieve snug, uniform tension on the mask. If only extreme tension on the straps will seal the respirator, report this to the Supervisor. A mask with uncomfortably tight straps rapidly becomes obnoxious to the wearer. 29 CFR 1910.134 (g)(1)(A) states: "Personnel with facial hair that comes between the sealing surface of the facepiece and the face or that interferes with valve function shall not be

permitted to wear tight-fitting respirators." Thus, respirator wearers with beards or side burns that interfere with the face seal are prohibited from wearing tight-fitting respirators on the job. Dental changes—loss of teeth, new dentures, braces, and so forth—may affect respirator fit and may require a new fitting with a different type mask. Any change to the face or mouth that may alter respirator fit must be brought to the immediate attention of the workcenter Supervisor.

VIII. Medical Surveillance

OSHA states that no one should be assigned a task requiring use of respirators unless they are found medically fit to wear a respirator by competent medical authorities. The company's Respiratory Protection Program includes a medical surveillance procedure:

A. *Pre-Employment Physical, Spirometric Evaluation, and 5-Year Follow-Up Physical Exam*
   1. All new and regular employees who fill job classifications that require respirator use in the performance of their duties are required to pass an initial medical examination to determine fitness to wear respiratory protection on the job. Annual spirometric evaluations will be conducted to ensure that employees covered under this program meet the OSHA requirements for fitness to wear respirators. On a continuous 5-year basis, all company employees covered under this program will be reexamined by competent medical authorities to ensure their continued fitness to wear respiratory protection on the job.
   2. Each department director will specify which job classifications require the employee to use respirators. A medical doctor will conduct pre-employment and 5-year follow-up medical evaluations. The Safety Division will conduct spirometric evaluations. The Safety Division will forward the employee's spirometry results to the Human Resources Manager for inclusion in the employee's personnel file.

B. *Annual Spirometric Evaluation*
   1. Annual spirometric evaluations will be conducted by the Safety Division on all employees filling job classifications requiring the use of respirators in the performance of their duties. Spirometry testing will be used to measure forced vital capacity (FVC) and forced expiratory volume–1 second (FEV_1). If FVC is less than 75% and/or FEV_1 is less than 70%, the employee will not be allowed to wear a respirator unless a written waiver is obtained from a medical

doctor. The supervisor determines whether the employee can be exempted from work functions that require wearing a respirator.

Note: The company will make reasonable accommodations to allow employees to retain their current positions with specified medical restrictions on respirator use.

The Safety Division will route annual results of spirometric testing to the Human Resources Manager for inclusion in each employee's personnel file and will notify appropriate supervisors of any employee who fails the test.

IX. Training

No worker may wear a respirator before spirometric evaluation, medical evaluation, fit testing, and training have all been completed and documented.

A. The Safety Division holds the responsibility for providing employee respirator training.
B. Supervisors are the day-to-day monitors of the program and have the responsibility to perform refresher training and to ensure that self fit-testing is accomplished by their employees as needed.

Available dates for Safety Division-administered training sessions will be published on a routine basis. Supervisors are responsible for scheduling their new employees for the next available session. Training on respiratory protection is also conducted at New Employee Safety Orientation sessions. This respiratory protection program is subject to changes and improvements as new regulations and technologies emerge. The Safety Division will train supervisors and employees as applicable on any new information.

X. Documentation Procedures

Documentation of safety training is very important. OSHA insists that certain records be maintained on all employees. All safety-training records should be considered legal records; the likelihood of having to use safety-training records in a court of law is real.

A. The following information will be maintained by the Safety Division:
   1. Date and location of initial employee training
   2. Inventory records of all company respirators

B. The following information will be processed by the Human Resources Manager for inclusion in the employee's personnel file:
   1. Results of annual employee fit-testing
   2. Results of new employee medical evaluation and annual spirometric testing (to remain on file for 5 years)

C. Supervisors will maintain:
   1. A file of respirator inspection records
   2. Respirator inventory records

Note: The maintenance and accuracy of all records specified in this will be evaluated quarterly by the Safety Division.

XI. Procedure for Safe Use of SCBA/Supplied-Air Respirators

To be in compliance with 1910.134(e)(3), the company is providing these written procedures covering the safe use of respirators (SCBA and supplied-air respirators only).

Note: Air-purifying or chemical cartridge respirators are to be used only for coating and sandblasting operations, and *never* for confined space entry or any other activity where oxygen deficiency or atmospheric contaminants are present. SCBA and/or supplied-air (with emergency escape bottles) respirators are to be used in all situations that involve chemical handling, confined space entry during normal operations, and in emergencies.

A. Safe Use Procedure in Dangerous Atmospheres

   This written procedure is prepared for safe respirator use in IDLH atmospheres that may occur in normal operations or emergencies. All Company personnel covered under this program are to be familiar with these procedures and respirators.
   1. Inspect all respirator equipment prior to use to ensure that it is complete and in good repair.
   2. Ensure that the respirator facepiece is the correct size for your face; perform a self fit-test.
   3. Ensure that available air is adequate for the expected time to be used.

Note: No company employee should use SCBA that is not 100% full.

   4. Test all alarms on the respirator to ensure that they work.
   5. Be sure that at least two fully trained and certified standby or rescue persons, equipped with proper rescue equipment (including SCBA), will be present in the nearest safe area for emergency rescue of those wearing respirators in an IDLH atmosphere.

6. Communications (visual, voice, signal line, telephone, radio, or other suitable type) must be maintained among all persons present (those in the IDLH atmosphere and the standby person or persons). The respirator wearers are to be equipped with safety harnesses and safety lines to permit their removal from the IDLH atmosphere if they are overcome.
7. The atmospheres in a confined space may be immediately dangerous to life or health (IDLH) because of toxic air contaminants or lack of oxygen. Before any company employee enters a confined space, tests must be performed to determine the presence and concentration of any flammable vapor or gas or any toxic airborne particulate, vapor, or gas and to determine the oxygen concentration (follow all procedures as outlined in the company's Confined Space Program).
8. No one is to enter if a flammable substance exceeds the lower explosive limit (LEL). No one should enter without wearing the proper type of respirator if any air contaminant exceeds the established permissible exposure limit (PEL), or if there is an oxygen deficiency. Be sure that the confined space is force ventilated to keep the flammable substance at a safe level.

Note: Even if the contaminant concentration is below the established breathing time-weighted average (TWA) limit and there is enough oxygen, the safest procedure is to ventilate the entire space continuously and to monitor the contaminant and oxygen concentrations continuously if people are to work in the confined space without respirators.

9. If the atmosphere in a confined space is IDLH due to a high concentration of an air contaminant or oxygen deficiency, those who must enter the space to perform work must wear a pressure-demand SCBA or a combination pressure-demand airline and SCBA that always maintains positive air pressure inside the respiratory inlet covering. Fully trained and equipped rescue must be onsite and ready to respond if needed. This is the best safety practice for confined space entry and *is required* at the company.

9.17.6.6.2 Respirator Program Evaluation

Safety officials must ensure that their organization's respiratory protection program complies with the elements covered in the sample program presented above, and they must also ensure that respirator program evaluations are also accomplished. Why? Because 29 CFR 1910.134 requires regular inspection and evaluation of the respirator program to determine its continued effectiveness in protecting employees. Remember that periodic air monitoring is also required, to determine if the workers are adequately protected. The overall program should be evaluated at least annually, and the written program or standard operating procedure (SOP) modified if necessary. Do you have questions about how to evaluate your respiratory protection program? Good. You should. Guidelines offered by the National Institute for Occupational Safety and Health (NIOSH, 1987) probably provide the best answers, including an evaluation checklist. Following is a sample from the NIOSH guide.

9.17.6.6.3 Sample Respiratory Protection Evaluation Checklist

Respirator Program Evaluation Checklist

In general, the respirator program should be evaluated for each job at least annually, with program adjustments made as appropriate to reflect the evaluation results. Program function can be separated into administration and operation.

A. *Program Administration*
1. Is there a written policy that acknowledges employer responsibility for providing a safe and healthful workplace and assigns program responsibility, accountability, and authority?
2. Is program responsibility vested in one individual who is knowledgeable and who can coordinate all aspects of the program at the jobsite?
3. Can feasible engineering controls or work practices eliminate the need for respirators?
4. Are there written procedures or statements covering the various aspects of the respirator program, including:
   - designation of an administrator;
   - respirator selection;
   - purchase of OSHA/NIOSH certified equipment;
   - medical aspects of respirator usage;
   - issuance of equipment;
   - fitting;
   - training;
   - maintenance, storage, and repair;
   - inspection;
   - use under special conditions; and
   - work area surveillance?

B. *Program Operation*
1. Respiratory protective equipment selection
   - Are work area conditions and worker exposures properly surveyed?
   - Are respirators selected on the basis of hazards to which the worker is exposed?

- Are selections made by individuals knowledgeable of proper selection procedures?

2. Are only certified respirators purchased and used; do they provide adequate protection for the specific hazard and concentration of the contaminant?
3. Has a medical evaluation of the prospective user been made to determine physical and psychological ability to wear the selected respiratory protective equipment?
4. Where practical, have respirators been issued to the users for their exclusive use, and are there records covering this issuance?
5. Respiratory protective equipment fitting
   - Are the users given the opportunity to try on several respirators to determine whether the respirator they will subsequently be wearing is the best fitting one?
   - Is the fit tested at appropriate intervals?
   - Are those users who require corrective lenses properly fitted?
   - Are users prohibited from wearing contact lenses when using respirators?
   - Is the facepiece-to-face seal tested in a test atmosphere?
   - Are workers prohibited from wearing respirators in contaminated work areas when they have facial hair or other characteristics may cause face seal leakage?
6. Respirator use in the work area
   - Are respirators being worn correctly (e.g., head covering over respirator straps)?
   - Are workers keeping respirators on all the time while in the work area?
7. Maintenance of respiratory protective equipment
   a. Cleaning and Disinfecting
      - Are respirators cleaned and disinfected after each use when different people use the same device, or as frequently as necessary for devices issued to individual users?
      - Are proper methods of cleaning and disinfecting utilized?

   b. Storage
      - Are respirators stored in a manner so as to protect them from dust, sunlight, heat, excessive cold or moisture, or damaging chemicals?
      - Are respirators stored properly in a storage facility so as to prevent them from deforming?
      - Is storage in lockers and toolboxes permitted only if the respirator is in a carrying case or carton?

c. Inspection
   - Are respirators inspected before and after each use and during cleaning?
   - Are qualified users instructed in inspection techniques?
   - Is respiratory protective equipment designated as "emergency use" inspected at least monthly (in addition to after each use)?
   - Is SCBA incorporating breathing gas containers inspected weekly for breathing gas pressure?
   - Is a record kept of the inspection of "emergency use" respiratory protective equipment?

d. Repair
   - Are replacement parts used in repair those of the manufacturer of the respirator?
   - Are repairs made by manufacturers or manufacturer-trained individuals?

e. Special use conditions
   - Is a procedure developed for respiratory protective equipment usage in atmospheres immediately dangerous to life or health?
   - Is a procedure developed for equipment usage for entry into confined spaces?

8. Training
   - Are users trained in proper respirator use, cleaning, and inspection?
   - Are users trained in the basis for selection of respirators?
   - Are users evaluated, using competency-based evaluation, before and after training?

9.17.6.6.4 Respiratory Protection: Summary

In previous eras, miners continuously tested the air in their underground worksites by keeping caged canaries with them. When the bird stopped singing, the miner knew the air was no longer fit to breathe and could take action to save himself. As an indicator of poor air quality, the canary was a primitive but necessary monitoring system. Today, of course, we have the technology to test and monitor the air quality in our worksites, as well as a measure of control over what enters our lungs through the use of respiratory equipment. To use these tools effectively, though, we must use them safely. Careless or improper use (for whatever reason) is pointless—and dangerous. Accordingly, a properly trained program administrator must administer the respiratory protection program. The employer's responsibilities include providing respirators, training, and medical evaluations at no cost to the employee.

9.17.7 Subpart J, General Environmental Controls

1910.141 Sanitation
1910.145 Specifications for Accident Prevention Signs and Tags
1910.146 Permit-Required Confined Spaces
1910.147 The Control of Hazardous Energy (Lockout/Tagout)
1910.151 Medical Services and First Aid

9.17.7.1 Personal and Sanitation Facilities

Note that lack of personnel sanitation facilities for workers is one of the top ten OSHA violations cited in Table 9.12. The site manager must factor into any workplace design several sanitation and personal hygiene requirements (i.e., provisions for potable water for drinking and washing; sewage, solid waste, and garbage disposal; sanitary food services; and drinking fountains, washrooms, locker rooms, toilets, and showers), in addition to providing a facility or plant site with easy-to-use and correct housekeeping activities. Housekeeping and sanitation are closely related. Control of health hazards requires sanitation, and control is usually enacted through good housekeeping practices. Disease transmission and ingestion of toxic or hazardous materials are controlled through a variety of sanitation practices, but if the workplace is not properly designed with appropriate sanitary and storm sewers, safe drinking water, and sanitary dispensing equipment, then sound sanitary practices are made much more difficult to implement within the workplace.

9.17.7.2 Recommended Color Codes for Accident-Prevention Tags

"DANGER"—Red, or predominately red, with lettering or symbols in a contrasting color

"CAUTION"—Yellow, or predominating yellow, with lettering or symbolism in a contrasting color

"WARNING"—Orange, or predominantly orange, with lettering or symbols in a contrasting color

"BIOLOGICAL HAZARD"—Fluorescent orange or orange-red, or predominantly so, with letters or symbols in contrasting colors

*9.17.7.3 OSHA's Confined Space Entry Program**

Note that in Table 9.12 the second most frequent OSHA citation issued for oil and gas drilling operations was for violation of 29 CFR 1910.146, the permit-required confined spaces standard. OSHA has a specific standard that mandates specific compliance with its requirements for making confined space

* Sections 9.17.7.3 through 9.17.7.8 are adapted from Spellman, F.R., *Confined Space Entry*, Technomic, Lancaster, PA, 1999.

entries; however, no matter how many standards and regulations OSHA and other regulators write, promulgate, and attempt to enforce, if employers and employees do not abide by their responsibilities under the act, the requirements are not worth the paper they are written on. OSHA's Confined Space Entry Program (CSEP) is a vital guideline to protect workers and others. CSEP was issued to protect workers who must enter confined spaces. It is designed and intended to protect workers from toxic, explosive, or asphyxiating atmospheres and from possible engulfment from small particles such as sawdust and grain (e.g., wheat, corn, and soybean normally contained in silos). It focuses on areas with immediate health or safety risks—areas with hazards that could potentially cause death or injury. These areas or spaces are classified as *permit-required* confined spaces. Under the standard, employers are required to identify all permit-required spaces in their workplaces, prevent unauthorized entry into them, and protect authorized workers from hazards through an entry-by-permit-only program. CSEP covers all of general industry, including agricultural services (the keyword here is "services" and not agriculture), manufacturing, chemical plants, refineries, transportation, utilities, wholesale and retail trade, and miscellaneous services. It applies to manholes, vaults, digesters, contact tanks, basins, clarifiers, boilers, storage vessels, furnaces, railroad tank cars, cooking and processing vessels, tanks, pipelines, and silos, among other spaces.

9.17.7.3.1 Confined Space Entry Definitions

Most rules, regulations, and standards have their own set of terms essential for communication between managers and the workers required to comply with the guidelines. Key terms that specifically pertain to OSHA's Confined Space Entry Program are defined here in alphabetical order. The definitions are from 29 CFR 1910.146 (Permit-Required Confined Spaces). Obviously, understanding any rule or regulation is difficult unless you have a clear and concise understanding of the terms used.

Acceptable entry conditions—The conditions that must exist in a permit space to allow entry and to ensure that employees involved with a permit-required confined space entry can safely enter into and work within the space.

Attendant—An individual stationed outside one or more permit spaces who monitors the authorized entrants and who performs all attendant duties assigned to the employer's permit space program.

Authorized entrant—An employee who is authorized by the employer to enter a permit space.

Blanking and blinding—The absolute closure of a pipe, line, or duct by the fastening of a solid plate (such as a spectacle blind or a skillet blind) that completely covers the bore and that is capable of withstanding the maximum pressure of the pipe, line, or duct with no leakage beyond the plate.

Confined space—A space that (1) is large enough and so configured that an employee can bodily enter and perform assigned work; (2) has limited or restricted means for entry or exit (tanks, vessels, silos, storage bins, hoppers, vaults, and pits are spaces that may have limited means of entry); and (3) is not designed for continuous employee occupancy.

Double block and bleed—The closure of a line, duct, or pipe by closing and locking or tagging (note interface with lockout/tagout) a drain or vent valve in the line between the two closed valves.

Emergency—Any occurrence or event (including any failure of hazard control or monitoring equipment) internal or external to the permit space that could endanger entrants.

Engulfment—The surrounding and effective capture of a person by a liquid or finely divided (flammable) solid substance that can be aspirated to cause death by filling or plugging the respiratory system, or that can exert enough force on the body to cause death by strangulation, constriction, or crushing.

Entry—The action by which a person passes through an opening into a permit-required confined space. Entry includes ensuing work activities in that space and is considered to have occurred as soon as any part of the entrant's body breaks the plane of an opening into the space.

Note: In the past, many workers thought that the actual entering of a confined space meant that they actually had to place their bodies in the space, that they had to actually physically enter the area. Many companies (and eventually OSHA) found that, under the old rules on confined space entry, the exact meaning of a confined space entry was not clearly spelled out. In fact, some of the old rules and regulations used in various industries were vague and ambiguous on this subject. OSHA, under a new 1993 rule, moved quickly to clear up the possibility of misunderstanding and to enhance the definition of exactly what constitutes a confined space entry. When OSHA initially issued its Final Rule on Confined Space Entry in early 1993, users in the field were confused about several items in the standard. One item of confusion, for example, was the original "point of entry" definition in the preamble, which stated that doorways and other portals through which a person can walk are not to be considered limited means for entry or exit. This was intended to limit the definition of confined spaces to those areas where an employee would be forced to enter or exit in a posture that not only might not be comfortable but that might also (and more importantly) slow self-rescue or make external rescue more difficult. Safety professionals in the field, however, knew from experience that even if a door or portal of a space is of sufficient size, obstructions could make entry into or exit from the space difficult. Even though OSHA's intent was that spaces that otherwise meet the definition of confined spaces and which have obstructed entry or exit (even though the portal is a standard-size doorway) should be classified as confined spaces, the problem was that OSHA's intent was not clear to the reader, to the supervisor, to other qualified or competent persons, or to the potential confined space entrant. Fortunately, OSHA recognized this ambiguity and changed this statement in its 1994 revision of the preamble to read, "OSHA notes that doorways and other portals through which a person can walk are not to be considered limited means for entry or exit. However, a space containing such a door or portal may still be deemed a confined space if an entrant's ability to escape in an emergency would be hindered."

Entry permit (permit)—The written or printed document provided by the employer to allow and control entry into a permit space and that contains the information shown in an approved entry permit.

Entry supervisor—The person (such as the employer, foreperson, or crew chief) responsible for determining whether acceptable entry conditions are present at a permit space where entry is planned, for authorizing entry and overseeing entry operations, and for terminating entry as required by the confined space entry standard.

Note: In practice (in the real world of performing confined space entry operations), common routine often designates the entry supervisor as the "competent" or "qualified" person. The competent or qualified person is that entry supervisor who has had the appropriate training and experience and possesses the knowledge required to supervise, and bring about, correct confined space entries. An entry supervisor may also serve as an attendant or as an authorized entrant, as long as that person is trained and equipped as required by the confined space entry standard for each role he or she plays. Also, the duties of entry supervisor may be passed from one qualified individual to another qualified individual during the course of an entry operation.

Hazardous atmosphere—An atmosphere that may expose employees to the risk of death, incapacitation, impairment of ability to self-rescue (i.e., to escape unaided from a permit space), injury, or acute illness from one or more of the following causes:

1. Flammable gas, vapors, or mist in excess of 10% of its lower explosive or lower flammable limit (LEL or LFL, which basically mean the same thing)
2. Airborne combustible dust at a concentration that meets or exceeds its LFL or LEL; this concentration may be approximated as a condition in which the dust obscures vision at a distance of 5 feet (1.52 meters) or less
3. Atmospheric oxygen concentration below 19.5% or above 23.5%

4. Atmospheric concentration of any substance for which a dose or a permissible exposure limit (PEL) is published in Subpart G, Occupational Health and Environmental Control, or in Subpart Z, Toxic and Hazardous Substances, which could result in employee exposure in excess of its dose or permissible exposure limit (PEL)

Note: An atmospheric concentration of any substance that is not capable of causing death, incapacitation, impairment of ability to self-rescue, injury, or acute illness due to its health effects is not covered by this provision.

5. Any other atmospheric condition that is immediately dangerous to life or health (IDLH)

Note: For air contaminants for which OSHA has not determined a dose or permissible exposure limit, other sources of information can provide guidance in establishing acceptable atmospheric conditions. These include Material Safety Data Sheets (MSDSs) that comply with the Hazard Communication Standard (commonly known as HazCom; 29 CFR 1910.1200), published information, and internal documents.

Hot work permit—The employer's written authorization to perform operations (e.g., riveting, welding, cutting, brazing, burning, heating) capable of providing a source of ignition.

Immediately dangerous to life or health (IDLH)—Any condition that poses an immediate or delayed threat to life that would cause irreversible adverse health effects or that would interfere with an individual's ability to escape unaided from a permit space.

Note: Some materials (e.g., hydrogen fluoride gas, cadmium vapor) may produce immediate transient effects that, even if severe, may pass without medical attention but are followed by sudden, possibly fatal collapse 12 to 72 hours after exposure. The victim feels normal after recovery from these transient effects until collapse. Such materials in hazardous quantities are still considered to be immediately dangerous to life or health.

Inerting—The displacement of the atmosphere in a permit space by a noncombustible gas (such as nitrogen) to such an extent that the resulting atmosphere is not combustible. This procedure produces an IDLH oxygen-deficient atmosphere.

Isolation—The process by which a permit space is removed from service and completely protected against the release of energy and material into the space by such means as blanking or blinding; realigning or removing sections of lines, pipes, or ducts; a double block and bleed system; lockout or tagout of all sources of energy; or blocking or disconnecting all mechanical linkages.

Line breaking—The intentional opening of a pipe, line, or duct that is or has been carrying flammable, corrosive, or toxic material, an inert gas, or any fluid at a volume, pressure, or temperature capable of causing injury.

Non-permit confined space—A confined space that does not contain or (with respect to atmospheric hazards) have the potential to contain any hazard capable of causing death or serious physical harm.

Oxygen-deficient atmosphere—An atmosphere containing less than 19.5% oxygen by volume.

Oxygen-enriched atmosphere—An atmosphere containing more than 23.5% oxygen by volume.

Permit-required confined space (permit space)—A confined space that has one or more of the following characteristics:

1. Contains or has a potential to contain a hazardous atmosphere;
2. Contains a material that has the potential for engulfing an entrant;
3. Has a configuration such that an entrant could be trapped or asphyxiated by inwardly converging walls, or by a floor that slopes downward and tapers to a smaller cross-section; or
4. Contains any other recognized serious safety or health hazard.

Permit-required confined space program (permit space program)—The employer's overall program for controlling (and where appropriate, for protecting employees from) permit space hazards, and for regulating employee entry into permit spaces.

Permit system—The employer's written procedure for preparing and issuing permits for entry and for returning the permit space to service following termination of entry.

Prohibited condition—Any condition in a permit space that is not allowed by the permit during the period when entry is authorized.

Rescue service—The personnel designated to rescue employees from permit spaces.

Retrieval system—The equipment, including a retrieval line, chest or full-body harness, wristlets (if appropriate), and a lifting device or anchor (usually a tripod and winch assembly), used for non-entry rescue of persons from permit spaces.

Testing—The process by which the hazards that may confront entrants in a permit space are identified and evaluated. Testing includes specifying the tests that are to be performed in the permit space.

Note: Testing allows employers both to develop adequate control measures for the protection of authorized entrants and to determine if acceptable entry conditions are present immediately prior to, and during, entry.

9.17.7.3.2 Evaluating the Workplace

> The employer shall evaluate the workplace to determine if any spaces are permit-required confined spaces.
>
> **—29 CFR 1910.146(c)(1)**

The organization's safety official needs to ask, "Does my organization need to comply with OSHA's confined space entry standard?" It depends, and OSHA wants all safety officials to make that determination by evaluating their workplaces. So, how do we go about evaluating our workplaces to determine if we must comply? Before we answer this question, a note of caution. In the evaluation procedure that you must follow to evaluate your workplace, you must take every care and caution that you do not walk into, climb into, or crawl into *any* space unless you are absolutely certain that it is safe to do so. In short, for safety, you must assume that any unfamiliar confined space presents hazards until you have determined by examination and testing that it does not. How do you go about evaluating a worksite for compliance? To determine if a particular worksite must comply with OSHA's confined space entry standard we must take certain steps. First, we must be familiar with what a confined space is. A confined space is defined as follows:

> A confined space is large enough and so configured that an employee can bodily enter and perform assigned work; has limited or restricted means for entry or exit (e.g., tanks, vessels, silos, storage bins, hoppers, vaults, and pits are spaces that may have limited means of entry); and is not designed for continuous employee occupancy.

The next step is to survey the plant site, the facility, the factory, or other type worksite to determine if any spaces or structures fall under OSHA's definition of a confined space. When performing such a survey, you must record on paper the name and location of each space or structure identified for evaluation later. You should also have a list of all worksite-confined spaces. This list should be distributed to all employees, placed in plain view on employee bulletin boards, and inserted into your site's written confined space program. One thing is certain—when OSHA audits your facility, they will want to see your list of confined spaces.

During the evaluation survey process, if confined spaces are identified, then the determination must be made whether or not they are "permit-required confined spaces" or "non-permit confined spaces." To do this, you must be familiar with OSHA's definitions for both. Recall from the definitions list that

1. *Non-permit confined space* is a confined space that does not contain or (with the respect to atmospheric hazards) have the potential to contain any hazard capable of causing death or serious physical harm.
2. *Permit-required confined space (permit space)* is a confined space that has one of more of the following characteristics:

a. Contains or has a potential to contain a hazardous atmosphere;
b. Contains a material that has the potential for engulfing an entrant;
c. Has an internal configuration such that an entrant could be trapped or asphyxiated by inwardly converging walls or by a floor that slopes downward and tapers to a smaller cross-section; or
d. Contains any other recognized safety or health hazard.

A permit-required confined space must be clearly labeled to inform employees of the location and the danger posed by the permit space.

After identifying and labeling all site permit-required confined spaces, the employer has two choices: (1) designate such spaces as off limits to entry by any employee (unauthorized entry must be prevented), or (2) develop a written confined space program. The requirements for a written confined space program are covered in the next section.

9.17.7.3.3 Permit-Required Confined Space Written Program

> If the employer decides that its employees will enter permit spaces, the employer shall develop and implement a written permit space program....The written program shall be available for inspection by employees and their authorized representatives.
>
> **—29 CFR 1910.146(c)(4)**

The first step the employer must take in implementing a permit-required confined space program is to take the measures necessary to prevent unauthorized entry. Typically, this is accomplished by identifying and labeling all confined spaces. The next step is to list all of those confined spaces and clearly communicate to employees that the listed spaces are not to be entered by organizational personnel under any circumstances. Remember that the *employer* is responsible for identifying, labeling, and listing all site permit-required confined spaces, in addition to identifying and evaluating the hazards of each confined space. Once the hazards have been identified and evaluated, the identity and hazards for each confined space must be listed in the organization's written confined space entry program (obviously, it is important that employees are made well aware of all the hazards). The next step is to develop written procedures and practices for those personnel who are required to enter, for any reason, permit-required confined spaces. The procedures and practices used for permit-required confined space entry must be *in writing* and at the very least must include the following:

- Specifying acceptable entry conditions
- Isolating the permit space
- Purging, inerting, flushing, or ventilating the permit space as necessary to protect entrants from external hazards

- Providing pedestrian, vehicle, or other barriers as necessary to protect entrants from external hazards
- Verifying that conditions in the permit space are acceptable for entry throughout the duration of an authorized entry

Under OSHA's program, the employer must also provide specified equipment to employees involved in confined space entry. The requirements under this specification and the required equipment are covered in the following section.

9.17.7.3.4 Permit-Required Confined Space Entry Equipment

In its permit-required confined spaces standard (1910.146), OSHA specifies the equipment required to make a safe and approved confined space entry into permit-required confined spaces. Note that the employer, at no cost, must provide this equipment to the employee. The employer is also required not only to procure this equipment at no cost to the employee but also to maintain the equipment properly. Most importantly, the employer is also required to ensure that employees use the equipment properly. We will come back to this important point later. For now, let's take a look at the type of equipment required for making a safe and legal permit-required confined space entry.

"Equipment" refers to equipment that has been approved, listed, labeled, or certified as conforming to applicable government or nationally recognized standards, or to applicable scientific principles. It does not mean jerry-rigged devices that might (or might not) be suitable for use by employees. It means that only safe and approved equipment in good condition is to be used. Period.

9.17.7.3.4.1 Testing and Monitoring Equipment Numerous makes and models of confined space air monitors (gas detectors or sniffers) are available on the market, and selection should be based on your facility's specific needs; for example, if the permit-required confined space to be entered is a sewer system, then the specific need is a multiple-gas monitor. This type of instrument is best suited for sewer systems, where toxic and combustible gases and oxygen-deficient atmospheres are prevalent. No matter what type of air monitor is selected for a specific use in a particular confined space, any user must be thoroughly trained on how to effectively use the device. Users must also understand the monitor's limitations and how to calibrate the device according to manufacturer's requirements. Having an approved air monitor is useless if workers are not trained in its operation or proper calibration. When choosing an air monitor for use in confined space entry, you must ensure that the monitor selected is suitable for the type of atmosphere to be entered and that it is equipped with audible and visual alarms that can be set, for example, at 19.5% or lower for oxygen and preset for levels of the combustible or toxic gases it is used to detect.

9.17.7.3.4.2 Ventilating Equipment In many cases, you can eliminate, reduce, or modify atmospheric hazards in confined spaces by ventilating—using a special fan or blower to displace the bad air inside a confined space with good air from outside the enclosure. Whatever blower or ventilator type you choose to use, a certain amount of common sense and consideration of the depth of the confined space, size of the enclosure, and number of openings available are required. Keep in mind that the blower must be equipped with a vaporproof, totally enclosed electrical motor or a non-sparking gas engine. Obviously, the size and configuration of the confined space dictate the size and capacity of the blower to be used. Typically, a blower with a large-diameter flexible hose (elephant trunk) is most effective.

9.17.7.3.4.3 Personal Protective Equipment Note that noncompliance with the requirements of the OSHA's personal protective equipment standard is listed as one of the top ten violations in Table 9.12. OSHA requires PPE for confined space entries. The entrant must be equipped with the standard personal protective equipment (PPE) required to make a vertical entry into a permit-required confined space (a full-body harness combined with a lanyard or lifeline), and also the PPE required to protect him or her from specific hazards. As an example, an employee who is to enter a manhole is typically equipped with (1) an approved hard hat to protect the head; (2) approved gloves to protect the hands; (3) approved footwear (safety shoes) to protect the feet; (4) approved safety eyewear or face protection to protect the eyes and face; (5) full body clothing (long-sleeved shirt and long trousers) to protect the trunk and extremities; and (6) a tight-fitting NIOSH-approved self-contained breathing apparatus (SCBA) or supplied-air hose mask with emergency escape bottle for IDLH atmospheres.

9.17.7.3.4.4 Lighting Many confined spaces could be described as nothing more than dark (and sometimes foreboding) holes in the ground—often a fitting description. As you might guess, typically many confined spaces are not equipped with installed lighting. To ensure safe entry into such a space, the entrant must be equipped with intrinsically safe lighting. Intrinsically safe? Absolutely. Think about it. The last thing you want to do is to send anyone into a dark space filled with methane with a torch in his or her hand, but a light source that emits sparks might as well be a torch. Confined spaces present enough dangers on their own without adding to the hazards. Even after the space has been properly ventilated (with copious and continuous amounts of outside fresh air) and, for example, the source of methane has been shut off, we still obviously have a space that has the potential for an extremely explosive atmosphere. Do not underestimate the hazards such a confined space presents! So, what do we do? If lighting is required in a confined space, we need to ensure that it is provided to the entrant—for his or her safety as well as to enable work to be done. For confined space entries, explosion-proof lanterns or flashlights (intrinsically safe devices) are recommended. Such devices, if NIOSH and OSHA approved, are equipped with spring-loaded bulbs that, upon breaking, eject themselves from the electrical circuit, preventing ignition of hazardous atmospheres. Another safe, low-cost, instant light source now readily available for confined space entry is lightsticks. They can be used safely near explosive materials because they contain no source of ignition. Lightsticks are available with illumination times ranging from 1/2 to 12 hours. Another common work light used for confined space entry is the droplight. UL-approved droplights that are vaporproof, explosion-proof, and equipped with ground-fault circuit interrupters (GFCIs) are the recommended type for confined space entry.

Note: If you have a confined space that has the potential for an explosive atmosphere and has light fixtures permanently installed in place, remember that these lights must be certified for use in hazardous locations and maintained in excellent condition.

9.17.7.3.4.5 Barriers and Shields We must be concerned with not only the safety of the confined space entrant but also the safety of those outside the confined space. An open manhole, for example, obviously presents a pedestrian and traffic hazard. To prevent accidents in areas where manhole work is in progress, we can use several safety devices, including manhole guard rail assemblies, guard rail tents, barrier tape, fences, and manhole shields. Remember that we want to prevent someone from falling into a manhole (or other type of confined space opening) and we also want to prevent unauthorized entry. Occasionally, manholes or ordinarily inaccessible areas, when open for work crews, present an attractive nuisance—even ordinary curiosity may lead people (especially children) to put themselves at risk by attempting to enter a confined space. Along with protecting the confined space opening from someone falling into it or entering it illegally, we must also control traffic around or near the opening. To do this we may need to employ the use of cones, signs, or stationed guard personnel. Don't forget the nighttime hours. After dark, it is obviously difficult to see an open confined space opening or guard device; these devices should be lighted with vehicle strobes or beacon lights.

9.17.7.3.4.6 Ingress and Egress Equipment: Ladders Have you ever peered down a 40-foot deep, 24-inch diameter vertical manhole? Not pleasant? Depends on your point of view. If the manhole has no lighting (as most do not) then you are peering into what appears to be a bottomless pit (and maybe it is). Have you been there? If so, no further explanation is needed. You know that at best, entering any manhole can be a perilous undertaking. If you have never faced entering a manhole, let's consider an important point. If you are tasked to enter such a confined space, you will obviously be interested in entering it (ingressing) safely (taking all required precautions) and returning (egressing) safely. Experience with assessing safety considerations in confined space areas has shown that many of the installed ladders (in place to allow entry and exit inside confined spaces) are not always in the best condition due to the environment to which they are constantly exposed year after year.

Confined spaces may be shrouded in moist, chemical-laden atmospheres—conditions excellent for corroding most metals. Most ladders installed in confined spaces are made of metal. In addition to requiring workers to enter dangerous permit-required confined spaces, if we do not properly evaluate all of the confined space's conditions then we may also be asking them to enter in a totally unsafe manner, on equipment that may fail. Installed ladders within confined spaces must be inspected on a periodic basic to ensure their integrity.

Don't forget about the devices used to hold the ladders in place—the securing or attachment bolts or screws. Most of these are also made of metal, which will corrode and weaken with time. How about those spaces that do not have installed ladders? For confined spaces not equipped with ladders, stairways, or some other installed means of ingress and egress, we often employ the use of portable ladders. One way or another, we are required to provide a safe way in and out of a confined space, and ladders often fit this need. Occasionally, though, ladders or stairways for safe entry or exit are not available, practical, or practicable. When such a situation arises, winches and hoisting devices are commonly used to raise and lower entrants. Remember that any lowering and lifting devices must be OSHA-approved as safe to use. Using a rope attached to the bumper of a vehicle to lower or raise an entrant, for example, is strictly prohibited. Only hand-operated lifting or hoisting devices should be employed. Motorized devices are unforgiving, especially when the entrant gets caught up in an obstruction (e.g., machinery, pipe, angle iron) that prevents him from moving. When a motorized device is used, a person stuck in a confined space could literally be pulled apart. OSHA regulations were created to prevent just such gruesome incidents from occurring, but gruesome and fatal events (sometimes involving multiple fatalities) do occur. When an entrant gets into trouble while inside a confined space, OSHA is quite specific on what should and should not be done in rescuing a confined space entrant who is in trouble.

9.17.7.3.4.7 Rescue Equipment When confined space rescue is to be effected by any agency other than the facility itself (e.g., emergency rescue service, fire department), the facility is not required to provide the rescue equipment; however, when confined space rescue is to be performed by facility personnel, proper rescue equipment is required. Proper rescue equipment consists of the equipment needed to remove personnel from confined spaces in a safe manner. "In a safe manner" means "to prevent further injury to the entrants and *any* injury to the rescuers." Confined space rescue equipment (commonly called *retrieval equipment*) typically consists of three components: *safety harness, rescue and retrieval line*, and a *means of retrieval*. Let's take a closer look at each of these components.

A full-body harness combined with a lanyard or lifeline evenly distributes the fall-arresting forces among the worker's shoulders, legs, and buttocks, reducing the chance of further internal injuries. A harness also keeps the worker upright and more comfortable while awaiting rescue. The full-body harness used for confined space rescue should consist of flexible straps that continually flex and give with movement, conforming to the wearer's body and eliminating the need to frequently stop and adjust the harness. Usually constructed of a combination of nylon, polyester, and specially formulated elastomer, the proper harness resists the effects of sun, heat, and moisture to maintain its performance on the job. The full-body harness should include a sliding back D-ring (to attach the retrieval line hook) and a non-slip adjustable chest strap.

The heavy-duty rescue and retrieval line is usually a component of a winch system. Both ends of the retrieval lines should be equipped with approved locking mechanisms of at least the same strength as the lines for attaching to the entrant's harness and anchor point. The winch systems used today are either an approved two-way system or three-way system. The two-way system is used for raising and lowering rescue operations whenever a retractable lifeline is not needed. Typical systems feature three independent braking systems, a tough two-speed gear drive, and approximately 60 feet of steel cable. Three-way systems offer additional protection when a self-retracting lifeline is used. The winch is usually a heavy-duty model (usually rated at 500 lb or 225 kg) with disc brakes to stop falls within inches, and it is equipped with a shock-absorption feature to minimize injuries. The proper winch should allow the user to raise and lower loads at an average speed of 10 to 32 feet per minute in an emergency.

The means of retrieval usually includes the proper winch with built-in fall protection attached to a 7- or 9-foot tripod. The tripod should be of sufficient height to allow the victim to be brought above the rim of the manhole or other opening and placed on the ground.

9.17.7.3.4.8 Other Equipment If tools are to be used during a confined space entry or rescue, it may be necessary to use non-sparking tools if flammable vapors or combustible residues are present. These non-sparking, non-magnetic, and corrosion resistant tools are usually fashioned from copper or aluminum. A fire extinguisher, additional radios for communication, spare oxygen bottles (for SCBA and cascade systems as needed), a first-aid kit, and any other equipment required for safe entry into and rescue from permit spaces may also be necessary.

9.17.7.3.5 Pre-Entry Requirements

Before anyone is allowed to enter a permit-required confined space, certain space conditions must first be evaluated. The first step taken should be to determine whether workers must enter the permit-required space to complete the task at hand. You should ask yourself, "Do we really need to enter the permit-required confined space?" If the answer is yes, then before initiating a confined space entry the space should be tested with a calibrated air monitor to determine if acceptable entry conditions exist before entry is authorized. If air monitoring indicates that entry can be made safely without respiratory protection or if appropriate respiratory protection must be worn, then the supervisor (qualified or competent person) must decide how to effect the entry in the safest manner possible. Whether the atmosphere is safe or unsafe without proper respiratory protection, monitoring must be continuous. Taking only one reading and basing decisions on that reading is not wise; in fact, it is unsafe. Conditions can change within a confined space at any time—it is critical to the wellbeing of the entrant to know when these changes take place and what the changes are. When conducting the air test for atmospheric hazards, a standard testing protocol should be followed:

1. Test for oxygen.
2. Test for combustible gases and vapors.
3. Test for toxic gases and vapors.

You should also test the atmosphere within a confined space at different levels. For example, if you are about to authorize the entry of workers into a manhole that is 30 feet deep, you should test top to bottom for a stratified atmosphere. Remember that some toxic gases (methane, for example) are lighter than air. They tend to accumulate at higher levels within the manhole. If the manhole may contain carbon monoxide (which has a vapor density similar to air) you should test at the middle level. Hydrogen sulfide (a deadly killer) is heavier than air; therefore, you should test close to the bottom of the manhole if hydrogen sulfide may be present. Along with testing at different levels for stratification of toxic gases, you should also check in all directions as much as possible.

The key point to remember is that atmospheric testing should be continuous, especially when entrants are inside the confined space. To ensure that continuous atmospheric testing is conducted while an entrant is inside the confined space, an attendant (at least one) must be stationed outside the space to conduct the testing. In addition to continuously monitoring the atmosphere of the permit-required confined space, the attendant or some other designated person must be familiar with the procedure for summoning rescue and emergency services. For those facilities having fully trained and equipped onsite rescue teams, it is common (and prudent) practice to have the rescue team standing outside the confined space to be immediately available if required.

Another important function of the attendant or other designated person involved in permit-required confined space entry is to ensure that unauthorized entry into the confined space is prevented. Before any permit-required confined space entry can be made, a proper confined space entry permit must be used. When employees from more than one work center (e.g., electricians, machinists, painters, and others from different work centers) or more than one employer are involved in confined space entry, an entry procedure to ensure the safety of all entrants must be developed and implemented.

After the confined space entry is completed, procedures must be in place and used to ensure that the space has been closed off and the permit canceled.

The final step that should be taken after any confined space entry has been made and is completed is to critique the procedure. Questions should be asked and answers given. Did anything go wrong during the entry procedure? Did an unauthorized person make an entry into the space? Did any of the equipment used fail? Was anyone injured? Were there any employee complaints about the procedure? If such questions do come up, steps must be taken to make sure they are answered or that corrections are made to ensure that the next entry into a permit-required confined space is a safer one. At least once each year, the permits accumulated during the year (confined space permits must be retained by the employer for one year) should be reviewed. If it is apparent from the review that the procedure should be changed, then it should be changed as needed.

9.17.7.3.6 Permit System

A permit system for permit-required confined space entry is required by the confined space entry standard. An entry supervisor (qualified or competent person) must authorize entry, prepare and sign written permits, order corrective measures if necessary, and cancel permits when work is completed. Permits must be available to all permit space entrants at the time of entry and should extend only for the duration of the task. They must be retained for a year to facilitate review of the confined space program.

The previous paragraph sums up OSHA's requirements under its confined space entry standard in 29 CFR 1910.146(e), Permits System, and 1910.146(f), Entry Permit. OSHA's requirements under these sections are intended to ensure that

1. A permit is actually used for entry into permit-required confined spaces.
2. An entry supervisor (the qualified or competent person) authorizes the entry.
3. The entry permit is signed.
4. Any corrective measures are taken if found necessary.
5. The permit is canceled when work is completed.

Confined space entry permits must be available to all permit space entrants at the time of entry and should extend only for the duration of the task. As we stated previously, the permits must be retained for a year to facilitate review of the confined space program.

9.17.7.3.6.1 Permit Requirements According to OSHA, an entry permit must include the following:

1. Identification of the permit space to be entered
2. The purpose of the entry
3. The date and authorized duration of the entry permit
4. The authorized entrants within the permit space by name, or by such other means as will enable the attendant to determine quickly and accurately, for the duration of the permit, which authorized entrants are inside the permit space
5. The personnel, by name, currently serving as attendants
6. The individual, by name, currently serving as the entry supervisor (qualified or competent person), with a space for the signature or initials of the entry supervisor who originally authorized entry
7. The hazards of the permit space to be entered
8. The measures used to isolate the permit space and to eliminate or control permit space hazards before entry (i.e., lockout/tagout must be completed)
9. The acceptable entry conditions
10. The results of initial and periodic tests performed, accompanied by the names or initials of the testers and by an indication of when the tests were performed
11. The rescue and emergency services that can be summoned and the means (such as the equipment to use and the numbers to call) for summoning those services
12. The communication procedures used by authorized entrants and attendants to maintain contact during the entry
13. Equipment, such as personal protective equipment, testing equipment, communications equipment, alarm systems, and rescue equipment
14. Any other information whose inclusion is necessary, given the circumstances of the particular confined space, to ensure employee safety
15. Any additional permits, such as for hot work, that has been issued to authorize work in the permit space

9.17.7.3.7 Confined Space Training

According to 29 CFR 1910.146(g), Training, the employer must provide training so that all employees whose work is regulated by the standard acquire the understanding, knowledge, and skills necessary for the safe performance of the duties assigned. Any work requirement is easier to perform if the person doing the task is fully trained on the proper way to accomplish it. Training offers another advantage as well—increased safety. In accomplishing any work task safely, proper training is critical. Confined space entry operations are extremely dangerous undertakings. We stated earlier that confined spaces are very unforgiving, and this is the case even for those workers who have been well trained; however, training helps to reduce the severity of any incident. When something goes wrong (as if often the case) it is better to have fully trained personnel standing by than to have people standing by who are not trained and do not know how to properly rescue an entrant, let alone how to rescue themselves. When you get right down to it, having fully trained workers for any job just makes good common sense.

OSHA is very clear on its requirement to train confined space entry personnel. Both initial and refresher training must be provided. This training must provide employees with the necessary understanding, skills, and knowledge to perform confined space entry safely. Refresher training must be provided and conducted whenever an employee's duties change, when hazards in the confined space change, or whenever an evaluation of the confined space entry program identifies inadequacies in the employee's knowledge. The training must establish employee proficiency in the duties required and introduce new or revised procedures as necessary for compliance with the standard.

OSHA also requires the employer to certify *in writing* that the employee has been trained. This certification must include the employee's name, the signature of the trainer, and the dates of training. Typically, employers certify this training by conducting written and practical examinations (including training dry runs or drills). When an employee meets the certification requirements, the employee is normally awarded a certificate stating that he or she has been trained and certified (by whatever means). These written certifications should be filed in the employee's personnel record and training records.

Any time you conduct safety training, you must keep accurate records of the training. OSHA will want to see these records when they audit your facility (for whatever reason). Any supervisor or training official that provides critically important and possibly life-saving training would be foolish not to keep and maintain accurate training records, as they may be needed in a legal action. To facilitate the recordkeeping process, a form or roster with a statement such as the one shown in Figure 9.2 is recommended.

Remember, not only does OSHA require training on its confined space entry standard and other associated standards (i.e., Lockout/Tagout, Respiratory Protection, and Hot Work Permits), but this training is also critically important to the wellbeing of workers. Making sure that they know that their work organization is taking all possible steps to ensure their safety should encourage them to buy into the required safe work practices themselves. You must be able to demonstrate that this training was actually conducted. The form shown in Figure 9.2 will aid in this effort. As stated before, all training must be documented.

ATTENDANCE ROSTER

Trainer ______________________________ **Date** ______________

Confined Space Training

In accordance with the recordkeeping and training requirements of the Confined Space Standard, I have received training on Confined Space Entry Procedures. I have agreed to verify my understanding and training on 29 CFR 1910.146, Confined Space Entry Standard, by signing this roster. This training meets the requirements as specified by 29 CFR 1910.146.

Employee's Name	**Work Center**

FIGURE 9.2
Typical training roster form.

9.17.7.3.7.1 Workplace Confined Space Training Program Any workplace training program on just about any OSHA requirement is somewhat site specific. Confined space training for wastewater workers, for example, might be different from the training given to telephone repair persons who have to enter underground vaults, because the hazards might not be the same. As a rule of thumb, it is hard to go wrong with any OSHA required training if the requirements spelled out in the applicable standard are explained to all workers involved. In addition, for confined space entry training it is important, at a minimum, to cover the following:

1. Explain and point out the requirements of OSHA's confined space entry standard, 29 CFR 1910.146.
2. Clearly explain who is responsible for what under the program.
3. Explain key definitions.
4. Inform each trainee of the exact location of the worksite's permit-required confined spaces.
5. Explain how to use the worksite's confined space permit.
6. Explain the potential for engulfment.
7. Explain and demonstrate how to use air-monitoring equipment.

8. Explain and demonstrate how to use required confined space entry equipment.
9. Explain the potential for hazardous atmospheres.
10. Explain the worksite's procedures for confined space rescue.
11. Explain the interface between confined space entry and lockout/tagout, respiratory protection, and hot work permits.
12. Explain how to properly use the worksite's pre-entry checklist.

9.17.7.3.8 Assignment of Onsite Personnel

Onsite personnel, including entrants, attendants, and entry supervisors, assigned to make permit-required confined entries must be fully aware of their duties under the OSHA standard. OSHA clearly defines these duties under 1910.146(h), Entrants; 1910.146(i), Attendants; and 1910.146(j), Entry Supervisors. Again, training is the key ingredient to effecting safe permit-required confined space entry. Obviously, assigning anyone specific duties is easy, but ensuring that these duties are performed in the correct manner—especially when training has not been conducted—is much more difficult. Supervisors and workers must know their duties and must know how to complete their duties in a safe and correct manner.

9.17.7.3.8.1 Duties of Authorized Entrants The key responsibility of *any* permit-required confined space entrant is to gain knowledge of the hazards that may be faced during entry. The entrant must also be knowledgeable enough to understand the mode, signs or symptoms, and consequences of exposure to hazards, whether they be immediate or potential hazards. This knowledge requirement is central to the critical role that training plays in achieving compliance with this program—or any safety program. Knowledge of the hazards or potential hazards is just part of the requirements involved in being a qualified entrant. Entrants must also know their equipment—how to use it, what it is to be used for, and its limitations. They must also know how to communicate with the attendant. Communication can be via radio/walkie-talkie (which must also be intrinsically safe and not capable of producing sparks), by hand signals (obviously, visual contact must be maintained), or by voice, whistle, or some other prearranged and practiced sound-making device. The entrant must alert the attendant whenever he or she recognizes any warning sign or symptom of exposure to a dangerous situation. He or she must also communicate to the attendant any changing condition that could make the entry more hazardous than it already is. The entrant must know when to exit the confined space—without hesitation, without prompting, without delay. He or she must maintain a position within the space whereby he or she can exit quickly if necessary. When ordered to exit by the attendant, the entrant must not delay, think about it, or pause for any reason. When ordered to exit, the entrant must exit *immediately*.

9.17.7.3.8.2 Duties of the Attendant The employer has the responsibility of ensuring that the permit-required confined space attendant is fully trained and knowledgeable. The attendant must know the hazards that may be faced during entry, including information on the mode, signs or symptoms, and consequences of the exposure. The attendant must be aware of the behavioral effects of hazard exposure to which the entrants may be subjected. That the attendant plays a critical role in confined space entry should be apparent. This critical role cannot be filled by just anyone; the attendant must be fully trained and qualified to perform his or her assigned responsibilities. The attendant is responsible for maintaining an accurate count of authorized entrants in the permit space and must ensure that the means used to identify authorized entrants accurately identifies who is in the permit space. The attendant remains outside the permit space until properly relieved by another qualified attendant. When the employer's permit entry program allows attendant entry for rescue, attendants may enter a permit space to attempt a rescue *only* if they have been trained and equipped for rescue operations. The attendant maintains communication between him- or herself and the entrants at all times. The attendant also monitors conditions within and outside the space that might endanger the entrants and orders the entrants to exit if necessary. Attendants who detect a hazardous situation or the behavioral effects of hazard exposure in an authorized entrant or who determine that they cannot (for whatever reason) perform their attendant duties must order the immediate evacuation of the permit space. The attendant is also responsible for summoning rescue and other emergency services as soon as he or she determines that authorized entrants may need assistance to escape from permit space hazards. The attendant prohibits unauthorized persons from entering a permit space or from interfering with an entry in progress. The attendant has one responsibility and one responsibility only: to perform the duties of a permit space attendant without allowing any distraction.

9.17.7.3.8.3 Duties of Entry Supervisors As with any other work activity, supervisors play a key role in permit-required confined space entry. In permit-required confined space entry, the supervisor is responsible for issuing confined space permits. To do this according to the standard, the entry supervisor must know the hazards of the confined spaces, verify that all tests have been conducted and all procedures and equipment are in place before endorsing a permit, terminate entry if necessary, cancel permits, and verify that rescue services are available and the means for summoning them are operable. In addition, entry supervisors are to remove unauthorized individuals who attempt to enter the confined space. They also must determine, at least when shifts and entry supervisors change, that acceptable conditions, as specified by the permit, continue. Remember, the entry supervisor signs off on the permit. Before signing any safety document, supervisors should use good judgment, along with care and caution. If and when anything goes wrong in a confined space entry, the first item that the OSHA investigator will want to see is the permit. When lawyers are involved, as is often the case when workers are killed or badly injured on the job, the permit becomes an important document that will end up in a court of law—along with the supervisor in charge of the confined space entry operation.

9.17.7.3.9 Confined Space Rescue

> Of the more than 1.6 million workers who enter confined spaces each year, approximately 63 die from asphyxiation, burns, electrocution, drowning and other tragedies related to confined space entry operations. But more alarming is the fact that 60 percent of those who die in confined spaces are untrained rescuers who not only fail to save a co-worker, but also are killed during the rescue attempt. OSHA requires that a trained, equipped rescue team be available whenever employees work in confined spaces.
>
> **—Coastal Video (1993)**

The employer who engages in permit-required confined space entry has the option of whether to use an offsite or in-plant rescue service. If the decision is made to use an offsite rescue service, a number of factors must be considered. The first factor to consider is whether such a rescue service is readily available. This may seem to be a logical, straightforward issue, but let's take a look at what typically occurs when the option to choose an offsite rescue service is chosen.

The natural inclination is to list dial 911 or another local emergency number on your confined space permit as your rescue service, but is such rescue service readily available to you from the local fire department or some other emergency service? You need to find out. In our experience, when we have called local fire departments to explain to them that we are about to make a confined space entry, that we are giving them a heads up to be aware of the operation, they are usually puzzled. "We fight fires and make some rescues, but confined space rescue? Sorry, we are not trained for that." If you try to locate an offsite rescue service, you probably will hear something similar. We cannot simply list 911 as the standby emergency rescue service and hope that whoever responds will be able to effect rescue. We must be absolutely certain that the service will respond in less than 4 minutes (remember, a victim in a confined space cannot live without air for more than 4 minutes) and are fully trained to effect the rescue.

The second factor that you must take into consideration after you have identified a rescue service that can respond in 4 minutes or less is whether or not that service is familiar with your facility. Have you invited the members of the service into your facility for familiarization with your facility? Another factor to consider is onsite training. Has the rescue service actually practiced making confined space rescues in your confined spaces? Are they willing to spend the time to acquire the information they need to handle a crisis situation at your facility? This is an important point, one that an OSHA auditor will be certain to verify if and when your facility is audited.

Onsite rescue teams have considerations, as well. If you decide to employ the services of an onsite rescue team, OSHA requires that

1. The employer shall ensure that each member of the rescue team is provided with, and is trained to use properly, the personal protective equipment and rescue equipment necessary for making rescues from permit spaces.
2. Each member of the rescue team shall be trained to perform the assigned rescue duties. Each member of the rescue team shall also receive the training required of authorized entrants.
3. Each member of the rescue team shall practice making permit space rescues at least once every 12 months by means of simulated rescue operations in which they remove dummies, mannequins, or actual persons from the actual permit spaces or from representative permit spaces. Representative permit spaces shall, with respect to opening size, configuration, and accessibility, simulate the types of permit spaces from which rescue is to be performed.
4. Each member of the rescue team shall be trained in basic first aid and in cardiopulmonary resuscitation (CPR). At least one member of the rescue team holding current certification in first aid and in CPR shall be available.

In the OSHA standard, the above requirements describe the rescue team as a "rescue service." From experience, calling this rescue service a rescue team is better (and more appropriate)—because a team is what it is. To properly effect confined space rescue, the rescue service must be a team—individuals who work together seamlessly. Each member must have good endurance, enthusiasm, a willingness to learn, and a team-oriented attitude.

9.17.7.3.9.1 Rescue Service Provided by Outside Contractors When an employer arranges to have persons other than the host employer's employees perform permit space rescue, the host employer shall

1. Inform the rescue service of the hazards they may confront when called on to perform rescue at the host employer's facility.
2. Provide the rescue service with access to all permit spaces from which rescue may be necessary so that the rescue service can develop appropriate rescue plans and practice rescue operations.

9.17.7.3.9.2 Non-Entry Rescue The rescue services we have discussed to this point all involve making external (non-entry rescue) confined space rescues—the preferred method of rescue recommended by this text, even though it may not be feasible on all occasions. The rule of thumb that we use is that if external rescue via a tripod, winch, retrieval line, and body harness cannot be made, then the confined space entry should not be made in the first place. When such retrieval systems are used, they must meet the following requirements:

1. Each authorized entrant shall use a chest or full body harness, with a retrieval line attached at the center of the entrant's back near shoulder level, or above the entrant's head. Wristlets may be used in lieu of the chest or full body harness if the employer can demonstrate that the use of a chest or full body harness is unfeasible or creates a greater hazard, and that the use of wristlets is the safest and most effective alternative.
2. The other end of the retrieval line shall be attached to a mechanical device or fixed point outside the permit space in such a manner that rescue can begin as soon as the rescuer becomes aware that rescue is necessary. A mechanical device (such as a tripod and winch assembly) must be available to retrieve personnel from vertical type permit spaces more than 5 feet (1.52 m) deep.

A final word on permit-required confined space rescue: In the event of a rescue where the entrant is exposed to a hazardous material for which a Material Data Safety Sheet (MSDS) or other similar written information is required to be kept at the worksite, that MSDS or written information must be made available to the medical facility treating the exposed entrant.

9.17.7.3.10 *Alternative Protection Methods*

Minimizing the amount of regulation that applies to spaces whose hazards have been eliminated encourages employers to actually remove all hazards. OSHA has specified alternative protection procedures that may be used for permit spaces where the only hazard is atmospheric and ventilation alone can control the hazard. Let's take a brief look at these alternative protection procedures.

9.17.7.3.10.1 Procedures for Atmospheric Testing You can never trust your senses to determine if the air in a confined space is safe! You cannot see or smell many toxic gases and vapors, nor can you determine the level of oxygen present. Personnel involved in permit-required confined space entry must understand that some vapors or gases are heavier than air and will settle to the bottom of a confined space. Other gases are lighter than air and will be found around the top of the confined space. Because of the behaviors of various toxic gases, you must test all areas (top, middle, bottom) of a confined space with properly calibrated testing instruments to determine what gases are present. Atmospheric testing is required for two distinct purposes: evaluation of the hazards of the permit space and verification that acceptable entry conditions exist for entry into that space. The requirements for atmospheric testing are summarized below:

1. *Evaluation testing*—The atmosphere of a confined space should be analyzed using equipment of sufficient sensitivity and specificity to identify and evaluate any hazardous atmospheres that may exist or arise, so that appropriate permit entry procedures can be developed and acceptable entry conditions stipulated for that space. Evaluation and interpretation of these data and development of the entry procedure should be done by, or reviewed by, a technically qualified professional, such as an OSHA consultation service, Certified Safety Professional (CSP), Certified Industrial Hygienist (CIH), or registered safety official, based on evaluation of all serious hazards.
2. *Verification testing*—The atmosphere of a permit space that may contain a hazardous atmosphere should be tested for residues of all contaminants identified by evaluation testing using permit-specified equipment to determine that residual concentrations at the time of testing and entry are within the range of acceptable entry conditions. Results of testing (e.g., actual concentration) should be recorded on the permit in the space provided adjacent to the stipulated acceptable entry condition.
3. *Duration of testing*—Measurement of values for each atmospheric parameter should be made for at least the minimum response time of the test instrument specified by the manufacturer.
4. *Testing stratified atmospheres*—When monitoring for entries involving a descent into atmosphere that may be stratified, the atmospheric envelope should be tested a distance of approximately 4 feet (1.22 m) in the direction of travel and to each side. If a sampling probe is used, the entrant's rate of progress should be slowed to accommodate the sampling speed and detector response.

9.17.7.3.10.2 Air Monitoring and OSHA When an OSHA compliance officer audits your facility, if you have permit-required confined spaces that are entered by your employees the auditor will pay particular attention to your air monitoring procedures. Typically, the OSHA auditor will want to see copies of your confined space permits for the past year. From these permits, the auditor will choose one and set it aside. Later, the auditor will ask to interview those involved in making that confined space entry. The auditor may ask the confined space personnel several different questions related to their knowledge of confined space entry. The auditor may desire to see these personnel perform the entry again.

During the OSHA auditor's interview process, air monitoring will be discussed. The auditor will want to see the instrument used during the confined space entry. The auditor will note the condition of the instrument, looking specifically for any damage, dirt, and battery condition (e.g., dead batteries, correct batteries being used rather than a jerry-rigged battery pack) and will test to determine if any sensors are malfunctioning. The

OSHA auditor almost always asks one of the confined space entry personnel to demonstrate both how to calibrate and how to use the instrument. In addition, the OSHA auditor typically asks several questions related to air monitoring to determine the knowledge level of the workers. These may include the following:

1. Have the operators been trained?
2. Who gave them the training? What was covered? How long did the training last? Any hands-on or on-the-job training?
3. What type of instruments are used?
4. Where is the manufacturer's instruction manual? Have they read the manual?
5. How often do they use the instrument?
6. Do they have calibration data, logbooks, etc.?
7. What calibration gas do they use? Why did they choose this gas?
8. Do they zero the instrument as part of the calibration?
9. Who calibrates the equipment? How often? How is it done?
10. Do they have a calibration curve or correction factor chart?
11. What are the interferences for the toxic sensors?
12. Is the meter intrinsically safe for the environment that they are monitoring?
13. Are they waiting long enough for the sensors to respond? (For remote sampling, some manufacturers suggest 1 second per foot of tubing.)
14. Are they testing all levels and areas where entrants will be working?
15. If using several individual instruments, are they testing in the right sequence (oxygen, flammables, toxics)?
16. Do they know what the numbers on the instrument mean? Are they exact?
17. What are they comparing the numbers to? Do they know what is considered safe for entry?
18. Have they replaced any sensors? Any batteries? Any other parts? Do they have maintenance logs?
19. Do they send the instrument back to the manufacturer on a regular basis for complete calibration and maintenance?
20. Do they field check?

9.17.7.3.11 Other OSHA Permit-Required Confined Space Audit Items

When an OSHA auditor audits your confined space entry program, you can be assured that they will look at most (if not all) of the items listed below:

1. Are aisles in the vicinity of the confined space marked?
2. Are aisles and passageways properly illuminated?
3. Are aisles kept clean and free of obstructions?
4. Are fire aisles, access to stairways, and fire equipment kept clear?
5. Is there safe clearance for equipment through aisles and doorways?
6. Have all confined spaces and permit-required confined spaces been identified?
7. Are danger signs posted (or other equally effective means of communication) to inform employees about the existence, location, and dangers of permit-required confined spaces?
8. Is the written permit-required confined space entry program available to employees?
9. Is the permit-required confined space sufficiently isolated? Have pedestrian, vehicle, or other necessary barriers been provided to protect entrants from external hazards?
10. When working in permit-required confined spaces, are environmental monitoring tests taken and means provided for quick removal of workers in case of an emergency?
11. Are confined spaces thoroughly emptied of any corrosive or hazardous substances (such as acids or caustics) before entry?
12. Are all lines to a confined space containing inert, toxic, flammable, or corrosive materials valved off and blanked or disconnected and separated before entry?
13. Is each confined space checked for decaying vegetation or animal matter that may produce methane?
14. Is the confined space checked for possible industrial waste that could contain toxic properties?
15. Before permit space entry operations begin, has the entry supervisor identified on the permit signed the entry permit to authorize entry?
16. Has the permit been made available at the time of entry to all authorized entrants (by being posted at the entry portal or by other equally effective means) so entrants can confirm that pre-entry preparations have been completed?
17. Is necessary personal protective equipment (PPE) available?
18. Has necessary lighting equipment been provided?
19. Has equipment (such as ladders) needed for safe ingress and egress by authorized entrants been provided?
20. Is rescue and emergency services equipment available?
21. Is it required that all agitators, impellers, or other rotating equipment inside confined spaces be locked out if they present a hazard?

22. Is all portable electrical equipment used inside confined spaces either grounded and insulated or equipped with ground fault protection?
23. Is at least one attendant stationed outside the confined space for the duration of the entry operation?
24. Is there at least one attendant whose sole responsibility is to watch the work in progress, sound an alarm if necessary, and render assistance?
25. Is the attendant trained and equipped to handle an emergency?
26. Are the attendant and other workers prohibited from entering the confined space without lifelines and respiratory equipment if there is any question as to the cause of an emergency?
27. Is communications equipment provided to allow the attendant to communicate with authorized entrants as necessary to monitor entrant status and to alert entrants of the need to evacuate the permit space?

9.17.7.4 Control of Hazardous Energy—Lockout/Tagout

> When maintenance and servicing are required on equipment and machines, the energy sources must be isolated and lockout/tagout procedures implemented. The terms *zero mechanical state* or *zero energy state* have often been used to describe machines with all energy sources neutralized. These terms have been incorporated in many standards. The current term indicating a machine at total rest is *energy isolation*. Machine energy can be electrical, pneumatic, steam, hydraulic, chemical, thermal, and others. Energy is also the potential energy from suspended parts or springs.
>
> **—National Safety Council (1992)**

OSHA's 29 CFR 1910.147 states that employers are required to develop, document, and utilize an energy control procedures program to control potentially hazardous energy. The energy control procedures must specifically outline the scope, purpose, authorization, rules, and techniques to be utilized for the control of hazardous energy and the means to enforce compliance including, but not limited to, the following:

- Specific statement of the intended use of the procedure
- Specific procedural steps for shutting down, isolating, blocking, and securing machines and equipment to control hazardous energy
- Specific procedural steps for the placement, removal, and transfer of lockout devices or tagout devices and the responsibility for them
- Specific requirements for testing a machine or equipment to determine and verify the effectiveness of lockout devices, tagout devices, and other energy control measures

It has been estimated by OSHA that full compliance with the lockout/tagout standard can prevent 120 accidental deaths, 29,000 serious injuries, and 32,000 minor injuries every year (Carney, 1991). Experience has shown that many workers mistake the results of atmospheric testing that show no hazard exists in a particular confined space as meaning that the space is totally safe for entry. Indeed, this might be the case; however, many other dangers inherent to confined spaces make entry into them hazardous. If the confined space has some type of open liquid stream flowing through it, the chance for engulfment exists. If the space has electrical devices and circuitry inside, an electrocution hazard exists. If hazardous chemicals are stored and taken into the space, the potential for a hazardous atmosphere exists. Many confined spaces contain physical hazards, including piping and other obstructions; for example, rotating machinery is often housed within confined spaces.

To ensure that the confined space is indeed safe, any and all sources of hazardous energy must be isolated before entry is made. The primary method employed to accomplish this is through lockout/tagout procedures; however, the intent of employing lockout/tagout procedures goes far beyond just providing for safe confined space entry. The control of hazardous energies by locking or tagging out also applies to most servicing, adjusting, or maintenance activities involving machines and processes that place personnel at elevated risk. In addition to the sources of machine energy mentioned earlier (electrical, pneumatic, steam, and so forth), of particular concern is inadvertent activation when personnel are in contact with the hazards.

Safety professionals employed in major industrial groups recognize that the need to incorporate a viable, fully compliant lockout/tagout program (one that includes all elements of 29 CFR 1910.147) cannot be overstated. Review the historical data. It has been estimated that 7% of all workplace deaths and nearly 10% of serious accidents in many major industrial groups are associated with the failure to properly restrain or de-energize equipment during maintenance. Maintenance workers account for one third of injuries, even though they are familiar with the machines they are working on. Statistical records show that most injuries involve machines that are still running or that have been accidentally activated. In the sawmill industry, start-ups and unwanted movements have been involved in about a third of accidents that occur, and, surprisingly, it has been found that no emergency shutoffs are available about 50% of the time.

9.17.7.4.1 Lockout/Tagout Key Definitions

Affected employee—An employee whose job requires him or her to operate or use a machine or equipment on which servicing or maintenance is being performed under lockout or tagout, or whose job requires him or her to work in an area where such servicing or maintenance is being performed.

Authorized employee—A person who locks out or tags out machines or equipment to perform servicing or maintenance on that machine or equipment. An affected employee becomes an authorized employee when that employee's duties include performing servicing or maintenance covered under the company's lockout/tagout program.

Capable of being locked out—An energy-isolating device is capable of being locked out if it has a hasp or other means of attachment to which (or through which) a lock can be affixed, or it has a locking mechanism built into it. Other energy-isolating devices are capable of being locked out if lockout can be achieved without the need to dismantle, rebuild, or replace the energy isolating device or permanently alter its energy control capability.

Energized—Refers to machines or equipment being connected to an energy source or containing residual or stored energy.

Energy isolating device—A mechanical device that physically prevents the transmission or release of energy, including (but not limited to) the following: a manually operated electrical circuit breaker; a disconnect switch; a manually operated switch by which the conductors of a circuit can be disconnected from all ungrounded supply conductors and, in addition, in which no pole can be operated independently; a line valve; a block; and any similar device used to block or isolate energy. Push buttons, selector switches, and other control circuit type devices are not energy-isolating devices.

Energy source—Any source of electrical, mechanical, hydraulic, pneumatic, chemical, thermal, or other energy.

Hot tap—A procedure used in repair, maintenance, and services activities that involves welding on a piece of equipment (pipelines, vessels, or tanks) under pressure, to install connections or appurtenances. Commonly used to replace or add sections of pipeline without the interruption of service for air, gas, water, steam, and petrochemical distribution systems.

Lockout—The placement of a lockout device on an energy-isolating device, in accordance with an established procedure, ensuring that the energy-isolating device and the equipment being controlled cannot be operated until the lockout device is removed.

Lockout device—A device that utilizes a positive means (such as a lock, either key or combination type) to hold an energy-isolating device in the safe position and prevent the energizing of a machine or equipment. Included are blank flanges and bolted slip blinds.

Normal production operation—The utilization of a machine or equipment to perform its intended production function.

Servicing and/or maintenance—Workplace activities such as constructing, installing, setting up, adjusting, inspecting, modifying, and maintaining or servicing machines or equipment. These activities include lubrication, cleaning or unjamming of machines or equipment, and making adjustments or tool changes where the employee may be exposed to the unexpected energization or startup of the equipment or release of hazardous energy.

Setting up—Any work performed to prepare a machine or equipment to perform its normal production operation.

Tagout—The placement of a tagout device on an energy-isolating device, in accordance with an established procedure, to indicate that the energy-isolating device and the equipment being controlled may not be operated until the tagout device is removed.

Tagout device—A prominent warning device, such as a tag and a means of attachment, that can be securely fastened to an energy-isolating device in accordance with an established procedure, to indicate that the energy-isolating device and the equipment being controlled may not be operated until the tagout device is removed.

9.17.7.4.2 Sample Lockout/Tagout Program

Lockout/Tagout Program

I. Purpose

This program has been developed to ensure protection of employees and to maintain compliance with OSHA standard 1910.147, Control of Hazardous Energy. These instructions establish the minimum requirements for the lockout or tagout of energy-isolating devices whenever maintenance or servicing is done on machines or equipment. It shall be used to ensure that the machine or equipment is stopped, isolated from all potentially hazardous energy sources, and locked out before employees perform any servicing or maintenance. These procedures apply:

1. Whenever an employee has to remove or bypass a guard or other safety device
2. Where such servicing results in employees placing all or part of their body in a danger zone
3. Where the unexpected energizing or start-up of the machine or equipment or release of stored energy could cause injury or death

II. Authorized Employees

Authorized company employees shall be trained in lockout/tagout procedure. Retraining will be done whenever an authorized employee's job responsibility changes or when the periodic inspection identifies procedural deficiencies. All *affected* (and other) employees whose work operations are or may be in the area shall be instructed in the purpose and use of lockout/tagout procedures.

III. Application and Compliance

This procedure applies to all company employees and shall be enforced. All employees must signify that they have read and understand each part of this procedure by signing the training record (see Section IX of this procedure). All employees are required to comply with the restrictions and limitations imposed upon them during the use of lockout or tagout. The employee, upon observing a machine that is locked or tagged out for servicing, *shall not* attempt to start or use that machine.

IV. Tags

Tags are used to prevent the unexpected energization of equipment and generally are restricted to those controls that cannot be locked or safeguarded by any other method.

1. When the energy-isolating devices cannot be locked out, use a tag to prevent inadvertent actuation.
2. Tags will be attached to equipment or machinery, at the control panel, and at other points of operation.
3. Tags must bear the words "DANGER" and "DO NOT OPERATE" or "DO NOT USE" printed on both sides of the tag.
4. Tags shall bear the name, department, date, and telephone number of the employee (or department) performing the work.

V. Locks

Locks are employee-identifiable, key-operated padlocks. Lockout devices are those openings in equipment control handles or switches that can accept a padlock. Multiple lockout devices are those devices that have spaces for the application of more than one lock. This type of device is locked into the lockout opening, and subsequent locks prevent removal of the device. Authorized employees will be assigned their own individual lock with one unique key. A master key will generally not be available except under very unusual circumstances. A master key is necessary; it will be under the strict control of the work center supervisor. *Locks will not be removed* by anyone other than the authorized employee who placed it on the control, unless the procedure described in the "Special Condition" section of this program is followed. Employees working on the same machine will each use their own lock. When work continues from one shift to the next, employees leaving must remove their locks and employees beginning work must apply their own locks.

VI. Energy Analysis

In 20__, the company's Safety Division surveyed each company facility and identified all machines or equipment covered by this program. All energy sources were identified, along with their methods of control. Each time a new machine is installed, the company Safety Division is responsible for performing an energy analysis. This energy analysis procedure is ongoing and must be updated as required.

Equipment Energy Analysis Survey

Starting in June of 1989, and any time a new machine or new configuration to an existing machine has been installed since then, the company's Safety Director has conducted a survey of all company properties and made the following findings:

(List all equipment that must be locked/tagged out.)

Equipment ______________________

Equipment ______________________

Equipment ______________________

Equipment ______________________

Equipment ______________________

with regard to various equipment and machinery identified in the company's Maintenance Management System (preventive maintenance program) that have specific lockout/tagout procedures. The company is not required to document the specific procedures for particular machines or pieces of equipment for which any of the following is true:

1. They have been found to have no potential for stored or residual energy or the re-accumulation of stored energy after shut down which could endanger employees.
2. The machine or equipment has a single energy source that can be readily identified and isolated.
3. The isolation and locking out of that energy source will completely de-energize and deactivate the machine or equipment.
4. The machine or equipment is isolated from the energy source and locked out during servicing or maintenance.
5. A single lockout device will achieve a locked-out condition.
6. The lockout device is under the exclusive control of the authorized employee performing the servicing or maintenance.
7. Servicing or maintenance does not create hazards for other employees.
8. The company, in utilizing this exception, has had no accidents involving the unexpected activation or re-energization of the machine or equipment during servicing or maintenance.

For those machines or equipment that do not require equipment-specific lockout/tagout procedures, the lockout/tagout procedures in Section VII is to be used at the company.

VII. Lockout/Tagout Procedures

Lockout/tagout procedures for company equipment other than equipment-specific lockout procedures for the incinerators and other covered equipment are as follows:

1. Notify appropriate operations and maintenance supervisors of lockout/tagout.
2. Place the main switch, valve, control, or operating lever in the "off," "closed," or "safe" position.
3. *Check and test to make certain* that the proper controls have been identified and deactivated.
4. Place a lock to secure the disconnection whenever possible. If a lock cannot be used on electrical equipment, an electrician shall remove fuses or disconnect circuit.
5. If a system cannot be locked out with a lock, attach a HOLD-OFF, DO NOT ENERGIZE, or other such tag to the switch, valve, or lever. If the company work center does not use employee-identifiable locks, a lock and tag must be used together.
6. When auxiliary equipment or machine controls are powered by separate supply sources, such equipment or controls shall also be locked or tagged to prevent any hazard that may be caused by operating the equipment or exposure to live circuits.
7. When equipment uses pneumatic or hydraulic power, pressure in lines or accumulators must be checked. Using whatever safe means possible, this pressure shall be disconnected or pressure lines disconnected.
8. When stored energy is a factor as a result of position, spring tension, or counterweighting, the equipment shall be placed in the bottom or closed position, or it shall be blocked to prevent movement.
9. When the work involves more than one person, additional employees must attach their locks and tags as they report.
10. When outside contractors are involved, the equipment shall be locked out and tagged in accordance with this procedure by the Project Manager supervising the work. Only in emergency cases is equipment to be shut down by other than a company representative.
11. When the servicing or maintenance is completed and the machine or equipment is ready to return to normal operating condition, the following steps will be used:

a. Check the machine or equipment and the immediate area around the machine to ensure that tools, materials, and other nonessential items have been removed and that the machine or equipment components are operationally intact. Ensure that all guards have been replaced.
b. Check the work area to ensure that all employees have been safely positioned or removed from the area.
c. Verify that the controls are in neutral.
d. Remove the lockout devices and re-energize the machine or equipment. The employee who applied the device will remove each lockout/tagout device from each energy-isolating device.
e. Notify affected employees that the servicing or maintenance is completed and that the machine or equipment is again ready for use.

VIII. Periodic Inspections

A periodic inspection of the energy control procedures will be conducted at least quarterly to ensure that the requirements of the program and the standard are being followed to ensure full employee protection. The work center supervisor and/or Safety Division personnel will perform the periodic inspection. The inspection will be conducted to identify any program inadequacies that require correcting. The inspection will review the employee's responsibilities under the procedure being inspected. The company will provide certification that the inspection of a lockout/tagout has been performed. The certification will identify the machine on which the lockout/tagout is being utilized, the date of the inspection, names of employees involved, and the name of the individual performing the inspection.

IX. Training

Training will be provided by the company's Safety Division to ensure that all employees understand the purpose and function of the lockout/tagout program and that employees acquire the knowledge and skills required for safe application, usage, and removal of energy controls. The training shall include the following:

1. Each *authorized employee* (person who actually performs the lockout/tagout) shall receive training in the recognition of hazardous energy sources, the type and magnitude of the energy available in the workplace, and the methods and means necessary for energy isolation and control.
2. Each *affected employee* (person affected by the lockout/tagout) shall be instructed in the purpose and use of the energy control procedure.

3. All other employees whose work operations are or may be in areas where energy control procedures may be used shall be instructed about the procedure and about the prohibitions relating to attempts to restart or re-energize machines or equipment that are locked out or tagged out.

When tagout is used, employees shall receive additional training on the limitations of tags.

A training record (see below) will be used to record each employee's training.

Sample Training Record

Training Roster

Lockout/Tagout Program

Date: ____________________

In accordance with the recording and training requirements of the Lockout/Tagout Program, OSHA 29 CFR 1910.147, I have received safety training on lockout/tagout requirements and procedures. I have agreed to verify my understanding and training of 1910.147 by signing and dating this form. This training meets requirements of 29 CFR 1910.147.

Name *Work Center*

____________________ ____________________

X. Special Conditions

Lockout/Tagout Removal When Authorized Employee Is Absent

When the authorized employee who applied the lockout or tagout device is not available to remove it, that device may be removed under the direction of the supervisor, provided that specific procedures and training for such removal have been developed, documented, and incorporated into the lockout/tagout program. Specific procedures include the following elements:

1. Verifying that the authorized employee who applied the device is not at the facility;
2. Making all reasonable efforts to contact the authorized employee to inform him/her that his/her lockout or tagout device has been removed;
3. Ensuring that the authorized employee has this knowledge before he/she resumes work at that facility; and
4. Completing the "Lockout/Tagout Removal When Authorized Employee Is Absent" form—the form shall be kept by the work center supervisor.

Lockout/Tagout Removal When Authorized Employee Is Absent

Department Machine/Equipment Lock Id #

_______________ ___________________________ ______________

Authorized Employee Type of Work Being Done

__________________________________ _________________________

Verification That Employee Is Not on Site

❑ Yes ❑ No

NOTE: ____

Verification That Employee Has Been Informed Before Coming Back to Work That His/Her Lockout/Tagout Device Has Been Removed

❑ Yes ❑ No

NOTE: ____

Supervisor Date

__________________________________ ______________________

XI. Methods of Informing Outside Contractors of Procedures

Whenever outside servicing personnel are to be engaged in lockout/tagout activities covered by this program, the company and the outside employer will inform each other of their respective lockout or tagout procedures. Company employees are to be trained to understand and comply with the restrictions and prohibitions of the outside contractor's lockout/tagout energy control program. The Safety Division will provide to the company work center supervisor a contractor briefing form (shown below) that will inform the contractor on precautionary measures involved in performing lockout/tagout on the site.

Record of Notification

Company and Outside Contractor Safety Arrival Conference

In accordance with the recordkeeping and training requirements under OSHA 1910.119 and other applicable standards, I have received a safety brief from the company Safety Division/plant personnel covering the company's lockout/tagout program. I further understand that the company expects all outside contractors, including subcontractors, to perform construction activities under OSHA required guidelines and company procedures. It is understood that all information regarding hazards, safety procedures, and lockout/tagout procedures shall be disseminated to all persons, subcontractors, and agents employed either directly or indirectly by us. Regarding Lockout/Tagout procedures, the company and the contractor must discuss these operations and work in a coordinated effort so as to prevent injury to anyone.

Signature	Printed Name	Company Name
________________	________________	________________
________________	________________	________________
________________	________________	________________

9.17.8 Subpart K, Medical and First Aid

1910.151 Medical Services and First Aid

Note that the lack of medical and first aid services at oil and gas drilling sites is listed as one of the top 10 OSHA citations (FY 2005) in Table 9.2. Subpart K of 29 CFR 1910 directly addresses eye-flushing capabilities in the workplace and indirectly the need to have medical personnel readily available. "Readily available" can mean that there is a clinic or hospital nearby. If such a facility is not located nearby, employers must have a person onsite that has had first-aid training. Because of these OSHA requirements, the organization's safety official must, as with all other regulatory requirements, ensure that the organization is in full compliance. First-aid awareness and training in the workplace usually require providing lectures, interactive video presentations, discussions, and hands-on training to teach participants how to

- Recognize emergency situations.
- Check the scene and call for help.
- Avoid bloodborne pathogen exposure.
- Care for wounds, bone and soft-tissue injuries, head and spinal injuries, burns, and heat and cold emergencies.
- Manage sudden illnesses, stroke, seizure, bites and poisoning.
- Minimize stroke.

First-aid services in the workplace typically include training and certification of selected individuals to perform CPR on workers when necessary. This training usually combines lectures, video demonstrations, and hands-on manikin training. This training teaches participants to

- Call and work with EMS.
- Recognize breathing and cardiac emergencies that call for CPR.
- Perform CPR and care for breathing and cardiac emergencies.
- Avoid bloodborne pathogen exposure.
- Know the role of AEDs in the cardiac chain of survival.

The American Red Cross points out that typical first-aid and CPR training for the workplace has been enhanced to include training on the Automated External Defibrillator (AED): "Although the idea of using a handheld device to deliver a shock directly into a coworker's heart may seem daunting, the American Red Cross hopes this life-saving practice becomes more common over the next year" (Orfinger, 2002). AED training focuses on typical AED equipment with hands-on simulation, lectures, and live and video demonstrations. Participants learn to

- Call and work with EMS.
- Care for conscious and unconscious choking victims.
- Perform rescue breathing and CPR.
- Use an AED safely on a victim of sudden cardiac arrest.

9.17.9 Subpart L, Fire Protection

1910.157 Portable Fire Extinguishers

1910.165 Employee Alarm Systems

9.17.9.1 Fire Safety

As shown in Table 9.12, noncompliance with OSHA's portable fire extinguisher standard was one of the top ten cited violations in the oil and gas drilling industry. Although technical knowledge about flame, heat, and smoke continues to grow, and although additional information continues to be acquired concerning the ignition, combustibility, and flame propagation of various solids, liquids, and gases, it still is not possible to predict with any degree of accuracy the probability of fire initiation or consequences of such initiation. Thus, while the study of controlled fires in laboratory situations provides much useful information, most unwanted fires happen and develop under widely varying conditions, making it virtually impossible to compile complete bodies of information from actual unwanted fire situations. This fact is further complicated because the progress of any unwanted fire varies from the time of discovery to the time when control measures are applied (Cote and Bugbee, 1991).

Industrial facilities are not immune to fire and its terrible consequences. Each year fire-related losses in the United States are considerable. According to conservative figures reported by Brauer (1994), about 1 million fires involving structures and about 8000 deaths occur each year. The total annual property loss is more than $7 billion. Complicating the fire problem is the point that Cote and Bugbee (1991) made above—the unpredictability of fire. Fortunately, facility safety officials are aided in their efforts in fire prevention and control by the authoritative and professional guidance readily available from the National Fire Protection Association (NFPA), the National Safety

Council (NSC), fire code agencies, local fire authorities, and OSHA regulations. In this section, we discuss fire prevention and control and fire protection provided by the use of fire extinguishers.

Along with providing fire prevention guidance, OSHA regulates several aspects of fire prevention and emergency response in the workplace. Emergency response and evacuation and fire prevention plans are required under 29 CFR 1910.38. The requirements for fire extinguishers and worker training are addressed in 1910.157. Along with state and municipal authorities, OSHA has listed several fire safety requirements for general industry.

All of the advisory and regulatory authorities approach fire safety in much the same manner; for example, they all agree that electrical short circuits or malfunctions usually start fires in the workplace. Other leading causes of workplace fires are friction heat, welding and cutting of metals, improperly stored flammable/combustible materials, open flames, and cigarette smoking.

For fire to start, three components must be present: *temperature (heat)*, *fuel*, and *oxygen*. Because oxygen is naturally present in most environments on Earth, fire hazards usually involve the mishandling of fuel or heat. The fire triangle helps us understand fire prevention, because the objective of fire prevention and firefighting is to separate any one of the fire ingredients from the other two. To prevent fires, it is necessary to keep fuel (combustible materials) away from heat (as in airtight containers), thus isolating the fuel from oxygen in the air. To gain a better perspective of the chemical reaction known as *fire*, remember that the combustion reaction normally occurs in the gas phase; generally, the oxidizer is air. If a flammable gas is mixed with air, there is a minimum gas concentration below which ignition will not occur. That concentration is known as the *lower flammable limit* (LFL). When trying to visualize the LFL and its counterpart, the *upper flammable limit* (UFL), it helps to use an example that most people are familiar with—the combustion process that occurs in the automobile engine. When an automobile engine has a gas/air mixture that is below the LFL, the engine will not start because the mixture is too lean. When the same engine has a gas/air mixture that is above the UFL, it will not start because the mixture is too rich (the engine is flooded). When the gas/air mixture is between the LFL and UFL levels, however, the engine should start (Spellman, 1996).

9.17.9.2 Fire Prevention and Control

The best way to prevent and control fires in the workplace is to institute a facility Fire Safety Program. Safety experts agree that the best way to reduce the possibility of fire in the workplace is prevention. For the facility safety official this begins with developing a fire prevention plan, which must be in writing and must list fire hazards and fire controls and specify the control jobs and personnel responsible and emergency actions to be taken. More specifically, in accordance with OSHA 29 CFR 1910.38, the elements that make up the plan must include the following:

1. A list of the major workplace fire hazards and their proper handling and storage procedures, potential ignition sources (such as welding, smoking, and others), their control procedures, and the type of fire protection equipment or systems that can control a fire involving them.
2. Names or regular job titles of those personnel responsible for maintenance of equipment and systems installed to prevent or control ignitions or fires.
3. Names or regular job titles of those personnel responsible for control of fuel source hazards.
4. Control of accumulation of flammable and combustible waste materials and residues so that they do not contribute to a fire emergency. These housekeeping procedures must be included in the written fire prevention plan.
5. All workplace employees must be apprised of the fire hazards of the materials and processes to which they are exposed.
6. All new employees must be made aware of those parts of the fire prevention plan that the employee must know to protect the employee in the event of an emergency. The written plan must be kept in the workplace and made available for employee review.
7. The employer is required to regularly and properly maintain, according to established procedures, equipment and systems installed on heat-producing equipment to prevent accidental ignition of combustible materials. The maintenance procedure must be included in the written fire prevention plan.

Fire prevention and control measures are those taken *before* fires start and include the following:

- Elimination of heat and ignition sources
- Separation of incompatible materials
- Adequate means of firefighting (e.g., sprinklers, extinguishers, hoses)
- Proper construction and choices of storage containers
- Proper ventilation systems for venting and reducing vapor buildup
- In the event of fire emergency, maintaining unobstructed means of egress for workers, as well as adequate aisle and fire-lane clearance for firefighters and equipment

In the event of a fire emergency, all employees need to know what to do; they need a plan to follow. The fire emergency plan normally is the protocol to follow for fire emergency response and evacuation. Typically, the facility safety official is charged with developing fire prevention and emergency response plans that spell out everyone's role. In this effort, the safety official's goal should be to make the plan as simple as possible. In addition to a fire emergency response plan, each facility needs to have a well-thought-out fire emergency evacuation plan.

9.17.9.3 Fire Protection Using Fire Extinguishers

OSHA, under 29 CFR 1910.157, requires employers to provide portable fire extinguishers that are mounted, located, and identified so they are readily accessible to employees without subjecting the employee to possible injury. OSHA also requires that each workplace institute a portable fire extinguisher maintenance plan. Fire extinguisher maintenance service must take place at least once a year, and a written record must be kept to show the maintenance or recharge date. *Note:* When the facility provides portable fire extinguishers for employee use in the facility, the employee must be provided with training to learn the general principles of fire extinguisher use and the hazards involved in firefighting. Employees who are expected to use fire extinguishers in the workplace must be trained on the types of fire extinguishers available to them, the different classes of fires, and where the fire extinguishers are located. The ABC type of fire extinguisher is probably best suited for most industrial applications because it can be used on Class A, B, and C fires. Class A is used for common combustibles (such as paper, wood, and most plastics); Class B is for flammable liquids (such as solvents, gasoline, and oils); and Class C is for fires in or near live electrical circuits. In areas such as electrical substations and switchgear rooms, only Class C (carbon dioxide, CO_2) should be used. Though combination Class A, B, and C extinguishers will extinguish most electrical fires, the chemical residue left behind can damage delicate electrical/electronic components; thus, the CO_2 type of extinguisher is more suitable for extinguishing electrical fires. Each employee must know how to use the fire extinguisher. Most importantly, employees must know when it is not safe to use fire extinguishers—that is, when the fire is beyond being extinguishable with a portable fire extinguisher. Emergency telephone numbers should be strategically placed throughout the workplace. Employees need to know where they are posted. Workers should be trained on the information they need to provide to the 911 operator (or other emergency service number) in case of fire.

9.17.9.4 Miscellaneous Fire Prevention Measures

In addition to basic fire prevention, emergency response training and fire extinguisher training, employees must be trained on the hazards involved with flammable and combustible liquids. 29 CFR 1910.106 addresses this area. Industrial facilities typically use all types of flammable and combustible liquids. These dangerous materials must be clearly labeled and stored safely when not in use. The safe handling of flammable and combustible liquids is a topic that needs to be fully addressed by the facility safety official and workplace supervisor. Worker awareness of the potential hazards that flammable and combustible liquids pose must be stressed. Employees need to know that flammable and combustible liquid fires burn extremely hot and can produce copious amounts of dense, black smoke. Explosion hazards exist under certain conditions in enclosed, poorly ventilated spaces where vapors can accumulate. A flame or spark can cause vapors to ignite, creating a flash fire with the terrible force of an explosion. One of the keys to reducing the potential spread of flammable and combustible fires is to provide adequate containment. All storage tanks should be surrounded by storage dikes or containment systems, for example. Correctly designed and built dikes will contain spilled liquid. Spilled flammable and combustible liquids that are contained are easier to manage than those that have free run of the workplace. Properly installed containment dikes can prevent environmental contamination of soil and groundwater.

9.17.10 Subpart N, Materials Handling and Storage

1910.176 Materials Handling and Storage

1910.178 Powered Industrial Trucks

1910.184 Slings

9.17.10.1 Rigging Safety Program

> In lifting the various materials and supplies, a number of standard chokers, slings, bridle hitches, and basket hitches can be used. Because loads vary in physical dimension, shape, and weight, the rigger needs to know what method of attachment can be safely used. It is estimated that 15% to 35% of crane accidents may involve improper rigging. The employer needs to train those employees who are responsible for rigging loads. They need to be able to (1) know the load, (2) judge distances, (3) properly select tackle and lifting gear, and (4) direct the operation. The single most important rigging precaution is to determine the weight of the load before attempting to lift it. The weight of the load will in turn determine the lifting device, such as a crane, and the rigging gear to be used. It is also important to rig a load so that it will be stable, that is, it does not move as it is lifted.
>
> **—National Safety Council (1992)**

The facility safety professional needs to realize that special safety precautions apply to rigging operations and to using and storing fiber ropes, rope slings, wire ropes, chains, and chain slings. The safety official should know the properties of the various types used, the precautions for use, and the maintenance required. In addition, the safety official must be familiar with the requirements of OSHA's rigging equipment for material handling standard (29 CFR 1926.251). Rigging operations are inherently dangerous. Any time any type of load is lifted, the operation is dangerous in itself. When heavy loads are lifted several feet and suspended in air while they are moved from one place to another, the dangers are increased exponentially. Although rigging and lifting operations include the use of several different types of mechanical devices such as cranes, winches, chain falls, and come-alongs, in this section we focus on those components that form the interface between the load and the lifting or hoisting equipment—the ropes, chains, and slings. We place our focus on these devices not only because they are the most commonly used rigging devices found in industrial applications but also because the safety professional is directly responsible for ensuring that they are safe to use—and are used safely.

9.17.10.1.1 Written Rigging Safety Program

We have said it before but we feel the need to say it again: Safety begins with written policies and procedures, because employees cannot be expected to perform their tasks safely and consistently, unless company safety policies and procedures are consistent and in writing. The same can be said about the Rigging Safety Program. Company policies, procedures, and responsibilities must be spelled out in straightforward, plain English. Anything less is unacceptable and clearly unworkable. Rigging is dangerous. Improper rigging is absolutely dangerous. Rigging procedures must be in writing and must be enforced, and employees must be trained on the requirements. The written Rigging Safety Program should also include a responsibilities section that clearly identifies those individuals responsible for the Rigging Safety Program and the names of all designated competent persons. In addition, definitions pertinent to rigging operations and equipment should be included. The definitions provided in 29 CFR 1910.184 provide a good example of the types of definitions that should be included in the written program:

Angle of loading—The inclination of a leg or branch of a sling measured from the horizontal or vertical plane.

Basket hitch—A sling configuration whereby the sling is passed under the load and has both ends, end attachments, eyes, or handles on the hook or a single master link.

Bridle wire rope sling—A wire rope formed by plaiting component wire ropes.

Cable laid endless sling-mechanical joint—A wire rope sling made endless by joining the ends of a single length of cable laid rope with one or more metallic fittings.

Cable laid grommet-hand tucked—An endless wire rope sling made from one length of rope wrapped six times around a core formed by hand tucking the ends of the rope inside the six wraps.

Cable laid rope—A wire rope composed of six wire ropes wrapped around a fiber or wire rope core.

Cable laid rope sling-mechanical joint—A wire rope sling made from a cable laid rope with eyes fabricated by pressing or swagging one or more metal sleeves over the rope junction.

Choker hitch—A sling configuration with one end of the sling passing under the load and through an end attachment, handle, or eye on the other end of the sling.

Coating—An elastomer or other suitable material applied to a sling or to a sling component to impact desirable properties.

Cross rod—A wire used to join spirals of metal mesh to form a complete fabric.

Designated—Selected or assigned by the employer or the employer's representative as being qualified to perform specific duties.

Equivalent entity—A person or organization (including an employer) which, by possession of equipment, technical knowledge, and skills, can perform with equal competence the same repairs and tests as the person or organization with which it is equated.

Fabric (metal mesh)—The flexible portion of a metal mesh sling consisting of a series of transverse coils and cross rods.

Female handle (choker)—A handle with a handle eye and a slot of such dimension as to permit passage of a male handle, thereby allowing the use of a metal mesh sling in a choker hitch.

Handle—A terminal fitting to which metal mesh fabric is attached.

Handle eye—An opening in a handle of a metal mesh sling shaped to accept a hook, shackle, or other lifting device.

Hitch—A sling configuration whereby the sling is fastened to an object or load, either directly to it or around it.

Link—A single ring of a chain.

Male handle (triangle)—A handle with a handle eye.

Master link or gathering ring—A forged or welded steel link used to support all members (legs) of an alloy steel chain sling or wire rope sling.

Mechanical coupling link—A nonwelded, mechanically closed steel link used to attach master links, hooks, etc. to alloy steel chain.

Proof load—The load applied in performance of a proof test.

Proof test—A nondestructive tension test performed by the sling manufacturer or an equivalent entity to verify construction and workmanship of a sling.

Rated capacity or working load limit—The maximum working load permitted.

Reach—The effective length of an alloy steel chain sling measured from the top bearing surface of the upper terminal component to the bottom bearing surface of the lower terminal component.

Selvage edge—The finished edge of synthetic webbing designed to prevent unraveling.

Sling—Connects the load to the material handling equipment.

Sling manufacturer—A person or organization that assembles sling components into their final form for sale to users.

Spiral—A single transverse coil that is the basic element from which metal mesh is fabricated.

Strand laid endless sling-mechanical joint—A wire rope sling made endless from one length of rope with the ends joined by one or more metallic fittings.

Strand laid grommet-hand tucked—An endless wire rope sling made from one length of strand wrapped six times around a core formed by hand tucking the ends of the strand inside the six wraps.

Strand laid rope—A wire rope made with strands (usually six or eight) wrapped around a fiber core, wire strand core, or independent wire rope core (IWRC).

Vertical hitch—A method of supporting a load by a single, vertical part or leg of the sling.

9.17.10.2 Safety: Ropes, Slings, Chains

Because of the dangers inherent in any rigging and lifting operation, the safety official must check out and ensure the safety of every element involved. This may seem like common sense to some, but others might be surprised to find out how often rigging mistakes are made, by assuming that the only factor that need be considered is the safe operation of hoisting equipment to lift a given load. Experience has shown that the attachments used to secure the hook to the load are often overlooked and thus the cause of failure and injuries. In this section, we discuss OSHA's general requirements and the main rigging attachments: ropes, slings, and chains.

9.17.10.2.1 Rigging Equipment and Attachments: General

In 29 CFR 1926.251, the point is made that rigging equipment for handling material must not be loaded in excess of its recommended safe working load (see Tables H-1 through H-20 in the standard). All such equipment must be inspected prior to use on each shift and as necessary during use to ensure safety. Any rigging equipment found to be defective must be immediately removed from service. Rigging equipment not in use that presents a hazard must be removed from the immediate working area to ensure the safety of employees. The safety official must ensure that all custom-designed grabs, hooks, clamps, or other lifting accessories are marked to indicate their safety working loads. Each device must be proof tested to 125% of its rated load before use. Whenever a sling is used, the following practices must be observed:

- Slings must not be shortened with knots, bolts, or other makeshift devices.
- Sling legs must not be kinked.
- Slings used in a basket hitch must have the loads balanced to prevent slippage.
- Slings must be padded or protected from the sharp edges of their loads.
- Shock loading is prohibited
- A sling must not be pulled from under a load when the load is resting on the sling.
- Hands or fingers must not be placed between the sling and its load while the sling is being tightened around the load.

9.17.10.2.2 Rope Slings

Ropes used in rigging (for slings) are usually divided into two main classes: fiber rope slings and wire rope slings. Fiber ropes are further divided into natural and synthetic fibers depending on their construction. There are many types of slings. Slings normally have a fixed length. They may be made from various materials and have the form of rope, belts, mesh, or fabric. Natural fiber ropes and slings are usually made from manila, sisal, or henequen fibers. Most natural fiber ropes and slings used in industry today are made from manila fibers because of its superior breaking strength, consistency between grades, excellent wear properties in both freshwater and saltwater atmospheres, and elasticity. The main advantages of natural fiber ropes are their price and their ability to form or bend around angles of the object being lifted. The disadvantages of using natural fiber ropes include increased susceptibility to cuts and abrasions, their reduced capability or inability to be used to lift materials at elevated temperatures, and that hot or humid conditions may reduce their service life. Fiber ropes should never be used in atmospheres where they may come in contact with acids and caustics, as these substances will degrade the fibers. Safe working loads of various sizes and classifications of natural fiber ropes can be determined from tables in 29 CFR 1926.251.

Synthetic fiber rope slings are made from synthetic fibers (such as nylon, polyester, polypropylene, polyethylene, or a combination of these) to obtain the desired properties. Synthetic fiber ropes have many of the same qualities as natural fiber rope slings, but are in much wider use throughout the industry because they can be engineered to fit a particular operation. Synthetic fiber ropes have many advantages, including increased strength and elasticity, over natural fiber rope. Synthetic fiber rope also stands up better to shock loading and has better resistance to abrasion than natural fiber rope. One of the key advantages of synthetic fiber rope is that it does not swell when wet. It is also more resistant to acids, caustics, alcohol-based solvents, bleaching solutions, and their atmospheres. As with the use of natural fiber rope, synthetic fiber rope also has some disadvantages, including damage from excessive heat (they can melt) or from alkalis and susceptibility to abrasion damage. They also cost more than natural ones.

9.17.10.2.3 Wire Rope

The most widely used type of rope sling in industry is the cable laid 6 × 19 and 6 × 37 wire rope. By definition, wire rope is a twisted bundle of cold-drawn steel wires, usually composed of wires, strands, and a core. When used in rope slings, wire ropes must have a minimum clear length of wire rope 10 times the component rope diameter between splices, sleeves, or end fittings. The main reasons for the wide usage of wire over fiber rope are its greater strength, durability, predictability of stretch characteristics when placed under heavy stresses, and stable physical characteristics over a wide variety of environmental conditions. The main advantages of wire rope that is preformed are its lessened tendency to unwind, set, kink, or generate sharp protruding wires.

9.17.10.2.4 Chains and Chain Slings

Steel and alloys (stainless steel, monel, bronze, and other metals) are commonly used for lifting slings made of chain. The safety official needs to know a number of facts related to chain slings and the type of chain that is authorized for use in slings; for example, the rated capacity (working load limit) for welded alloy steel chain slings must conform to the values in the appropriate tables in 29 CFR 1926.251. Whenever wear at any point of any chain link exceeds that specified, the assembly must be removed from service. All such slings have permanently affixed durable identification that states size, grade, rated capacity, and the sling manufacturer. Finally, regular hardware chain or other chain not specifically designed for use in slings should not be used for load lifting.

9.17.10.2.5 Rigging Training

Along with emphasizing and reemphasizing the need for written programs and procedures, we have gone to lengths to reinforce the idea that employee training is critically important. Many experienced safety professionals have stated that training is at the very heart of safety—and we agree. Training is especially important in work practices that involve rigging operations. One false move, one mistaken perception in distance, one careless mistake can yield devastating consequences, not only to expensive equipment and machinery but more importantly to humans. Rigging is a dangerous enterprise. We cannot make this point strongly enough. No employee can or should be expected to have the gift of innate knowledge. Knowledge has to be learned through training. Rigging is dangerous, but training lessens the danger. As always, any time you conduct training, do not forget to put it on paper. Training that is not documented is training that … well, by now you should have gotten this picture in Technicolor.

9.17.10.2.6 *Proof Testing Rigging Equipment*

One of the safety official's primary duties involving rigging operations is to ensure that the equipment used is safe to use. Ropes, slings, and chains, and other lifting devices must be certified via proof testing to verify their soundness and safety for use. Proof testing is a nondestructive tension test performed by the sling manufacturer or an equivalent entity to verify construction and workmanship of a sling or other lifting device. During proof testing, a *proof load* is applied to test the lifting device. The safety official is responsible for ensuring that, before each use, each new, repaired, or reconditioned lifting device (rope, chain, or sling)—including all welded components in the sling assembly—is proof tested by the sling manufacturer or equivalent entity, in accordance with American Society of Testing and Materials Specification A391-65 (ANSI G61.1-1968). The safety official should ensure that a written certification of the proof test is provided and that such records are available for review by regulatory auditors. Typically, sling proof test or load test results are stamped, marked, or labeled right on the sling itself. In addition to verifying the satisfactory condition of each sling or other rigging component, the safety official should ensure that certification labels and identification tags are attached and visible and that test data (e.g., load rating) are current.

9.17.10.2.7 *Rigging Inspections*

Each day before being used, the sling and all rigging fastenings and attachments must be inspected for damage or defects by a competent person designated by the employer. A few of the kinds of items that should be inspected to ensure that slings are safe to use include the following:

1. Alloy steel chain slings must have permanently affixed, durable identification stating size, grade, rated capacity, and reach.
2. A thorough periodic inspection of alloy steel chain slings in use must be made on a regular basis (at least once every 12 months).
3. A record must be maintained of the most recent month in which each alloy steel chain sling was thoroughly inspected.
4. Alloy steel chains slings must be permanently removed from service if they are heated above 1000°F.
5. Worn or damaged alloy steel chain slings and attachments must be taken out of service until repaired.
6. Wire rope slings must be used only with loads that do not exceed the rated capacities.
7. Fiber core wire rope slings of all grades must be permanently removed from service if they are exposed to temperatures in excess of 200°F.
8. Welding of end attachments, except covers to thimbles, must be performed prior to the assembly of the sling.

9. Welded end attachments must be proof tested by the manufacturer or equivalent entity at twice their rated capacity prior to initial use.
10. All synthetic web slings must be marked or coded to show the rated capacities for each type of hitch and type of synthetic web material.

Additional inspection must also be performed during sling use where service conditions warrant. Damaged or defective slings must be immediately removed from service. Make them unusable by burning or cutting them before they are discarded; otherwise, they may mysteriously reappear and be used again.

9.17.10.2.8 Rigging Safe Work Practices

Written safe work practices are an important element of any effective Rigging Safety Program. The purpose of such safe work practices, rules, or regulations is, of course, to reduce the chances of employee injury and property damage. The organization safety person should include those safe work practices that apply specifically to the kinds of operations and rigging practices that the employees of the company perform and are responsible for. Most organization will have different safe work practices, because the kind of work each does will be different. The following text provides a sample safe work practice for using fiber and synthetic rope slings. Though it specifically targets fiber and synthetic rope slings, this sample is a guide that can be used to write a safe work practice procedure for most rigging equipment.

9.17.10.2.8.1 Safe Work Practice: Fiber and Synthetic Rope Slings

1. Do not attempt to lift loads which exceed the rated load capacity of the rope.
2. Fiber rope slings should have a diameter of curvature meeting at least minimum OSHA or manufacturer's specifications.
3. Natural fiber and synthetic fiber rope slings, except for wet frozen slings, may be used in a temperature range from minimum 20°F to plus 180°F without decreasing the work load limit. For operations outside this temperature range and for wet frozen ropes, the sling manufacturer's recommendations should be followed.
4. Spliced fiber rope slings should not be used unless they have been spliced in accordance with the requirements of the manufacturer.
5. Natural and synthetic fiber rope slings should be immediately removed from service if any of the following conditions are present:
   - Abnormal wear
   - Powdered fibers between strands
   - Broken or cut fibers

- Variations in the size or roundness of strands
- Discoloration or rotting
- Distortion of hardware in the sling.

6. Only fiber rope slings made from new rope should be used. Law prohibits use of repaired or reconditioned fiber rope slings.
7. When synthetic web slings are used, certain precautions should be taken:
   - Nylon web slings must not be used where fumes, vapors, sprays, mists, acids, or phenolics are present.
   - Polyester and polypropylene web slings must not be used where fumes, vapors, sprays, mists, or liquid forms of caustics are present.
   - Web slings with aluminum fittings must not be used where fumes, vapors, sprays, mists, or liquid forms of caustics are present.
   - Synthetic web slings of polyester and nylon must not be used at temperatures in excess of 180°F; slings of polypropylene must not be used at temperatures in excess of 200°F.
   - Synthetic web slings must be immediately removed from service if there exists any of the following conditions: acid or caustic burns; melting or charring of any part of the sling surface; snags, punctures, tears, or cuts; broken or worn stitches; or distortion of fittings.
8. Sling legs should not be kinked.
9. Slings should be securely attached to their loads.
10. All employees must keep clear of loads about to be lifted and of suspended loads.

9.17.10.3 Forklift Operator Safety

While workers usually have received their standard license training somewhere else (most come equipped with their license, of course), they may or may not have experience driving any motorized vehicle other than a car or truck. Regardless of their stated prior experience with forklifts, to gain their company license to drive such equipment they should undergo the complete company training on driving and handling such equipment. This protects themselves, their co-workers, company property, possibly the local environment, and especially the company's liability in case of accident. The 1998 version of OSHA's forklift standard, 29 CFR 1910.178, came about because the Industrial Truck Association (ITA) requested that OSHA include training requirements for forklift operators. OSHA added them in performance-oriented language, permitting the employer flexibility in teaching methods and to an extent in training content. Full compliance with the revised regulation prevents up to 22 deaths and 14,000 injuries each year. Effective forklift training includes classroom work with examples and testing, as well as hands-on experience.

9.17.10.4 Forklift Operator Training

9.17.10.4.1 Prior to Training Session

Instructor inspects the forklift, checking each item on the Driver's Daily Forklift Maintenance/Safety Checklist (Figure 9.3) for the type of forklift to be used in the training and signs the form. Instructor tests brakes and steering, then parks the forklift.

9.17.10.4.2 Training Session: Safe Work Practices for Forklift Operations

Instructor explains company requirements for all employees who operate forklifts on the job and lists each safe work practice for the trainees, ensuring that they copy the practices for themselves and explaining what each listed item means and its importance. This training method ensures that trainees process the information in at least three different ways—they see it, hear it, and generate it through writing, reinforcing the lesson. This repetition increases how much information the trainee retains. Practices and explanations include the following:

- Only company certified personnel are permitted to drive forklifts on company property in the performance of company business.

  Explanation: Certified means that the employee has a company forklift driver's license. The procedure to obtain a license includes classroom training, viewing a safety forklift video, passing the written examination, and passing the forklift driver's test.

- Use care and caution, and wear seatbelts when operating a forklift.

  Explanation: Operating the forklift without a seatbelt, "hot-rodding," or performing any unsafe act while operating the forklift is grounds for dismissal.

- Complete all daily maintenance and visual checks, and fill out and initial the Daily Checklist Form prior to operation.

  Explanation: Before daily use of any forklift, drivers fill out a Driver's Daily Checklist Form (Figure 9.3). If anything is wrong, broken, or unsatisfactory during the inspection, the driver must report it immediately to the supervisor. Under no circumstances should drivers operate a forklift that does not pass inspection until the supervisor clears it.

Electric Forklift Driver's Daily Checklist

Check at the start of each shift.

Truck No. ____________ Date ____________

Operator ____________ Supervisor's OK ____________

Hour meter reading ____________ Start of day ____________

Check (✓) appropriate boxes:

OK	Needs Attention or Repair		OK	Needs Attention or Repair	
❑	❑	Obvious damage and leaks (report to foreman immediately)	❑	❑	Service brakes (report problems to foreman immediately)
❑	❑	Tire condition	❑	❑	Battery load test (Watch battery indicator while holding tilt lever on full back tilt. If needle falls to red area, battery does not have sufficient charge to operate truck properly.)
❑	❑	Battery plug connection (be sure battery plug connection is tight)			
❑	❑	Head and tail lights	❑	❑	Parking brake (report problems to foreman immediately)
❑	❑	Warning lights	❑	❑	Seat brake (report problems to foreman immediately)
❑	❑	Hour meter	❑	❑	Hydraulic controls
❑	❑	Other gauges and instruments			
❑	❑	Battery discharge indicator (key on needle should indicate in green area)			
❑	❑	Horn			
❑	❑	Steering (report problems to foreman immediately)			

Gas, LPG, or Diesel Forklift Driver's Daily Checklist

Check at the start of each shift.

Truck No. ____________ Date ____________

Operator ____________ Supervisor's OK ____________

Hour meter reading ____________ Start of day ____________

Check (✓) appropriate boxes:

OK	Needs Attention or Repair		OK	Needs Attention or Repair	
❑	❑	Engine oil level (driver replenish)	❑	❑	Hour meter
❑	❑	Radiator water level (driver replenish)	❑	❑	Other gauges and instruments
❑	❑	Fuel level (driver replenish)	❑	❑	Horn
❑	❑	Battery water level	❑	❑	Steering (report problems to foreman immediately)
❑	❑	Obvious damage or leaks (report to foreman immediately)	❑	❑	Service brakes (report problems to foreman immediately)
			❑	❑	Parking brake (report problems to foreman immediately)
❑	❑	Tire condition	❑	❑	Hydraulic controls
❑	❑	Head and tail lights			
❑	❑	Warning lights			

Remarks (explain all items requiring attention or repair):

FIGURE 9.3
Forklift driver's daily checklist.

Forklift safe work practices that relate to driving and handling skills include the following:

- Mount and dismount carefully.
- Sit on the seat and keep arms and legs inside the cab at all times.
- Observe traffic and keep to the right.
- Do not allow passengers on the forklift.
- Slow down or stop at all blind intersections; yield at all others.
- Always observe all traffic rules, warning load limits, and overhead clearance.
- Sound warning device at all cross aisles, exits, elevators, sharp corners, ramps, and blind corners and when approaching pedestrians.
- Face direction of travel. *Never* back up without looking.
- Do not exceed rated capacity. Check unit capacity if attachments are installed.
- Keep forks 4 to 6 inches above the ground when traveling.
- Travel with the load facing uphill on inclines and downgrades.
- Operate in reverse when carrying bulky loads and whenever you cannot see over the load.
- Travel at speeds that allow safe stops.
- Keep forklift and forks clear of any pedestrians.
- Position the load evenly on both forks.
- Ensure that awkward loads are secured.
- Never permit anyone to stand or pass under the elevated portion of the mast or attachment.
- Lower forks, put forklift in neutral, shut off the forklift, set the brake, and, if parked on an incline, block the wheels before leaving the forklift. Park the forklift in authorized areas only.
- Refuel or recharge only at safe locations. Fully lower the load engaging means, neutralize controls, set brakes, shut off power, remove key, block wheels on an incline, and *do not smoke* while refueling or recharging.
- Do not operate unit in areas without an overhead guard in place.
- Do not operate any forklift that is in need of repair, defective, or in any way unsafe.
- Never raise anyone up in the air while they stand on the forks. Use an appropriate safety platform.

Forklift safe work practices that relate to hazardous materials include the following:

- Be alert at all times, as hazardous materials containers are stored throughout the warehouse area.
- No one is authorized to operate a forklift on company grounds without first have been trained on the company's Hazard Communication Standard and Emergency Response Plan for Chemical Spills.
- Report all accidents involving chemicals to the supervisor immediately. *Do not* attempt to clean up a chemical spill without first alerting the company's chemical response team.

  Explanation: Careless forklift drivers cause many chemical spills within the warehouse. The company's policy on chemical spills is in place to protect workers, the workplace, the community, and the surrounding environment. The company's policies work to maintain and protect the company's excellent reputation in preventing and mitigating spills, and workers are expected to follow those policies and practices to further the same goals.

Instructor asks for questions, answering any that arise. Instructor has trainees date and sign their handwritten copies of their training form, makes duplicate copies, and gives the copy to the trainee, keeping the original for the training record file. Instructor presents the Forklift Safety Training video, then again asks for questions. Instructor presents trainees with the Forklift Operation Exam and allows trainees 15 minutes for completion.

9.17.10.4.3 Sample Forklift Operation Exam

1. Only trained, tested, licensed, and authorized personnel should operate a forklift within the company.

   True *or* False

2. A daily maintenance check that includes the visual check of fuel, air, water, hydraulics, transmission, battery, brakes, tires, and controls shall be performed and documented on the driver's daily checklist.

   True *or* False

3. While operating a forklift, a ____________ and a hardhat are required.

   a. Two-way radio

   b. A tool belt

   c. Safety belt

4. Never permit anyone to ____________ or ____________ under the elevated portion of the mast or attachment.
   a. Smoke or drink
   b. Talk or use hand signals
   c. Stand or pass
5. When carrying bulky loads, operate in reverse for better visibility.
   True *or* False
6. Refuel while the forklift is running.
   True *or* False
7. Travel with the load facing uphill on inclines and ____________.
8. If a forklift requires repair, the person trained to operate it should try to repair it.
   True *or* False
9. Always know the capacity of your forklift and the weight of the materials to be lifted—never lift more than the rated capacity of your equipment.
   True *or* False
10. The forklift should be positioned so that the load is lifted evenly on both forks, then the load should be carefully and smoothly lifted.
    True *or* False
11. The company has responsibility for making sure forklift drivers read all product warning labels.
    True *or* False
12. If, while operating the forklift, you run into and damage any container of chemicals, what should you do first?
    a. Clean up the spill.
    b. Call your lawyer.
    c. Admit nothing.
    d. Report the incident immediately to the environmental health and safety manager.

Answer Key: (1) True; (2) True; (3) C; (4) C; (5) True; (6) False; (7) Downgrades; (8) False; (9) True; (10) True; (11) False; (12) D.

After the allotted time is over, the instructor goes over each question to make sure that trainees understand the correct answers. The instructor makes sure each trainee replaces any wrong answers with correct answers to ensure that the trainee has only correct information on their study materials.

9.17.10.4.4 Hands-on Driver Training

Onsite, the instructor reinspects the forklift, demonstrates how to perform and fill out the Driver's Daily Checklist, and explains what each checklist item is for and what to look for. After the inspection is complete, the instructor mounts the forklift, demonstrating and explaining each operating control and answering any questions. The instructor starts up the forklift, demonstrating and explaining the various functions. The instructor maneuvers the truck through an obstacle course, lifts and stacks empty pallets, and picks up and moves a fully loaded pallet. The instructor directs the trainee to mount and start up the machine, making sure he or she is familiar with all of the controls. When the trainee is ready to operate the forklift, the instructor videorecords the trainee's forklift operation. The trainee practices operating the control levers, picking up pallets, moving away from the stack, and building a new stack one pallet at a time, and then practices moving a stack. The trainee then drives through the obstacle course as many times as necessary for the instructor to be satisfied with the performance. The instructor asks for final questions, then has the trainee sign and date the attendance roster form, telling the trainee that the operator's license will be generated within a few days and not to operate a company forklift until the license is in his or her possession.

9.17.10.4.5 Post-Training Session

The instructor inventories the training materials, including the attendance roster sheet, filled-out student handouts, examinations, and videorecordings of the trainee driving exercises, and enters the pertinent data in the employee's training file. The instructor checks the video for quality, labels it, and shelves it according to the filing system for trainee video records. The instructor fills out a Driver's Operator License form for Human Resources and routes it through company mail. Effective employee training, retraining, and proper documentation are the best regulatory insurance. They ensure that mishaps, incidents, and accidents happen infrequently in a facility. Also, in court or at an OSHA hearing, you have proof that your company took the proper and necessary steps to provide workers with access to the required information and with effective training.

9.17.11 Subpart O, Machinery and Machine Guarding

1910.212 General requirements for all machines

1910.215 Abrasive wheel machinery

1910.219 Mechanical power-transmission apparatus

9.17.11.1 Machine Guarding

> [S]afety and health on the job begin with sound engineering and design. The engineer and designer will be familiar with most of the common hazards to be dealt with in the design phase. For the senior manager,

> however, highlighting the most common hazards found in equipment and the ones requiring particular alertness [is called for here]. The most common sources of mechanical hazards are unguarded shafting, shaft ends, belt drives, gear trains, and projections on rotating parts. Where a moving part passes a stationary part or another moving part, there can be a scissor-like effect on anything caught between the parts. A machine component which moves rapidly with power or a point of operation where the machine performs its work are also typical hazard sources.
>
> There are probably over 2 million metalworking machines and half that many woodworking machines in use that are at least 10 years old. Most are poorly guarded, if at all. Even the newer ones may have substandard guards, in spite of OSHA requirements. …The basic objective of machine guarding is to prevent personnel from coming in contact with revolving or moving parts such as belts, chains, pulleys, gears, flywheels, shafts, spindles, and any working part that creates a shearing or crushing action or that may entangle the worker.
>
> Machine guarding is visible evidence of management's interest in the worker and its commitment to a safe work environment. It is also to management's benefit, as unguarded machinery is a principal source of costly accidents, waste, compensation claims, and lost time.
>
> **—Ferry (1990)**

The basic purpose of machine guarding is to prevent contact of the human body with dangerous parts of machines. Moving machine parts have the potential for causing severe workplace injuries, such as crushed fingers or hands, amputations, burns, and blindness, just to name a few. Machine guards are essential for protecting workers from these needless and preventable injuries. Any machine part, function, or process that may cause injury must be safeguarded. When the operation of a machine or accidental contact with it can injure the operator or others in the vicinity, the hazards must be either eliminated or controlled (OSHA, 2003). Our experience has clearly (and much too frequently) demonstrated that when an arm, finger, hair, or any body part enters into or makes contact with moving machinery, the results can be not only gory, bloody, and disastrous but also sometimes fatal.

Depending on the machine and the types of hazards it presents, methods of machine guarding vary greatly. The intent of this section is to familiarize safety professionals with the hazards of unguarded machines, common safeguarding methods, and the safeguarding of machines—all of which, if followed, combine to ensure that Ferry's main point—"Machine guarding is visible evidence of management's interest in the worker and its commitment to a safe work environment"—becomes a reality. It logically follows that if the employer provides a safe workplace then all sides benefit from the results.

Complying with 29 CFR 1910.212–219 regarding safeguarding machines is an important step the safety official takes to ensure control of workplace hazards and to protect the safety of employees; however, making sure that machines are safeguarded with the types of guards and devices discussed in the previous section is only part of the compliance effort. *Safe work practices* are an important element of any machine guarding safety program (and most other specialized safety programs). Experience has clearly demonstrated that, if written safe work practices are not in place and employees do not have a written protocol to follow to safeguard themselves from the hazards presented by many machines, then the machine safeguarding safety program is incomplete—and less than fully effective. Consider the following safe work practices provided by Hoover et al. (1989), which are designed to be employed in addition to the machine safeguarding guards and devices, as well as other practices:

1. Guards should not be removed unless:
   a. Permission is given by a supervisor.
   b. The person concerned is trained.
   c. Machine adjustment is a normal part of his or her job.
2. Do not start machinery unless guards are in place and in good condition.
3. Report missing or defective guards immediately to your supervisor.
4. When removing safeguards for repair, adjustment, or service, turn off power and lock and tag the main switch.
5. Do not permit employees to work on or around equipment while wearing ties, loose clothing, watches, rings, etc.
6. Inspect and conduct a maintenance program of guards on a regularly scheduled basis.
7. Instruct operators of mechanical equipment in all safe work practices for operation of that machine.

9.17.11.1.1 Basics of Safeguarding Machines

Any mechanical motion that threatens a worker's safety should not remain unguarded. OSHA's reasoning behind this point is quite clear and is reinforced often—anytime the safety professional investigates on-the-job injuries involving crushed hands and arms, severed fingers, blindness, and other horrifying machinery-related injuries. For the safety official, the goal is quite clear; when the operation of a machine or accidental contact with it can injure the operator or others in the vicinity, the hazards must be either controlled or eliminated.

Application of appropriate safeguards keeps people and their clothing from coming into contact with hazardous parts of machines and equipment. They also keep flying particles from an operation and broken machine parts from striking or injuring people. Guards may also serve to enclose noise or dust hazards. Machine safeguarding is intended to minimize the risk of accidents of machine–operator contact (NSC, 1987). Such contact can occur as follows:

1. An individual makes contact with the machine—usually the moving part—because of inattention caused by fatigue, distraction, curiosity, or deliberate chance taking.
2. An individual is exposed to flying metal chips, chemical and hot metal splashes, and circular saw kickbacks, to name a few;
3. An individual makes contact as a direct result of a machine malfunction, including mechanical and electrical failure.

Guards should have certain characteristics (Brauer, 1994). They should be a permanent part of the machine or equipment, must prevent access to the danger zone during operation, and must be durable and constructed strongly enough to resist the wear and abuse expected in the environment where machines are used. Guards should not interfere with the operation of the machine; that is, guards must not create hazards. Finally, machine guards should be designed to allow the more frequently performed maintenance tasks to be accomplished without removal of the guards.

9.17.11.1.2 Types of Machine Safeguards Required

Dangerous moving parts in three basic areas require safeguarding:

1. *The point of operation*, where work is performed on the material, such as cutting, shaping, boring, or forming of stock.
2. Power transmission apparatus, including all components of the mechanical system that transmit energy to the part of the machine performing the work. These components include flywheels, pulleys, belts, connecting rods, couplings, cams, spindles, chains, cranks, and gears.
3. *Other moving parts*, including all parts of the machine that move while the machine is working. These can include reciprocating, rotating, and transverse moving parts, as well as feed mechanisms and auxiliary parts of the machine.

9.17.11.1.3 Mechanical Hazards: Motions and Actions

Three types of machine motion and four types of actions may present hazards to the worker. These can include the movement of rotating members, reciprocating arms, moving belts, meshing gears, cutting teeth, and any parts that impact or shear. These different types of hazardous mechanical motions and actions are basic in varying combinations to nearly all machines, and recognizing them is the first step toward protecting workers from the danger they present. The basic types of hazardous mechanical motions and actions are as follows:

Motions

- Rotating (including in-running nip points)
- Reciprocating
- Transversing

Actions

- Cutting
- Punching
- Shearing
- Bending

9.17.11.1.4 Common Safeguarding Methods

The safety official has several safeguarding methods to consider when he or she has determined that machine guarding is needed. The type of operation, the size or shape of stock, the method of handling, the physical layout of the work area, the type of material, as well as production requirements or limitations will help to determine the appropriate safeguarding method for the individual machine. As a general rule, power transmission apparatus is best protected by fixed guards that enclose the danger areas. For hazards at the point of operation, where moving parts actually perform work on stock, several kinds of safeguarding may be possible. The safety official must always choose the most effective and practical means available. Safeguards include guards, devices, automatic and semiautomatic feeding and ejecting methods, location and distance, and miscellaneous safeguarding accessories.

9.17.11.1.4.1 Guards Guards are barriers that prevent access to danger areas. Guards can be of several types. These include fixed, interlocked, adjustable, and self-adjusting. *Fixed guards,* as the names implies, are a permanent part of the machine. Unlike other types of guards, these do not move to accommodate the work being performed. They are not dependent upon moving parts to perform their intended function. They may be constructed of sheet metal, screen, wire cloth, bars, plastic, or any other material that is substantial enough to withstand the impact they may receive, and to endure prolonged use. If feasible, these guards are usually preferable to all other types, because of their relative simplicity and permanence. Limitations include interference with visibility, that they are limited to specific operations, and that machine adjustment and repair may require removal, thereby necessitating other means of protection for maintenance personnel.

Interlocked guards shut off or disengage power and prevent starting of the machine when the guard is open. An interlocked guard may use electrical, mechanical, hydraulic or pneumatic power, or any combination of these. Interlocked guards have the advantage of providing the maximum protection, and they allow access to the machine for setup, adjustment, or maintenance purposes. This type of guard, however, requires careful adjustment and maintenance and can be made inoperable.

Adjustable guards provide a barrier that may be adjusted to facilitate a wide variety of production operations. Advantages include their ability to be constructed to suit many specific applications and that they can be adjusted to admit varying sizes of stock. Protection may not be complete at all times because hands may enter the danger area, and they often require frequent adjustment and maintenance.

Self-adjusting guards also accommodate different sizes of stock, but the movement of the stock determines the openings of these barriers. As the operator moves the stock into the danger area, the guard is pushed away, providing an opening that is only large enough to admit the stock. After the stock is removed, the guard returns to the rest position. This guard protects the operator by placing a barrier between the danger area and the operator. The guards may be constructed of plastic, metal, or other substantial material. Self-adjusting guards offer different degrees of protection and are often easier to purchase and fit to the machine. This type of guard, however, does not always provide maximum protection, can limit visibility, and requires frequent adjustment and maintenance.

9.17.11.1.4.2 Devices Devices can also be used to safeguard machinery. A safety device may perform many functions. It may stop the machine if a hand or any part of the body is inadvertently placed in the danger area; restrain or withdraw the operator's hands from the danger area during operation; require the operator to use both hands on machine controls, thus keeping both hands and body out of danger; or provide a barrier synchronized with the operating cycle of the machine to prevent entry to the danger area during the hazardous part of the cycle. This category includes presence-sensing devices, pullback mechanisms, restraints, safety controls, and gates.

Presence-sensing devices commonly operate on photoelectric, radiofrequency, or electromagnetic principles to disengage the machine when something is detected in the zone of concern. The photoelectric (optical) presence-sensing device uses a system of light sources and controls to interrupt the operating cycle of the machine. If the light field is broken, the machine stops and will not cycle. This device must be used only on machines that can be stopped before the worker can reach the danger area. The design and placement of the guard depend upon the time it takes to stop the mechanism and the speed at which the employee's hand can reach across the distance from the guard to the danger zone. This type of device allows freer movement for the operator, is simple to use, can be used by multiple operators, provides passerby protection, and requires no adjustment. It does not protect against mechanical failure and is limited to machines that can be stopped.

The *radiofrequency (capacitance) presence-sensing device* uses a radio beam as part of the machine control circuit. When the capacitance field is broken, the machine stops or will not activate. Like the photoelectric device, this device is only to be used on machines that can be stopped before the worker can reach the danger area. This requires the machine to have a friction clutch or other reliable means for stopping. This device allows freer movement for the operator but does not protect against mechanical failure. In addition, antennae sensitivity must be properly adjusted; this adjustment must be maintained properly.

The *electromechanical presence-sensing device* has a probe or contact bar that descends to a predetermined distance when the operator initiates the machine cycle. If an obstruction prevents it from descending its full predetermined distance, the control circuit does not actuate the machine cycle. This device allows for access at point of operation, but the contact bar or probe must be properly adjusted for each application. This adjustment must be maintained properly.

Pullback devices use cables attached to the operator's hands, wrists, or arms to prevent hands from entering the point of operation. This type of device is primarily used on machines with stroking action. When the slide or ram is up between cycles, the operator is allowed access to the point of operation. When the slide or ram begins to cycle by starting its descent, a mechanical linkage automatically ensures withdrawal of the hands from the point of operation. This type of device eliminates the need for auxiliary barriers or other interference at the danger area; however, it limits movement of the operator and may obstruct workspace around the operator.

The *restraint (holdback) device* uses cables or straps that are attached to the operator's hands at a fixed point. The cables or straps are adjusted to let the operator's hands travel within a predetermined safe area, with no extending or retracting action involved; consequently, hand-feeding tools are often necessary if the operation involves placing material into the danger area. Because restraints prevent the operator from reaching into the danger area there is little risk of danger. Adjustments must be made, however, for specific operations and for each individual. Also, frequent inspections and regular maintenance are required, close supervision of the operator's use of the equipment is required, movement of the operator is limited, work space may be obstructed, and adjustments must be made for specific operations and each individual.

Safety controls use involvement of the operator as a safeguarding method, and include safety trip controls, two-hand controls, and two-hand trips. For safety trip controls, if the operator or anyone trips, loses their balance, or is drawn toward a machine, the resulting pressure applied to a bar installed on the machine will stop the operation. The positioning of the bar, therefore, is critical. It must stop the machine before a part of the employee's body reaches the danger area. Whereas safety trip controls offer simplicity of use, they must still be manually activated, which may be difficult because of their location. Safety trip controls work to protect only the operator. They may require special fixtures to hold work and often require a machine braking mechanism.

Another type of safety control is the *two-hand control*, which requires constant concurrent pressure by the operator to activate the machine. This kind of control requires a part-revolution clutch, brake, and a brake monitor if used on a power press. With this type of device, the operator's hands must be at a safe location (on control buttons) and at a safe distance from the danger area while the machine completes its closing cycle. The advantages of this type of safety control are that the operator's hands are at a predetermined location and that the operator's hands are free to pick up a new part after the first half of a cycle is completed. Some two-handed controls, however, can be rendered unsafe by using an arm on the controls or blocking, thereby permitting one-hand operation. The safety control only protects the operator.

The *two-hand trip* requires concurrent application of both the operator's control buttons to activate the machine cycle, after which the hands are free. This device is usually used with machines equipped with full-revolution clutches. The trips must be placed far enough from the point of operation to make it impossible for operators to move their hands from the trip buttons or handles into the point of operation before the first half of the cycle is completed. The distance from the trip button depends upon the speed of the cycle and the band speed constant, so the operator's hands are kept far enough away to prevent them from being placed in the danger area prior to the slide/ram or blade reaching the full "down" position. The two-hand trip offers the advantages of keeping the operator's hands away from the danger area, its ability to be adapted to multiple operations, that it presents no obstruction to hand feeding, and that it does not require adjustment for each operation. The operator, however, may try to reach into a danger area after tripping the machine, and some trips can be rendered unsafe by holding with an arm or blocking, thereby permitting one-hand operation. To be effective, both two-hand controls and trips must be located so that the operator cannot use one hand and another part of his or her body to trip the machine.

Gates can also provide a high degree of protection to both the operator and other workers in the area. A gate is a movable barrier that protects the operator at the point of operation before the machine cycle can be started. Gates, in many instances, are designed to be operated with each machine cycle. To be effective, a gated horizontal injection-molding machine must be interlocked so the machine will not begin a cycle unless the gate guard is in place. It must be in the closed position before the machine can function. The main advantage of using gates is that they prevent reaching into or walking into the danger area; however, gates may require frequent inspection and regular maintenance, and they may interfere with the operator's ability to see the work.

9.17.11.1.4.3 Feeding and Ejection Methods Automatic and semiautomatic feeding and ejection of parts are other ways of safeguarding machine processes. These methods eliminate the need for the operator to work at the point of operation. In some situations, no operator involvement is necessary after the machine is set up. In other cases, operators can manually feed the stock with the assistance of a feeding mechanism. Properly designed ejection methods do not require any operator involvement after the machine starts to function. Note that using these feeding and ejection methods does not eliminate the need for guards and devices. Guards and devices must be used wherever they are necessary and possible to provide protection from exposure to hazards.

9.17.11.1.4.4 Safeguarding by Location and Distance Location and distance can also be used to safeguard machinery. A thorough hazard analysis of each machine and particular situation is absolutely essential before attempting this safeguarding technique. To consider a part of a machine to be safeguarded by location, the dangerous moving part of a machine must be so positioned that those areas are not accessible or do not present a hazard to a worker during the normal operation of the machine. This may be accomplished by positioning a machine so that the hazardous parts of the machine are located away from operator workstations or areas where employees walk or work; for example, a machine could be positioned with its power transmission apparatus against a wall, leaving all routine operations to be conducted on the other side of the machine. Enclosure walls or fences could restrict access to machines. Another possible solution is to have dangerous parts located high enough to be out of the normal reach of any worker. The feeding process can be safeguarded by location, if a safe distance can be maintained to protect the worker's hands.

The dimensions of the stock being worked on may provide adequate safety; for example, if the stock is several feet long and only one end of the stock is being worked on, the operator may be able to hold the opposite end while the work is being performed. An example of this would be a single-end punching machine. Depending on the machine, protection might still be required for other personnel. The positioning of the operator's control station provides another potential approach to safeguarding by location. Operator controls may be located at a safe distance from the machine if there is no reason for the operator to tend it.

9.17.11.1.4.5 Miscellaneous Safeguarding Accessories A variety of methods and tools can be used to help lower the hazard potential created by certain machines, even though they do not provide full or complete machine safeguarding. Note that sound judgment is necessary for their application and usage:

- *Awareness barriers* may be used. Though the barrier does not physically prevent a person from entering the danger area, it calls attention to it. For an employee to enter the danger area an overt act must take place; the employee must reach or step over, under, or through the barrier.
- *Shields* may be used to provide protection from flying particles, splashing cutting oils, or coolants.
- *Special devices or hand tools* for placing objects in power presses allow the operator's hands and arms to remain away from the point of operation.
- *Push sticks/blocks and jigs* allow employees to keep their hands at a safe location when guiding wood or other materials during joiner and shaper operations.
- *Spreaders and nonkickback devices* help prevent work from being thrown back at the operator, particularly with woodworking machines such as circular and radial saws.

9.17.11.1.4.6 Machine Guarding: Training, Enforcement, and Inspections As with all other safety programs, training is at the heart of the safety effort, because even the most elaborate safeguarding system and precise step-by-step safe work practices cannot offer effective protection unless the worker knows how to use it and why. Specific and detailed training is therefore a crucial element of any effort to provide safeguarding against machine-related hazards. Thorough operator training should involve instructions and hands-on training should include the following:

1. Description and identification of the hazards associated with particular machines
2. The various safeguards, how they provide protection, and the hazards for which they are intended
3. How to use the safeguards and why
4. How and under what circumstances safeguards can be removed and by whom (in most cases, repair or maintenance personnel only)
5. What to do (e.g., contact the supervisor) if a safeguard is damaged, missing, or unable to provide adequate protection

This kind of safety training is necessary for new operators and maintenance or setup personnel, when any new or altered safeguards are put in service, or when workers are assigned to a new machine or operation. Properly installed machinery safeguards, well-written safe work practices, and a strong training program are all important elements of a company's Machine

Guarding Safety Program; however, if employees are allowed to overtly disregard company safe work practices and rules, the Machine Guarding Safety Program is worthless. Enforcement of safety rules and safe work practices is required. Though the safety official is not normally associated with disciplinary action, he or she must take an active role in enforcing company safety policies; likewise, the safety official must ensure that supervisors and workers alike understand the importance of company safety policies, rules, regulations, and safe work practices—and, more importantly, that they will be strictly enforced.

Machinery safety guards must be periodically inspected and maintained to ensure their integrity—to ensure that they are in place and working as designed, to ensure that they are continually effective, and to ensure that they have not been tampered with or bypassed in any way. Generally, machinery safety guards are inspected throughout the company's preventive maintenance program checks. Whether discovered through a scheduled maintenance or while in operation, broken or inoperable parts must be replaced; however, good engineering practice dictates that machine safety guards should be inspected before and after each use—to ensure their operability. To aid the safety official in inspecting his or her workplace machinery to determine the safeguarding needs of his or her own workplace, OSHA (1992) has provided a Machine Guarding Checklist. Answers to the following questions should help the interested reader determine the safeguarding needs of his or her own workplace, by drawing attention to hazardous conditions or practices requiring correction.

Requirements for All Safeguards

Yes	No	
❑	❑	1. Do the safeguards provided meet the minimum requirements?
❑	❑	2. Do the safeguards prevent workers' hands, arms, and other body parts from making contact with dangerous moving parts?
❑	❑	3. Are the safeguards firmly secured and not easily removable?
❑	❑	4. Do the safeguards ensure that no object will fall into the moving parts?
❑	❑	5. Do the safeguards permit safe, comfortable, and relatively easy operation of the machine?
❑	❑	6. Can the machine be oiled without removing the safeguard?
❑	❑	7. Is there a system for shutting down the machinery before safeguards are removed?
❑	❑	8. Can the existing safeguards be improved?

Point-of-Operation Hazards

Yes No

❑ ❑ 1. Does the machine have a point-of-operation safeguard?

❑ ❑ 2. Does it keep the operator's hands, fingers, and body out of the danger area?

❑ ❑ 3. Is there evidence that the safeguards have been tampered with or removed?

❑ ❑ 4. Could you suggest a more practical, effective safeguard?

❑ ❑ 5. Could changes be made on the machine to eliminate the point-of-operation hazard entirely?

Power Transmission Apparatus

❑ ❑ 1. Are there any unguarded gears, sprockets, pulleys, or flywheels on the apparatus?

❑ ❑ 2. Are there any exposed belts or chain drives?

❑ ❑ 3. Are there any exposed set screws, key ways, collars, etc.?

❑ ❑ 4. Are starting and stopping controls within easy reach of the operator?

❑ ❑ 5. If there is more than one operator, are separate controls provided?

Other Moving Parts

❑ ❑ 1. Are safeguards provided for all hazardous moving parts of the machine, including auxiliary parts?

Nonmechanical Hazards

❑ ❑ 1. Have appropriate measures been taken to safeguard workers against noise hazards?

❑ ❑ 2. Have special guards, enclosures, or personal protective equipment been provided where necessary to protect workers from exposure to harmful substances used in machine operation?

Electrical Hazards

❑ ❑ 1. Is the machine installed in accordance with National Fire Protection Association and National Electrical Code requirements?

❑ ❑ 2. Are there loose conduit fittings?

❑ ❑ 3. Is the machine properly grounded?

❑ ❑ 4. Is the power supply correctly fused and protected?

❑ ❑ 5. Do workers occasionally receive minor shocks while operating any of the machines?

Training

Yes No

❑ ❑ 1. Do operators and maintenance workers have the necessary training in how to use the safeguards and why they are used?

❑ ❑ 2. Have operators and maintenance workers been trained in where the safeguards are located, how they provide protection, and what hazards they protect against?

❑ ❑ 3. Have operators and maintenance workers been trained in how and under what circumstances guards can be removed?

❑ ❑ 4. Have workers been trained in the procedures to follow if they notice guards that are damaged, missing, or inadequate?

Protective Equipment and Proper Clothing

❑ ❑ 1. Is protective equipment required?

❑ ❑ 2. If protective equipment is required, is it appropriate for the job, in good condition, kept clean and sanitary, and stored carefully when not in use?

❑ ❑ 3. Where several maintenance persons work on the same machine, are multiple lockout devices used?

❑ ❑ 4. Do maintenance persons use appropriate and safe equipment in their repair work?

❑ ❑ 5. Is the maintenance equipment itself properly guarded?

❑ ❑ 6. Are maintenance and servicing workers trained in the requirements of 29 CFR 1910.147, and do the procedures for lockout/tagout exist *before* they attempt their tasks?

9.17.11.1.4.7 Machine Hazard Warnings One or more warnings are needed on a machine to communicate hazards that may be present. Machine hazard warnings are of several different types. Hazard signs and/or labels use *signal words* such as DANGER, WARNING, or CAUTION. Danger signs indicate an imminently hazardous situation, which if not avoided could result in death or serious injury. Warning signs indicate a potentially hazardous situation, which if not avoided could result in death or serious injury. Caution signs indicate that a hazard may result in moderate or minor injury. Warning signs with the appropriate signal words are often used to indicate dangerous or hazardous conditions:

- KEEP HANDS OUT OF MACHINERY
- EYE PROTECTION REQUIRED IN THIS AREA

Danger signs include the following:

- DANGER: PINCH POINTS! WATCH YOUR HANDS.
- DANGER: THIS MACHINE HAS NO BRAIN. USE YOUR OWN.
- DANGER: THIS MACHINE CYCLES.
- DANGER: THIS MACHINE STARTS AUTOMATICALLY.

Another kind of sign is often used—the notice sign; however, notice signs are used to state a company policy and should not be associated directly with a hazard or hazardous situation. They must not be used in place of DANGER, WARNING, or CAUTION. If a machine has guards, the warnings should include a notice to keep guards in place and not to operate the machine without them. On guard devices, the warnings should state any hazards, any limitations the device may have, and protective actions the operator must take.

9.17.11.1.4.8 Employee Clothing and Jewelry Engineering controls that eliminate the hazard at the source and do not rely on the worker's behavior for their effectiveness offer the best and most reliable means of safeguarding. Engineering controls must be the employer's first choice for eliminating machine hazards, but whenever engineering controls are not available or are not fully capable of protecting the employee then an extra measure of protection is necessary. Operators must wear protective clothing or personal protective equipment. Note that it is management's responsibility to ensure that employees wear appropriate clothing when operating or working around hazardous machines. If it is to provide adequate protection, the protective clothing and equipment selected must always be

- Appropriate for the particular hazards
- Maintained in good condition
- Properly stored to prevent damage or loss when not in use
- Kept clean, fully functional, and sanitary

Protective clothing and equipment can create their own hazards. Protective gloves can become caught between rotating parts, or a respirator facepiece can hinder the wearer's vision. The use of such protective equipment requires alertness and continued attentiveness.

Other parts of the worker's clothing may present additional safety hazards; for example, loose-fitting, oversized clothing might possibly become entangled in rotating spindles or other kinds of moving machinery. Rings, bracelets, or watchbands can catch on machine parts or stock and lead to serious injury by pulling a hand into the danger area. Employees with long hair may need to wear hats or hair nets if the long hair represents a hazard because of proximity to moving machinery.

9.17.11.1.4.9 Lockout/Tagout Setup, maintenance, and servicing of machinery often requires that existing safeguarding be removed or disengaged to provide access to machine parts. At such times, the machine should be locked out and tagged out of service to prevent anyone from activating it while someone else expects it to be de-energized.

9.17.12 Subpart P, Hand and Portable Powered Tools and Other Hand-Held Equipment

1910.242 Hand and portable power tools and equipment

9.17.12.1 Safe Work Practice for Hand Tools, Power Tools and Portable Power Equipment

1. Use care and caution when using hand tools, power tools, and portable power equipment.
2. Do not use tools and equipment unless trained and experienced in the proper use and operation of the tools and equipment.
3. Use the proper tools and equipment for the required task. *Never* use tools or equipment in a misapplication.
4. Inspect tools carefully before using them and discard any tool that appears unsafe.
5. Use care and caution when using tools with sharp points or edges such as saws, knives, chisels, punches, and screwdrivers. Hand tools of this type are not to be set down on surfaces where they can be tripped over, stepped on, or bumped.

DID YOU KNOW?

OSHA 29 CFR 1910.215, Abrasive Wheel Machinery, addresses the danger of and the need for safe work practices associated with all abrasive wheel machinery. An abrasive wheel is a cutting tool consisting of abrasive grains held together by organic or inorganic bonds. Diamond and reinforced wheels are included. *Note*: Immediately before an abrasive wheel is mounted to any machine normally equipped with such devices, the user should thoroughly inspect it and perform a ring test before it is mounted to make sure no damage occurred in transit, storage, or otherwise. Before using the ring test to sound an abrasive wheel, be sure that the wheel is dry and free from sawdust; otherwise, the sound will be deadened. Organic bonded wheels do not make the same clear metallic ring when struck by a handheld metallic tool as do vitrified and silicate wheels.

6. Use equipment guards and other safety devices at all times when operating tools and equipment. *Never* bypass a safety guard or switch.
7. Use safety glasses, goggles, and face shields as appropriate.
8. Inspect tools on a regular basis and before each use to ensure that tools and equipment are in good working order.
9. Keep tools and power equipment clean and in good operating condition. *Never* use broken hand tools and power tools.
10. Replace worn-out tools and equipment.
11. Use only grounded or double-insulated electrical tools.
12. Never use electrical tools in or near water without a ground fault interrupter circuit. Never stand in water when using an electrical tool or equipment.
13. Have frayed or broken electrical cords repaired or replaced immediately.
14. Shut off gasoline or diesel engines before refueling whenever possible.
15. Direct exhaust fumes from gasoline or diesel engines away from work areas.
16. Apply working force away from the body to minimize the chance for injury if the hand tool slips.
17. Be sure tool handles are fitted to tools and free of grease and other slippery substances.
18. Dress cold chisels, punches, hammers, drift pins, and other similar tools that have a tendency to mushroom from repeated poundings. As soon as they begin to crack and curl, grind a slight bevel (approximately 3/16 inch or 4.7 mm) around the head to prevent it from mushrooming.
19. Do not carry sharp edges or pointed tools in clothing pockets.
20. Do not use defective wrenches, such as open-end and adjustable wrenches with spur jaws or pipe wrenches with dull teeth.
21. Do not apply hand tools to moving machinery except tools designed for the purpose and necessary in the operation.
22. Do not throw tools and material from one employee to another, or from one location to another. Use a suitable container to raise or lower small equipment or tools between elevations.

9.17.13 Subpart Q, Welding, Cutting, and Brazing

1910.252 General Requirements (Hot Work Permit)

1910.253 Oxygen–Fuel Gas Welding and Cutting

9.17.13.1 Hot Work Permit Procedure

> Many organizations use a permit procedure for all hot work, except that involving normal operations or processes. Hot work is any kind of welding, cutting, burning, or activity that involves or generates sparks or open flame. It includes heated equipment that may provide an ignition source for a fire. Hot work often involves people from a maintenance department going to other departments to perform activities. The main idea in a hot work permit procedure is to ensure that supervisors of all departments involved and workers who may be involved in any way in the work participate in the decision to start work and to conduct it safely.
>
> **—Brauer (1996)**

In this section, we discuss an important procedure applicable to ensuring that confined space operations are conducted safely: hot work permit procedures. In addition to ensuring that any type of hot work to be performed in confined spaces is accomplished in a safe manner by utilizing hot work permit requirements, other workplace operations might require the use of hot work permit procedures.

In the performance of hot work in the workplace, OSHA 29 CFR 1910.119, Process Safety Management of Highly Hazardous Chemicals, requires the following: The employer shall issue a hot work permit for hot work operations conducted on or near a covered process (including confined spaces). The permit should document that the fire prevention and protection requirements in 29 CFR 1910.252(a), Fire Prevention and Protection, have been implemented prior to beginning the hot work operations; it should indicate the dates authorized for the hot work; and it should identify the object on which hot work is to be performed. The permit must be kept on file until completion of the hot work operations.

Under 29 CFR 1910.119, any time hot work is to be performed on, near, or around covered chemical processes, a hot work permit must be used. Many companies require the use of hot work permits any time hot work is to be performed anywhere within the organization, outside normal operations and processes. "Normal operations and processes" might be defined as work normally performed in a welding, brazing, or hot torch cutting shop or hot work performed as part of a assembly line process, such as that conducted by robots on automobile assembly lines. "Outside normal operations and processes" might be described as performing hot work in work areas where hot work is not typically performed—for example, in office, storage, or production areas.

Typically, the organizational safety official is responsible for implementing and managing the hot work permitting procedure. The primary elements required to be incorporated into a viable hot work permit system include a standard operating procedure consisting of (1) a written procedure, (2) a permit, (3) worker training, and (4) fire watch provisions.

Exactly what is accomplished by employing the use of a hot work permitting system? A hot work permitting procedure works primarily to ensure that work areas and all adjacent areas to which sparks and heat might be spread (including floors above and below and on opposite sides of walls) are inspected during the work and again 30 minutes after the work is completed, to ensure they are firesafe. During the inspection, work areas and surrounding areas should be inspected to ensure that

- Sprinklers are in service.
- Cutting and welding equipment is in good repair.
- Floors are swept clean of combustibles.
- Combustible floors are wetted down and covered with damp sand, metal, or other shields.
- No combustible material or flammable liquids are within 35 feet of the work.
- Combustibles and flammable liquids within 35 feet of work are protected with covers, guards, or metal shields.
- All wall and floor openings within 35 feet of work are covered.
- Covers are suspended beneath the work to collect sparks.
- For work on walls or ceilings, construction is of noncombustible materials.
- Combustibles are moved away from the opposite side of the wall.
- For work on or in enclosed tanks, containers, ducts, etc., equipment is cleaned of all combustibles and purged of flammable vapors.
- Fire watch is provided during and 30 minutes after operation.
- The assigned fire watch is properly trained and equipped.

9.17.13.1.1 Fire Watch Requirements

A fire watch must be assigned whenever hot work operations are being performed around hazardous materials, in confined spaces, and other times when there is the danger of fire or explosion from such work. OSHA has specific requirements regarding fire watch duties. Fire watchers are required whenever welding or cutting is performed in locations where other than a minor fire might develop or where any of the following conditions exist:

1. Appreciable combustible material, in building construction or contents, are closer than 35 feet (10.7 m) to the point of operation.
2. Appreciable combustibles are more than 35 feet (10.7 m) away but are easily ignited by sparks.

3. Wall or floor openings within a 35-foot (10.7-m) radius expose combustible materials in adjacent areas, including concealed spaces in walls or floors.
4. Combustible materials are adjacent to the opposite side of metal partitions, walls, ceilings, or roofs and are likely to be ignited by conduction or radiation.

Fire watchers must have fire-extinguishing equipment readily available and be trained in its use. They should be familiar with facilities for sounding an alarm in the event of a fire. They should watch for fires in all exposed areas and try to extinguish them only when obviously within the capacity of the equipment available; otherwise, they should sound the alarm. A fire watch should be maintained for at least a half hour after completion of welding or cutting operations to detect and extinguish possible smoldering fires.

9.17.13.2 Welding and Cutting Safety

Welding is typically thought of as the electric arc and gas (fuel gas/oxygen) welding process; however, welding can involve many types of processes. Some of these other processes include inductive welding, thermite welding, flash welding, percussive welding, plasma welding, and others. The most common type of electric arc welding also has many variants, including gas shielded welding, metal arc welding, gas–metal arc welding, gas–tungsten arc welding, and flux cored arc welding (McElroy, 1980). Welding, cutting, and brazing are widely used processes. 29 CFR 1910 Subpart Q contains the standards relating to these processes in all of their various forms. The primary health and safety concerns are fire protection, employee personal protection, and ventilation. The standards contained in this subpart are as follows:

1910.251 Definitions
1910.252 General Requirements
1910.253 Oxygen–Fuel Gas Welding and Cutting
1910.254 Arc Welding and Cutting
1910.255 Resistance Welding
1910.256 Sources of Standards
1910.257 Standards Organization

A study on deaths related to welding/cutting incidents (OSHA, 1989) revealed that of 200 deaths over an 11-year period, 80% were caused by failure to practice safe work procedures. Surprisingly, only 11% of deaths involved malfunctioning or failed equipment, and only 4% were related to environmental factors. The implications of this study should be obvious:

Equipment malfunctions or failures are not the primary causal factor of hazards presented to workers. Instead, the safety official's emphasis should be on establishing and ensuring safe work practices for welding tasks.

9.17.13.2.1 General Welding Safety

A viable Welding Safety Program should consist of the elements provided in 29 CFR 1910.252, Welding, Cutting, and Brazing. The fire prevention and protection element of any welding safety program begins with basic precautions, including the following:

1. *Fire hazards*—If the material or object cannot be readily moved, all movable fire hazards in the area must be moved to a safe location.
2. *Guards*—If the object to be welded or cut cannot be moved, and if all the fire hazards cannot be removed, then guards are to be used to confine the heat, sparks, and slag and to protect the immovable fire hazards.
3. *Restrictions*—If the welding or cutting cannot be performed without removing or guarding against fire hazards, then the welding and cutting should not be performed.
4. *Combustible material*—Wherever floor openings or cracks in the flooring cannot be closed, precautions must be taken so that no readily combustible materials on the floor below will be exposed to sparks that might drop through the floor. The same precautions should be taken with cracks or holes in walls, open doorways, and open or broken windows.
5. *Fire extinguishers*—Suitable fire extinguishing equipment must be maintained in a state of readiness for instant use. Such equipment may consist of pails of water, buckets of sand, hoses, or portable extinguishers, depending on the nature and quantity of the combustible material exposed.
6. *Fire watch*—Fire watchers are required whenever welding or cutting is performed in locations where other than a minor fire might develop. Fire watchers are required to have fire-extinguishing equipment readily available and must be trained in its use. They must be familiar with facilities for sounding an alarm in the event of fire. They must watch for fires in all exposed areas, try to extinguish them only when obviously within the capacity of the equipment available, or otherwise sound the alarm. A fire watch must be maintained for at least a half-hour after completion of welding or cutting operations to detect and extinguish possible smoldering fires.
7. *Authorization*—Before cutting or welding is permitted, the individual responsible for authorizing cutting and welding operations must inspect the area. The responsible individual must designate

precautions to be followed in granting authorization to proceed preferably in the form of a written permit (hot work permit).

8. *Floors*—Where combustible materials such as paper clippings, wood shavings, or textile fibers are on the floor, the floor must be swept clean for a radius of at least 35 feet (OSHA requirement). Combustible floors must be kept wet, covered with damp sand, or protected by fire-resistant shields. Where floors have been wet down, personnel operating arc welding or cutting equipment must be protected from possible shock.
9. *Prohibited areas*—Welding or cutting must not be permitted in areas that are not authorized by management. Such areas include in sprinklered buildings while such protection is impaired; in the presence of explosive atmospheres, or explosive atmospheres that may develop inside uncleaned or improperly prepared tanks or equipment that have previously contained such materials, or that may develop in areas with an accumulation of combustible dusts; and in areas near the storage of large quantities of exposed, readily ignitable materials such as bulk sulfur, baled paper, or cotton.
10. *Relocation of combustibles*—Where practicable, all combustibles must be relocated at least 35 feet from the work site. Where relocation is impracticable, combustibles must be protected with fireproofed covers, or otherwise shielded with metal of fire-resistant guards or curtains.
11. *Ducts*—Ducts and conveyor systems that might carry sparks to distant combustibles must be suitably protected or shut down.
12. *Combustible walls*—Where cutting or welding is done near walls, partitions, ceilings, or roofs of combustible construction, fire-resistant shields or guards must be provided to prevent ignition.
13. *Noncombustible walls*—If welding is to be done on a metal wall, partition, ceiling, or roof, precautions must be taken to prevent ignition of combustibles on the other side from conduction or radiation, preferably by relocating the combustibles. Where combustibles are not relocated, a fire watch on the opposite side from the work must be provided.
14. *Combustible cover*—Welding must not be attempted on a metal partition wall, ceiling, or roof that has combustible coverings, nor on any walls or partitions, ceilings, or roofs that have combustible coverings or on walls or partitions of combustible sandwich-type panel construction.
15. *Pipes*—Cutting or welding on pipes or other metal in contact with combustible walls, partitions, ceilings, or roofs must not be undertaken if the work is close enough to cause ignition by conduction.

16. *Management*—Management must recognize its responsibility for the safe usage of cutting and welding equipment on its property, must establish areas for cutting and welding, and must establish procedures for cutting and welding in other areas. Management must also designate an individual responsible for authorizing cutting and welding operations in areas not specifically designed for such processes. Management must also insist that cutters or welders and their supervisors are suitably trained in the safe operation of their equipment, and the safe use of the process. Management has a duty to inform contractors about flammable materials or hazardous conditions of which they may not be aware.
17. *Supervisor*—The supervisor has many responsibilities in welding and cutting operations, including the following:
    - Is responsible for the safe handling of the cutting or welding equipment and the safe use of the cutting or welding process.
    - Must determine the combustible materials and hazardous area present or likely to be present in the work location.
    - Must protect combustibles from ignition by whatever means necessary.
    - Must secure authorization for the cutting or welding operations from the designated management representative.
    - Must ensure that the welder or cutter secures his or her approval that conditions are safe before going ahead.
    - Must determine that fire protection and extinguishing equipment are properly located at the site.
    - Where fire watches are required, must ensure that they are available at the site.
18. *Fire prevention precautions*—Cutting and welding must be restricted to areas that are or have been made fire safe. When work cannot be move practically, as in most construction work, the area must be made safe by removing combustibles or protecting combustibles from ignition sources.
19. *Welding and cutting used containers*—No welding, cutting, or other hot work is to be performed on used drums, barrels, tanks, or other containers until they have been cleaned so thoroughly as to make absolutely certain that no flammable materials are present, or any substances such as greases, tars, acids, or other materials that when subjected to heat might produce flammable or toxic vapors. Any pipelines or connections to the drum or vessel must be disconnected or blanked.

20. *Venting and purging*—All hollow spaces, cavities, or containers must be vented to permit the escape of air or gases before preheating, cutting, or welding. Purging with inert gas (e.g., nitrogen) is recommended.
21. *Confined spaces*—To prevent accidental contact in confined space operations involving hot work, when arc welding is to be suspended for any substantial period of time (such as during breaks or overnight), all electrodes are to be removed from the holders and the holders carefully located so that accidental contact cannot occur. The machine must be disconnected from the power source. To eliminate the possibility of gas escaping through leaks or improperly closed valves, when gas welding or cutting, the torch valves must be closed and the gas supply to the torch positively shut off at some point outside the confined area whenever the torch is not to be used for a substantial period of time (such as during breaks or overnight). Where practicable, the torch and hose must also be removed from the confined space.

Note: The safety official should use the proceeding information as guidance in preparing the organizational Welding Safety Program.

9.17.13.2.2 Personal Protective Equipment and Other Protection

Personnel involved in welding or cutting operations not only must learn and abide by safe work practices but must also be aware of possible bodily dangers during such operations. They must learn about the personal protective equipment (PPE) and other protective devices and measures designed to protect them.

1. *Railing and welding cable*—A welder or helper working on platforms, scaffolds, or runways must be protected against falling. This may be accomplished by the use of railings, safety harnesses, lifelines, or other equally effective safeguards. Welders must place welding cable and other equipment so that it is clear of passageways, ladders, and stairways.
2. *Eye protection*—Helmets or hand shields must be used during all arc welding or arc cutting operations (excluding submerged operations). Helpers or attendants must be provided with the same level of proper eye protection. Goggles or other suitable eye protection must be used during all gas welding or oxygen cutting operations. Spectacles without side shields with suitable filter lenses are permitted for use during gas welding operations on light work, for torch brazing, or for inspection. Operators and attendants of resistance welding or resistance brazing equipment must use transparent face shields or goggles (depending on the particular job) to protect their

faces or eyes as required. Helmets and hand shields must meet certain specifications, including being made of a material that is an insulator for heat and electricity. Helmets, shields, and goggles must not be readily flammable and must be capable of sterilization. Helmets and hand shields must be so arranged as to protect the face, neck, and ears from direct radiant energy from the arc. Helmets must be provided with filter plates and cover plates designed for easy removal. All parts must be constructed of a material that will not readily corrode or discolor the skin. Goggles must be ventilated to prevent fogging of the lenses as much as possible. All glass for lenses must be tempered and substantially free from striae, air bubbles, waves, and other flaws. Except when a lens is ground to provide proper optical correction for defective vision, the front and rear surfaces of lenses and windows must be smooth and parallel. Lenses must also bear some permanent distinctive marking by which the source and shade may be readily identified. Table 9.16 provides a guide for the selection of proper shade numbers. These recommendations may be

TABLE 9.16

Construction Industry Requirements for Filter Lens Shade Numbers for Protection Against Radiant Energy (29 CFR 1910.252)

Welding Operation	Shade Number
Shielded metal-arc welding 1/16-, 3/32-, 1/8-, 5/32-inch-diameter electrodes	10
Gas-shielded arc welding (nonferrous) 1/16-, 3/32-, 1/8-, 5/32-inch-diameter electrodes	11
Gas-shielded arc welding (ferrous) 1/16-, 3/32-, 1/8-, 5/32-inch-diameter electrodes	12
Shielded metal-arc welding 3/16-, 7/32-, 1/4-inch-diameter electrodes	12
Shielded metal-arc welding 5/16-, 3/8-inch-diameter electrodes	14
Atomic hydrogen welding	10–14
Carbon arc welding	14
Soldering	14
Torch brazing	2
Light cutting, up to 1 inch	3 or 4
Medium cutting, 1 inch to 6 inches	4 or 5
Heavy cutting, 6 inches and over	5 or 6
Gas welding (light) up to 1/8 inch	4 or 5
Gas welding (medium) 1/8 to 1/2 inch	5 or 6
Gas welding (heavy) 1/2 inch and over	6 or 8

Note: In gas welding or oxygen cutting where the torch produces a high yellow light, use a filter or lens that absorbs the yellow or sodium line in the visible light of the operation.

varied to suit the individual's needs. All filter lenses and plates must meet the test for transmission of radiant energy prescribed in ANSI Z87.1-1968, American National Standard Practice for Occupational and Educational Eye and Face Protection. Where the work permits, the welder should be enclosed in an individual booth painted with a finish of low reflectivity (such as zinc oxide and lamp black) or must be enclosed with noncombustible screens similarly painted. Booths and screens must permit circulation of air at floor level. Workers or other persons adjacent to the welding areas must be protected from the rays by noncombustible or flameproof screens or shields or must be required to wear appropriate eye protection.

3. *Protective clothing*—Employees exposed to the hazards created by welding, cutting, or brazing operations must be protected by personal protective equipment, including appropriate protective clothing required for any welding operation.
4. *Confined spaces*—For welding or cutting operations conducted in confined spaces (i.e., in spaces that are relatively small or restricted spaces, such as tanks, boilers, pressure vessels, or small compartments of a ship), personal protective and other safety equipment must be provided. Protection of personnel performing hot work in confined spaces includes adhering to the following:
   - Gas cylinders and welding machines must be left on the outside and secured to prevent movement.
   - Where a welder must enter a confined space through a manhole or other small opening, means (e.g., lifelines) must be provided for quickly removing that person in case of emergency.
   - When arc welding is to be suspended for any substantial period of time, all electrodes must be removed from the holds, the holders carefully located so that accidental contact cannot occur, and the machine disconnected from the power source
   - To eliminate the possibility of gas escaping through leaks of improperly closed valves, when performing gas welding or cutting, the torch valves must be closed and the fuel-gas and oxygen supply to the torch positively shut off at some point outside the confined area whenever the torch is not to be used for a substantial period of time.
   - After welding operations are completed, the welder must mark the hot metal or provide some other means of warning others.

9.17.13.2.3 Ventilation and Health Protection

All welding should be accomplished in well-ventilated areas. There must be sufficient movement of air to prevent accumulation of toxic fumes or possible oxygen deficiency. Adequate ventilation becomes extremely critical in confined spaces where dangerous fumes, smoke, and dust are likely to collect. Where considerable hot work is to be performed, an exhaust system is necessary to keep toxic gases below the prescribed health limits. An adequate exhaust system is especially necessary when hot work is performed on zinc, brass, bronze, lead, cadmium, or beryllium-bearing metals. This also includes galvanized steel and metal painted with lead-bearing paint. Fumes from these materials are toxic—they are very hazardous to health. What does OSHA require for ventilation for hot work operations?

Ventilation must be provided when:

- Hot work is performed in a space of less than 10,000 cubic feet per welder.
- Hot work is performed in a room having a ceiling height of less than 16 feet.
- Hot work is performed in confined spaces where the hot work space contains partitions, balconies, or other structural barriers to the extent that they significantly obstruct cross ventilation.

The minimum rate of ventilation must be as follows:

- The minimum rate of 2000 cubic feet per minute per welder, except where local exhaust hoods and booths are provided, or where approved airline respirators are provided.

9.17.13.2.4 Arc Welding Safety

In 29 CFR 1910.254, Arc Welding and Cutting, OSHA specifically lists various safety requirements that must be followed when arc welding; for example, in equipment selection, OSHA stipulates that welding equipment must be chosen for safe application to the work to be done. Welding equipment must also be installed safely as per the manufacturer's guidelines and recommendations. Finally, OSHA specifies that workpersons designated to operate arc-welding equipment must have been properly trained and qualified to operate such equipment. Training and qualification procedures are important elements that must be included in any Welding Safety Program. Along with OSHA's requirements above, the safety official must ensure that the facility's Welding Safety Program includes written safe work practices detailing and explaining safety requirements that must be followed whenever arc welding is performed. In the following section, we summarize OSHA and industry requirements and recommendations for performing arc-welding operations safely.

9.17.13.2.4.1 Arc Welding Safe Work Practices Arc welding includes shielded metal arc, inert gas shielded arc, and resistance welding. In the following safe work practices, only general safety measures are indicated for these areas, because arc-welding equipment varies considerably in size and type. Equipment may range from a small portable shielded metal-arc welder to highly mechanized production spot or gas shielded arc welders. In each instance, specific manufacturer's recommendations should be followed. Along with OSHA requirements, the following work practices include safety practices that are generally common to all types of arc welding operations:

1. Be sure that all welding equipment is installed according to provisions of the National Electrical Code (NEC) and regulatory bodies.
2. Be sure that the welding machine is equipped with a power disconnect switch, conveniently located at or near the machine so the power can be shut off quickly.
3. Be sure that the range switch is not operated under load. The range switch, which provides the current setting, should be operated only while the machine is idling and the current is open. Switching the current while the machine is under a load will cause an arc to form between the contact surfaces.
4. Be sure that repairs to welding equipment are not made unless the power to the machine is shut OFF. The high voltage used for arc welding machines can inflict severe and fatal injuries.
5. Be sure that welding machines are properly grounded in accordance with the NEC. Stray current may develop which can cause severe shock when ungrounded parts are touched. Be sure that the ground to your work is securely attached. Grounds are not to be attached to pipelines carrying gases or flammable liquids.
6. Be sure that electrode holds do not have loose cable connections. Keep connections tight at all times. Avoid using electrode holders with defective jaws or poor insulation.
7. Be sure that the polarity switch is not changed when the machine is under a load. Wait until the machine idles and the circuit is open; otherwise, the contact surface of the switch may be burned and the person throwing the switch may receive a severe burn from the arcing.
8. Be sure that welding cables are not overloaded, and do not operate a machine with poor connections.
9. Be sure that welding is conducted in dry areas and that hands and clothing are dry.
10. Be sure that an arc is not struck whenever someone without proper eye protection is nearby.
11. Be sure that pieces of metal that have just been welded or heated are allowed to cool before picking them up.
12. Be sure to always wear protective safety glasses.
13. Be sure that hollow (cored) castings have been properly vented before welding.

14. Be sure that press-type welding machines are effectively guarded.
15. Be sure that suitable spark shields are used around equipment in flash welding.
16. When welding is completed, be sure to turn OFF the machine, pull the power disconnect switch, and hang the electrode holder in its designated place.
17. Be sure to inspect cables for cuts, nicks, or abrasion.

9.17.13.2.5 Gas Welding and Cutting

Specific safety requirements for oxygen–fuel gas welding and cutting are covered under 29 CFR 1910.253 and are listed in the units involving oxyacetylene welding. These safety requirements (precautions) cover proper handling of cylinders, operation of regulators, use of oxygen and acetylene, welding hose, testing for leaks, and lighting a torch. All of these safety requirements are extremely important and should be followed with the utmost care and regularity. Along with the normal precautions to be observed in gas welding operations, a very important safety procedure involves the piping of gas. All piping and fittings used to convey gases from a central supply system to work stations must withstand a minimum pressure of 150 psi. Oxygen piping can be of black steel, wrought iron, copper, or brass. Only oil-free compounds should be used on oxygen threaded connections. Piping for acetylene must be of wrought iron. (*Note:* Acetylene gas must never come into contact with unalloyed copper, except in a torch; any such contact could result in a violent explosion.) After assembly, all piping must be blown out with air or nitrogen to remove foreign materials. Five basic rules contribute to the safe handling of oxyacetylene equipment (Giachino and Weeks, 1985):

1. Keep oxyacetylene equipment clean, free of oil, and in good condition.
2. Avoid oxygen and acetylene leaks.
3. Open cylinder valves slowly.
4. Purge oxygen and acetylene lines before lighting a torch.
5. Keep heat, flame, and sparks away from combustibles.

9.17.13.2.6 Torch-Cutting Safety

Whenever torch-cutting operations are conducted, the possibility of fire is very real, because proper precautions are often not taken. Torch cutting is particularly dangerous because sparks and slag can travel several feet and can pass through cracks out of sight of the operator. The safety official must ensure that the persons responsible for supervising or performing cutting of any kind follow accepted safe work practices. Accepted safe work practices for torch cutting operations typically include the following:

1. Use of a cutting torch where sparks will be a hazard is prohibited.
2. If cutting is to be over a wooden floor, the floor must be swept clean and wet down before starting the cutting.
3. A fire extinguisher must be kept in reach any time torch-cutting operations are conducted.
4. Cutting operations should be performed in wide-open areas so sparks and slag will not become lodged in crevices or cracks.
5. In areas where flammable materials are stored and cannot be removed, suitable fire-resistant guards, partitions, or screens must be used.
6. Sparks and flame must be kept away from oxygen cylinders and hoses.
7. Never perform cutting near ventilators.
8. Fire watchers with fire extinguishers should be used.
9. Never use oxygen to dust off clothing or work.
10. Never substitute oxygen for compressed air.

9.17.14 Subpart S, Electrical

1910.302 Electric Utilization Systems
1910.303 General Requirements
1910.304 Wiring Design and Protection
1910.305 Wiring Methods, Components, and Equipment for General Use
1910.307 Hazardous (Classified) Locations
1910.332 Training
1910.333 Selection and Use of Work Practices
1910.334 Use of Equipment
1010.335 Safeguards to Personnel Protection

9.17.14.1 Electrical Safety

Note that noncompliance with various OSHA-required electrical safe work practices is one of the top ten violations listed in Table 9.12. The common use of electricity and electrical equipment and appliances has resulted in failure of most persons to appreciate the hazards involved. These hazards can be divided into five principal categories: (1) shock to personnel, (2) ignition of combustible (or explosive) materials, (3) overheating and damage to equipment, (4) electrical explosions, and (5) inadvertent activation of equipment (Hammer, 1989).

If you were to take a look at the annual on-the-job injury statistics for all employers in the United States, you would quickly notice that many of these injuries are typically the result of electrical shock, injuries received during electrical fires, or injuries received when some electrical component fails due to faulty installation, faulty maintenance conducted on electrical equipment, or equipment malfunction caused by manufacturer errors.

Although it is normally true that most workers fear electricity and its power, or at least have a healthy respect for electricity, it is also true that on-the-job electrocutions do occur and that the number one cause of fire in the workplace is electrical. For the organization safety official, electrical safety in the workplace not only is an important priority but also requires constant vigilance on his or her part and on the part of all supervisors and workers to ensure that safe work practices are followed when working with or around electrical circuits and components. All company employees should be made aware that maintaining the integrity of all electrical equipment and systems requires constant vigilance. An organization standing order should be that any discovered electrical discrepancy is to be reported to responsible parties immediately.

Per NEC and all local code requirements, safety officials must ensure that any planned electrical equipment is suitable for installation in the proposed installation areas. If a new electrical motor and controller are to be installed in an area that contains explosive vapors, the proper class of electrical motor and control equipment must be installed in such a space to prevent the possibility of explosion, based on NFPA recommendations.

OSHA standards relating to electricity can be found in 29 CFR 1910, Subpart S. They are extracted from the National Electrical Code. Subpart S is divided into the following two categories of standards: (1) Design of Electrical Systems, and (2) Safety-Related Work Practices.

9.17.14.1.1 Control of Electrical Hazards

We have stated consistently throughout this text that when the object is to control hazards, the goal should first be to engineer-out the hazard—any hazard—whenever possible. This, of course, is also the case with electrical hazards; for example, a company policy that insists that only intrinsically safe electrical equipment and tools (i.e., double- and triple-insulated hand tools) will be purchased and used within the organization is a type of engineering control. Another type of electrical engineering control is the installation of low-voltage systems. Other types of controls can reduce or eliminate electrical hazards, including switching devices, grounding and bonding, ground fault circuit interrupters and procedures, and lessening the hazardous effects of static electricity. Facility safety officials must be fully aware of the hazards of electricity, electrical circuits, and components and must also be familiar with the common means of electrical hazard control. This includes knowledge of applicable codes, regulations, and standards that provide detailed specifications and procedures for safeguarding electrical equipment and systems.

Because safety officials need to have some knowledge of electricity, electrical equipment and systems, and electrical hazard control methodologies, they also need to have some basic understanding of electricity itself, its uses, and the potential hazards it presents to all who might come into contact with it. In hazard control, the facility safety official must have fundamental knowledge of the electrical materials used, design of components, and placement of electrical equipment. An understanding of shielding methods and the enclosing and positioning of electrical devices can reduce contact by employees.

Note: This is not to say that the safety official must be an electrical engineer. Instead, we recommend some training in the fundamentals of electricity—this training should be included in the safety official's formal college or advanced short-school training. The minimum electrical system and component operation knowledge that safety officials should have is covered in the following discussion.

The facility safety official should understand that *overcurrent devices*, which limit the current that can flow through a circuit or electrical device, should be included in any electrical system design. Such a device cuts off power if current exceeds a given limit. The two most common overcurrent devices in use at present are fuses and circuit breakers. *Fuses* are composed of materials (usually lead or a lead alloy) that are designed to limit the current flow in the circuit. When current in the circuit exceeds some limiting value, the lead or lead alloy material heats above its melting point and separates, opening the circuit, thereby stopping the flow of current. Safety officials must understand that fuses are rated at certain design levels. In other words, not every fuse is suited for every electrical circuit. In fact, the danger with fused circuits is when the fuses are replaced with fuses that are too large for the circuit they are designed to protect. When this occurs, so does the danger that too much current will be allowed to flow in a circuit not designed to handle the high level of current flow, which could lead to electrical fires and other problems.

Circuit breakers are actually a form of switch designed to open when current passing through them exceeds a designed limit. Circuit breakers are designed to limit current flow in two different ways. One type is designed to open when the temperature of the breaker reaches a predetermined level. A common problem with this type of breaker is that the temperature of the environment around it can affect its operation. The second type is magnetic and opens when a predetermined current level is reached. The advantage of this type of breaker is that environmental conditions have little impact on its operation.

In addition to overcurrent devices, certain *switching devices* can reduce or eliminate electrical hazards. These include interlocks, lockouts, and thermal or overspeed switches. *Interlocks* are switches that prevent access to an energized or dangerous location. Often attached to access doors, panels, and gates, interlocks act to shut off power to the equipment whenever these devices are opened. Probably the most commonly used and most familiar interlock device is the one installed in most washing machine lids, which shuts down the machine when the lid is opened. A lockout procedure involves placing a lock on a switch, circuit breaker, or other device to prevent the switch, circuit breaker, or equipment from being turned on or energized.

Thermal and overspeed cutout devices are commonly used to protect electrical equipment (and thus the operator). A *thermal cutout* is simply a temperature-sensitive switch with a preset limit designed to interrupt power when the temperature exceeds a certain value. As its name implies, an *overspeed switch* operates when it senses that a motor or other device is operating at too fast a speed. Obviously, excessive speed may create dangerous conditions and indicate failure of equipment. The overspeed switch operates to shut down an overspeeding device by interrupting power to it.

Grounding and *bonding* control the electrical potential between two bodies. If there is a difference of potential between two bodies, a conductor between them will allow charge or current to flow. That flow may be dangerous, particularly as a source of ignition. Important information on grounding and bonding can be found in Lee (1969) and Hammer (1989). *Note:* The information provided in the following assumes that the reader has some fundamental knowledge of electrical terms and their meaning.

The Earth acts as an infinite store from which electrons (current flow) can be drawn or to which they can return. Providing a path for that flow can eliminate any undesirable excess or deficiency. Gaining electrons can neutralize positive ions in a system, and electrons can be conducted to Earth (called *earthing* in some countries). In the United States, the term *grounding* is preferred, and the path to Earth or the Earth itself is a *ground*. In some instances (such as in electronic equipment), a massive metallic body acts as the reservoir of electrons and ions (the ground) in place of the Earth.

Grounds can be designed and installed into a system or they can be accidental. Unless noted otherwise, the word *ground* used here indicates one of design. Installed grounds are basically safety mechanisms to prevent (1) overloading of circuits and equipment that would destroy them or shorten their lives, (2) shock to personnel, and (3) arcing or sparking that might act as an ignition source.

Grounds may protect a system, equipment, or personnel. Certain designs used on high-voltage transmission lines are sophisticated types that follow the standards set by the American Institute of Electrical Engineers or other codes. The ground systems and standards of the National Electrical Code, which apply to buildings and related facilities, are more common.

Safety officials should know several terms used in the NEC that are related to grounding and bonding. *System ground* refers to an electric circuit and is designed to protect conductors (wires/wiring) for a transmission, distribution, or wiring system. The term *voltage to ground* is often used in electrical codes. It indicates the maximum voltage in a grounded circuit measured between the ground wire and a wire that is not grounded. Where a ground is not used, voltage to ground indicates the maximum voltage between any two wires. The wire that connects the circuit to Earth is the *grounding wire* or *ground*; the wire to which it is connected is the *grounded wire*.

Probably the simplest way in which to illustrate the principles of grounding is to use a typical three-wire system as an example. In a three-wire system, current generally flows along two wires; the third is neutral. In distribution systems for buildings and related facilities, the neutral wire is always the one grounded when grounding is installed. High-voltage transmission lines sometimes ground all three wires, but this is less common. The types of grounding systems that have been used on transmission lines include the following:

1. *Solid grounds*—The neutral wire is grounded without any impedance, which might restrict current flow.
2. *Resistance grounds*—The neutral wire is connected to ground through a high resistance at a transformer.
3. *Reactance grounds*—The neutral wire is connected to ground through an impedance which is principally reactance.
4. *Capacitance grounds*—Each line of a circuit is connected to a capacitor; the other side of each capacitor is grounded.
5. *Resonant grounds*—This is a tuned, parallel system that uses capacitance grounds and a ground from a transformer neutral through an induction coil.

Solid grounds are the most commonly used, especially in interior electrical systems of buildings. Resistance and capacitance grounds are designed into most electronic equipment. These types of grounds involve circuitry comparable to two-wire systems in which it is necessary to maintain potentials within prescribed limits.

One purpose of grounding the neutral in a three-wire system is to activate overcurrent protection devices before damage is done when a fault occurs. Should one of the two wires that normally carries current be broken or accidentally grounded, current will flow through the neutral, through the installed ground, and back to the power source. This short circuit will open the protection devices and de-energize the affected portion of the system.

Where the neutral is not grounded, accidental grounding of one of the other wires will cause an increase in voltage to ground of the remaining system. The definitions of *voltage to ground* for grounded and ungrounded systems will illustrate this point. A 220-volt, three-wire grounded neutral system will have a voltage between any two wires. The excessively high voltages may cause burnout of equipment, burning or breakdown of insulation, arcing and sparking, and shock to personnel who come in contact with metal energized through the breaks.

Other possibilities exist by which an excessively high voltage can be produced, which would create similar hazards if the system is not grounded. A fault in a step-down transformer could result in the distribution system potential, or part of it greater than normal, being applied to a building wiring system. An accidental connection between the two systems would produce the same result. Where grounds existed, the overcurrent protection devices would de-energize and safeguard the system.

Equipment grounds may be used on the metal parts of a wiring system, such as the conduit, armor, switch boxes, and connected apparatus other than the wire, cable, or other circuit components. They may also be provided for equipment such as metal tables and cabinets that might come in contact with an energized circuit or source of electrical charges. Equipment on which undesirable charges may be induced or generated should also be grounded.

Metal of electrical equipment may come in contact with an energized circuit whose insulation is deteriorated or cut or through which arcing can take place. A person may then touch the metal surface inadvertently, receiving a shock. The degree of shock would depend on whether the equipment was grounded. If it was not, the person in contact with the metal would act as a ground, the current passing through his body. If the equipment was grounded, the person might or might not receive a shock at all. If current did pass through his body, the amount would be inversely proportional to the resistance of his body compared to that of the equipment ground. If the resistance of his body were high enough, no current would pass.

Bonding ensures that all major parts of a piece of equipment are linked to provide a continuous path to ground. A bond is a mechanical connection that provides a low-resistance path to current flow between two surfaces that are physically separated or may become separated. A bond can be permanent, such as one in which the connection is welded or brazed to the two surfaces, or it may be semipermanent, bolted or clamped where required.

Where permanent types are used, the parts themselves can be joined and narrow gaps filled with weld or brazing metal. Where separation is wider, a strip of metal can be welded or brazed at both ends across the gap. Bonds connecting one vibrating part to another part that may or may not vibrate should be of a flexible material that will not fail under vibration. Corrosion because of the joining of dissimilar metals may cause the electrical resistance across the bond to increase. This is especially noticeable in humid or corrosive atmospheres. The types of metal for the bond and its fastenings must therefore be selected with care. Grounds and bonds should

- Be permanent wherever possible.
- Have ample capacity to conduct any possible current flow (a ground should not normally be designed to be part of a current-carrying circuit).
- Have as low impedance as possible.

- Be continuous and, wherever possible, be made directly to the basic structure rather than through other bonded parts.
- Be secured so that vibration, expansion, contraction, or other movement will not break the connection or loosen it so the resistance varies.
- Have connections located in protected areas and where accessible for inspection or replacement.
- Not impede movement of movable components.
- Not be compression-fastened through nonmetallic materials.
- Not have dissimilar metals in contact.
- Have metals selected to minimize corrosion.

Grounding is not always advantageous in all cases; some electrical systems are safer ungrounded. As Lee (1969, p. 162) pointed out: "Some electrical systems (necessarily of limited extent), must be left ungrounded for safety reasons. For example, the electrical system of a hospital operating room is purposely ungrounded because a spark from an insulation failure would otherwise ignite the anesthesia-permeated atmosphere. When ungrounded, an insulation failure 'to ground' produces no current flow and hence no spark, no ignition and no explosion. Electric blasting caps present a similar condition; a short-circuit current returning through the earth could fire the caps if their two connecting wires touched the earth more than a few inches apart."

Ground-fault circuit interrupters (GFCIs) are designed to open the circuit before a fault path through the operator can cause harm, at levels as low as 5 milliamps (Ma). A GFCI compares current normally flowing through the power distribution wire and the grounded neutral wire of a circuit. The current flowing through one must pass through the other for the circuit to work. If current is not equal, some electrical energy is flowing to ground through other than the normal route, perhaps through a person. When the current is not equal, the GFCI detects this current differential and shuts off the current. Though GFCIs protect normal 115-volt circuits where users can form a ground with energized equipment, they do not work on line-to-line connections found in distributions of 220 volts and higher. GFCIs are required by the NEC for outdoor receptacles or circuits and for bathrooms and other locations.

Static electricity is a workplace hazard because of its potential to ignite (by arc) certain vapor or dust mixtures in air. Various controls are available to minimize the effects of static charges, dependent on the individual case:

- Selection of suitable materials (i.e., avoiding the use of materials such as clothing composed of synthetic fabrics that generate static electricity) is often the simplest method.

- Modifying a material by spraying its surface to make it conductive frequently can reduce or eliminate the static electricity problem.
- Bonding and grounding can be utilized to provide a path by which various surfaces on which charges could accumulate can be neutralized.
- Electrostatic neutralizers can be used to neutralize charges on materials.
- Humidification (raising the relative humidity above 65%) permits static charges to leak off and dissipate.

9.17.14.1.2 Summary of Safety Precautions for Electrical Circuits

- Ensure that power has been disconnected from the system before working with it. Test the system for de-energization. Capacitors can store current after power has been shut off.
- Allow only fully qualified and trained personnel to work on electrical systems.
- Do not wear conductive material such as metal jewelry when working with electricity.
- Screw bulbs securely into their sockets. Ensure that bulbs are matched to the circuit by the correct voltage rating.
- Periodically inspect insulation.
- If working on a hot circuit, use the buddy system and wear protective clothing.
- Do not use a fuse with a greater capacity than was prescribed for the circuit.
- Verify circuit voltages before performing work.
- Do not use water to put out an electrical fire.
- Check the entire length of electrical cord before using it.
- Use only explosion-proof devices and nonsparking switches in flammable liquid storage areas.
- Enclose uninsulated conductors in protective areas.
- Discharge capacitors before working on the equipment.
- Use fuses and circuit breakers for protection against excessive current.
- Provide lightning protection on all structures.
- Train people working with electrical equipment on a routine basis in first aid and cardiopulmonary resuscitation (CPR).

9.17.15 Subpart Z, Toxic and Hazardous Substances

1910.1000 Air Contaminants

1910.1030 Bloodborne Pathogens

1900.1200 Hazard Communication

9.17.15.1 Bloodborne Pathogens

A fairly recent peril has been added to the list of work-related hazards: exposure to *bloodborne pathogens.* When accidents, injuries, or illnesses occur in the workplace, employees need to know how to respond safely and correctly. This is particularly the case for employees rendering first aid. The safety official must look at this potential life-threatening area with particular attention. Specifically, the safety official should ensure that a proper accident response scenario is in writing and in place to detail exact response procedures. A worksite procedure entailing the procedures to be followed when rendering first aid services must be in place. Another important aspect cannot be overlooked: *accident reporting.* Employees must be trained to report all on-the-job accidents, no matter what their level of severity. Accidents that involve the release of body fluids that other employees come into contact with must be reported and proper medical response effected (the victim and the employees who came in contact should be medically evaluated and offered hepatitis B vaccinations). Employees must be thoroughly trained on avoiding bloodborne pathogens.

9.17.15.2 Hazard Communication (29 CFR 1910.1200)

Note that the number one OSHA citation listed in Table 9.12 for the oil and gas drilling industry is for noncompliance with 29 CFR 1910.1200, Hazard Communication. The 1984 Bhopal Incident, the ensuing chemical spill, and the resulting tragic deaths and injuries are well known; however, the repercussions—the lessons learned—from this incident are not as well known. After Bhopal, a worldwide outcry arose: "How could such an incident occur? Why wasn't something done to protect the inhabitants? Weren't safety measures taken or in place to prevent such a disaster from occurring?"

In the United States, these questions and others were bandied about and about by the media and Congress, and Congress took the first major step to prevent such incidents from occurring in the United States. Congress directed OSHA to take a close look at chemical manufacturing in the United States to see if a Bhopal-type incident could occur in this country. OSHA did a study and reported to Congress that a Bhopal-type incident in the United States was very unlikely. Within a few months of OSHA's report to Congress, however, a chemical spill occurred in Institute, West Virginia, which was similar to Bhopal, but fortunately not as deadly (although over 100 people became ill).

Needless to say, Congress was upset. Because of Bhopal and the Institute, West Virginia, incidents, OSHA mandated its Hazard Communication Program, 29 CFR 1910.1200, in 1984. Later, other programs, such as SARA (Superfund) Title III reporting requirements for all chemical users, producers, suppliers, storage entities, were mandated by USEPA.

There is no all-inclusive list of chemicals covered by the Hazard Communication Standard; however, the regulation refers to "any chemical which is a physical or health hazard." Those specifically deemed hazardous include the following:

- Chemicals regulated by OSHA in 29 CFR Part 1910, Subpart Z, Toxic and Hazardous Substances
- Chemicals included in the latest edition of the American Conference of Governmental Industrial Hygienists (ACGIH) *Threshold Limit Values (TLVs®) for Chemical Substances and Physical Agents in the Work Environment*
- Chemicals found to be suspected or confirmed carcinogens by the National Toxicology Program (NTP) and published in NIOSH's *Registry of Toxic Effects of Chemical Substances*, chemicals appearing in the latest edition of the NTP's *Annual Report on Carcinogens*, or chemicals appearing in the latest editions of IARC *Monographs*

Congress decided that those personnel involved with working with or around hazardous materials had a right to know about the hazards near them or the ones they worked with; thus, OSHA's Hazard Communication Standard was created. The Hazard Communication Standard is, without a doubt, the regulation most important to the communication of hazards to employees. Under its Hazard Communication Standard (more commonly known as "HazCom" or the "Right to Know Law"), OSHA requires employers who use or produce chemicals on the worksite to inform all employees of the hazards that might be involved with those chemicals. HazCom states that employees have the right to know what chemicals they are handling or could be exposed to. HazCom's intent is to make the workplace safer. Under the HazCom Standard, the employer is required to fully evaluate all chemicals on the worksite for possible physical and health hazards. All information relating to these hazards must be made available to the employee 24 hours each day. The standard is written in a performance manner, meaning that the specifics are left to the employer to develop. HazCom also requires the *employer* to ensure proper labeling of each chemical, including chemicals that might be produced by a *process*. In the wastewater industry, for example, deadly methane gas is generated in the waste stream. Another common wastewater hazard is the generation of hydrogen sulfide (which produces the characteristic rotten-egg odor) during degradation of organic substances in the wastestream; it can kill quickly. HazCom requires the employer to label methane and hydrogen sulfide hazards so that workers are warned, and safety precautions are followed.

Labels must be designed to be clearly understood by all workers. Employers are required to provide both training and written materials to make workers aware of what they are working with and what hazards they might be exposed to.

9.17.15.2.1 Material Safety Data Sheet (MSDS)

Employers are also required to make Material Safety Data Sheets (MSDSs) available to all employees. An MSDS is a fact sheet for a chemical posing a physical or health hazard at work. MSDSs must be in English. Blank spaces are not permitted on an MSDS. If relevant information in any one of the categories is unavailable at the time of preparation, the MSDS must indicate no information was available. Your facility must have an MSDS for each hazardous chemical it uses. Copies must be made available to other companies working on your worksite (outside contractors, for example), and they must do the same for you. The facility Hazard Communication Program must be in writing and, along with MSDSs, made available to all workers 24 hours each day, each shift. The following information is required on an MSDS:

1. Specific identity of each hazardous chemical or mixture ingredient and common product names
2. Physical and chemical characteristics of the hazardous material, such as:
    a. Density or specific gravity of liquid or solid
    b. Density of gas or vapor relative to air
    c. Boiling point
    d. Melting point
    e. Flash point
    f. Flammability range
    g. Vapor pressure
3. Physical hazard data such as stability, reactivity, flammability, corrosivity, explosivity
4. Health hazard data including acute and chronic health effects and target organ effects
5. Exposure limits such as OSHA permissible exposure limits (PELs)
6. Carcinogenicity of material
7. Precautions to be taken, including use of PPE
8. Emergency and first aid procedures (including spill cleanup information and USEPA spill reportability information)

9. Supplier or manufacturer data:
   a. Name
   b. Address
   c. Telephone number
   d. Data prepared.

9.17.15.2.2 Hazard Communication Program Definition of Terms

The Hazard Communication Program defines various terms as follows:

Chemical—Any element, compound, or mixture of elements and/or compounds.

Chemical name—The scientific designation of a chemical in accordance with the nomenclature system developed by the International Union of Pure and Applied Chemistry (IUPAC) or the Chemical Abstracts Service (CAS) Rules of Nomenclature, or a name that will clearly identify the chemical for the purpose of conducting a hazard evaluation.

Combustible liquid—Any liquid having a flashpoint at or above 100°F (37.8°C) but below 200°F (93.3°C).

Common name—Any designation or identification, such as code name, code number, trade name, brand name, or generic name used to identify a chemical other than its chemical name.

Compressed gas—A compressed gas is

- A gas or mixture of gases in a container having an absolute pressure exceeding 40 psi at 70°F (21.1°C).
- A gas or mixture of gases in a container having an absolute pressure exceeding 104 psi at 130°F (54.4°C) regardless of the pressure at 70°F (21.1°C).
- A liquid having a vapor pressure exceeding 10 psi at 100°F (37.8°C), as determined by ASTM D-323-72.

Container—Any bag, barrel, bottle, box, can, cylinder, drum, reaction vessel, storage tank, or the like that contains a hazardous chemical.

Explosive—A chemical that causes a sudden, almost instantaneous release of pressure, gas and heat when subjected to sudden shock, pressure, or high temperature.

Exposure—The actual or potential subjection of an employee to a hazardous chemical through any route of entry in the course of employment.

Flammable aerosol—An aerosol that, when tested by the method described in 16 CFR 1500.45, yields a flame projection exceeding 18 inches at full valve opening, or a flashback (flame extending back to the valve) at any degree of valve opening.

Flammable gas—A gas that at ambient temperature and pressure forms a flammable mixture with air at a concentration of 13% by volume or less, or a gas that at ambient temperature and pressure forms a range of flammable mixtures with air wider than 12% by volume regardless of the lower limit.

Flammable liquid—A liquid having a flashpoint of 100°F (37.8°C).

Flammable solid—A solid, other than a blasting agent or explosive as defined in 29 CFR 1910.109(a), that is likely to cause fire through friction, absorption of moisture, spontaneous chemical change or retained heat from manufacturing or processing, or which can be ignited and that, when ignited, burns so vigorously and persistently as to create a serious hazard. A chemical shall be considered to be a flammable solid if, when tested by the method described in 16 CFR 1500.44, it ignites and burns with a self-sustained flame at a rate greater than 1/10 of an inch per second along its major axis.

Flashpoint—The minimum temperature at which a liquid gives off a vapor in sufficient concentration to ignite.

Hazard warning—Any words, pictures, symbols, or combination thereof appearing on a label or other appropriate form of warning that convey the hazards of chemicals in containers.

Hazardous chemical—Any chemical that is a health or physical hazard.

Hazardous Chemical Inventory List—An inventory list of all hazardous chemicals used at the site and containing the date of each chemical's MSDS insertion.

Health hazard—A chemical for which there is statistically significant evidence based on at least one study conducted in accordance with established scientific principles that acute or chronic health effects may occur in exposed employees.

Immediate use—The use under the control of the person who transfers the hazardous chemical from a labeled container and only within the work shift in which it is transferred.

Label—Any written, printed, or graphic material displayed on or affixed to containers or hazardous chemicals.

Material Safety Data Sheet—The written or printed material concerning a hazardous chemical, developed in accordance with 29 CFR 1910.

Mixture—Any combination of two or more chemicals if the combination is not, in whole or in part, the result of a chemical reaction.

NFPA Hazardous Chemical Label—A color-coded labeling system developed by the National Fire Protection Association (NFPA) that rates the severity of the health hazard, fire hazard, reactivity hazard, and special hazard of the chemical.

Organic peroxide—An organic compound that contains the bivalent O–O structure and which may be considered to be a structural derivative of hydrogen peroxide, where one or both of the hydrogen atoms has been replaced by an organic radical.

Oxidizer—A chemical (other than a blasting agent or explosive as defined in 29 CFR 1910.198(a)) that initiates or promotes combustion in other materials, thereby causing fire either of itself or through the release of oxygen of other gases.

Physical hazard—A chemical for which there is scientifically valid evidence that it is a combustible liquid, a compressed gas explosive, flammable, an organic peroxide, an oxidizer, pyrophoric, unstable (reactive), or water reactive.

Portable container—A storage vessel that is mobile, such as a drum, side-mounted tank, tank truck, or vehicle fuel tank.

Primary route of entry—Primary means (e.g., inhalation, ingestion, skin contact) whereby an employee is subjected to a hazardous chemical.

"Right to know" station binder—Located in the "Right to Know" work station, it contains the company's Hazard Communication Program, the Hazardous Chemicals Inventory List and corresponding MSDSs, and the Hazard Communication Program Review and Signature Forms.

"Right to know" work station—Central area where employees can access MSDS sheets, Hazardous Chemical Inventory Lists, and the company's written Hazard Communication Program.

Pyrophoric—A chemical that will ignite spontaneously in air at a temperature of 130°F (54.4°C) or below.

Stationary container—A permanently mounted chemical storage tank.

Unstable (reactive) chemical—A chemical which in its pure state or as produced or transported will vigorously polymerize, decompose, condense, or become self-reactive under conditions or shock, pressure, or temperature.

Water-reactive chemical—A chemical that reacts with water to release a gas that is either flammable or presents a health hazard.

Work center—Any convenient or logical grouping of designated unit processes or related maintenance actions.

9.18 Thought-Provoking Discussion Questions

1. Write your own definitions of *pollution*, *contaminant*, *pollutant*, and *environment*. Include a list of the elements you consider of particular concern for the definitions.

2. Discuss the perception of pollution as individual judgment.
3. Discuss the idea of balance and tradeoffs for environmental pollution control.
4. Discuss how environmental pollution remediation can be affected by politics.
5. Discuss how environmental pollutants and remediation are affected by scientific and technical levels of certainty and uncertainty.
6. Look at environmental pollution from a historical perspective. What has changed in how we view environmental pollution, and what has changed in our levels of concern and action?
7. Why is finding a solution to an environmental conflict so complex? Explain your answer.
8. Discuss extinction as it relates to habitat, maximum sustainable yield, and natural capital.
9. Do we need environmental regulations?
10. Have we become a Regulation Nation?
11. Do you believe the society you live in is on an unsustainable path? Explain.
12. Explain why you agree or disagree with the following proposition: The world will run out of renewable resources because we cannot use technology to find substitutes.

References and Recommended Reading

AAA. (2003). AAA encourages motorists to prepare now for safe winter driving. *PR Newswire*, October 1.

AAAdvantage. (2003). Nine vital driving tips for mature operators. *Home & Away Magazine*, November/December.

ALL Consulting and Montana Board of Oil and Gas Conservation (MBOGC). (2002). *Handbook on Best Management Practices and Mitigation Strategies for Coal Bed Methane in the Montana Portion of the Powder River Basin*, prepared for National Energy Technology Laboratory, National Petroleum Technology Office, U.S. Department of Energy, Tulsa, OK, 44 pp.

ALL Consulting and Montana Board of Oil and Gas Conservation (MBOGC). (2004). *Coal Bed Natural Gas Handbook: Resources for the Preparation and Review of Project Planning Elements and Environmental Documents*, prepared for National Petroleum Technology Office, U.S. Department of Energy, Washington, DC, 182 pp.

Allaby, A. and Allaby, M. (1991). *The Concise Dictionary of Earth Sciences*. Oxford University Press, Oxford, UK.

Archea, J., Collins, B.L., and Stahl, F.I. (1976). *Guidelines for Stair Safety*, Building Sciences Series 120. National Bureau of Standards, Washington, DC.

Arms, K. (1994). *Environmental Science*, 2nd ed. Saunders College Publishing, Fort Worth, TX.

Associated Press. (1998). Town evacuated after acid spill. *Lancaster New Era*, September 6.

AWWA. (1995). *Water Quality*, 2nd ed. American Water Works Association, Denver, CO.

Baden, J. and Stroup, R.C., Eds. (1981). *Bureaucracy vs. Environment*. University of Michigan Press, Ann Arbor.

Bangs, R. and Kallen, C. (1985). *Rivergods: Exploring the World's Great Wild Rivers*. Sierra Club Books, San Francisco, CA.

Barker, J.R. and Tingey, D.T. (1991). *Air Pollution Effects on Biodiversity*, Van Nostrand Reinhold, New York.

BLM. (2001). *Sundry Notices and Reports on Wells*, Form 3160-5. Bureau of Land Management, U.S. Department of the Interior, Washington, DC (www.blm.gov/ca/pdfs/bakersfield_pdfs/minerals/Form_3160-5.pdf).

BLS. (1984). *Injuries Resulting from Falls from Elevation*, Bulletin 2195. Bureau of Labor Statistics, Washington, DC.

BNA. (1971). *The Job Safety and Health Act of 1970*, 1st ed. Bureau of National Affairs, Washington, DC.

Botkin, D.B. (1995). *Environmental Science: Earth as a Living Planet*. Wiley, New York.

Boyce, A. (1997). *Introduction to Environmental Technology*. Van Nostrand Reinhold, New York.

Brahic, C. (2009). Fish "an ally" against climate change. *New Scientist*, January 16 (www.newscientist.com/article/dn16432-fish-an-ally-against-clmate-cahnge.html).

Brauer, R.L. (1994). *Safety and Health for Engineers*. Van Nostrand Reinhold, New York.

Brauer, R.L. (1996). *Safety and Health for Engineers*, 2nd ed. Wiley-Interscience, New York.

Caldeira, K. and Wickett, M.E. (2003). Anthropogenic carbon and ocean pH. *Nature*, 425(6956):365.

Carney, A. (1991). Lock out the chance for injury. *Safety & Health*, 143(5):46.

Carson, J.E. and Moses, H. (1969). The validity of several plume rise formulas. *Journal of the Air Pollution and Control Association*, 19(11):862–866.

CDPHE. (2007). *Pit Monitoring Data for Air Quality*. Colorado Department of Public Health and Environment, Denver.

CH2M HILL. (2007). *Review of Oil and Gas Operation Emissions and Control Options, Final Report*, prepared for Air Pollution Control Division, Colorado Department of Public Health and Environment, Denver, CO.

CICA. (2008). *The Construction Industry Compliance Assistance Center*, www.cicacenter.org/index.cfm.

Coastal Video. (1993). *Confined Space Rescue Booklet*. Coastal Video Communication Corporation, Virginia Beach, VA.

Cobb, R.W. and Elder, C.D. (1983). *Participation in American Politics*, 2nd ed. The Johns Hopkins University Press, Baltimore, MD.

Cohen, H.H. and Compton, D.M.J. (1982). Fall accident patterns. *Professional Safety*, 27(6):16–22.

Cote, A. and Bugbee, P. (1991). *Principles of Fire Protection*. National Fire Protection Association, Batterymarch Park, MA.

CoVan, J. (1995). *Safety Engineering*. John Wiley & Sons, New York.

Davis, M.L. and Cornwell, D.A. (1991). *Introduction to Environmental Engineering*. McGraw-Hill, New York.

Diamond, J. (2006). *Guns, Germs, and Steel: The Fates of Human Societies*. Norton, New York.

Downing, P.B. (1984). *Environmental Economics and Policy*. Little, Brown, Boston.

DRBC. (2010). *Administrative Manual—Part III: Water Quality Regulations: 18 CFR Part 410*. Delaware River Basin Commission, West Trenton, NJ.

Easterbrook, G. (1995). *A Moment on the Earth: The Coming Age of Environmental Optimism*. Viking Penguin, Bergenfield, NJ.

Eisma, T.L. (1990). Rules change: worker training helps simplify fall protection. *Occupational Health & Safety*. 59(3):52–55.

Ellis, J.E. (1988). *Introduction to Fall Protection*. American Society of Safety Engineers, Des Plaines, IL.

FEMA. (1981). *Planning Guide and Checklist for Hazardous Materials Contingency Plans*, FEMA-10. Federal Emergency Management Agency, Washington, DC.

Ferry, T. (1990). *Safety and Health Management Planning*. Van Nostrand Reinhold, New York.

Field, B.C. (1996). *Environmental Economics: An Introduction*, 2nd ed. McGraw-Hill, New York.

Franck, I. and Brownstone, D. (1992). *The Green Encyclopedia*. Prentice Hall, New York.

Freedman, B. (1989). *Environmental Ecology*. Academic Press, New York.

Gasaway, D.C. (1985). *Hearing Conservation: A Practical Manual and Guide*. Prentice Hall, Englewood Cliffs, NJ.

Giachino, J. and Weeks, W. (1985). *Welding Skills*. American Technical Publications, Homewood, IL.

Godish, T. (1997). *Air Quality*, 3rd ed. Lewis Publishers, Boca Raton, FL.

Hammer, W. (1989). *Occupational Safety Management and Engineering*. Prentice Hall, Englewood Cliffs, NJ.

Hardin, G. (1968). The tragedy of the commons. *Science*, 162:1243–1248.

Healy, R.J. (1969). *Emergency and Disaster Planning*. John Wiley & Sons, New York.

Henry, J.G. and Heinke, G.W. (1995). *Environmental Science and Engineering*, 2nd ed. Prentice Hall, New York.

Holmes, G., Singh, B.R., and Theodore, L. (1993). *Handbook of Environmental Management and Technology*. John Wiley & Sons, New York.

Hoover, R.L., Hancock, R.L., Hylton, K.L., Dickerson, O.B., and Harris, G.E. (1989). *Health, Safety and Environmental Control*. Van Nostrand Reinhold, New York.

ICAS. (1975). *The Possible Impact of Fluorocarbons and Halocarbons on Ozone*. Interdepartmental Committee for Atmospheric Sciences, Washington, DC, 75 pp.

Ingram, C. (1991). *The Drinking Water Book*. Ten Speed Press, Berkeley, CA.

IOGCC. (1996). *Review of Existing Reporting Requirements for Oil and Gas Exploration and Production Operators in Five Key States*. Interstate Oil and Gas Compact Commission, Washington, DC.

IOGCC. (2008). *Issues: States' Rights*. Interstate Oil and Gas Compact Commission, Washington, DC.

IPAA. (2000). *IPAA Opposes EPA's Possible Expansion of TRI*, Environment and Safety Fact Sheets. Independent Petroleum Association of America, Washington, DC (www.ipaa.org/issues/factsheets/environment_safety/tri.php).

IPAA. (2004). *Reasonable and Prudent Practices for Stabilization (RAPPS) of Oil and Gas Construction Sites*, Guidance Document. Independent Petroleum Association of America, Washington, DC.

Jackson, A.R. and Jackson, J.M. (1996). *Environmental Science: The Natural Environment and Human Impact*. Longman, New York.

Jacobson, J. (1998). *The Supervisor's Tough Job: Dealing with Drug and Alcohol Abusers*, Supervisors' Safety Update 97. Eagle Insurance Group, Largo, FL.

Jacus, J.R. (2011). Air quality constraints on shale development activities, in *Proceedings of The Institute for Energy Law Second Conference on the Law of Shale Plays*, Fort Worth, TX, September 7–8.

Karliner, J. (1998). Earth predators. *Dollars and Sense*, 218:7.

Keller, E.A. (1988). *Environmental Geology*. Merrill, Columbus, OH.

Kimberlin, J. (2009). That reeking paper mill keeps Franklin running. *The Virginian-Pilot*, February 1.

Kohr, R.L. (1989). Slip slidin' away. *Safety & Health*, 140(5):52.

LaBar, G. (1989). Sound policies for protecting workers' hearing. *Occupational Hazards*, July, p. 46.

Lave, L.B. (1981). *The Strategy of Social Regulations: Decision Frameworks for Policy*. Brookings Institution, Washington, DC.

Lee, R.H. (1969). Electrical grounding: safe or hazardous? *Chemical Engineering*, 76:158.

Leopold, A. (1970). *A Sand County Almanac*. Ballentine Books, New York.

Lewis, S.A. (1996). *Safe Drinking Water*. Sierra Club Books, San Francisco, CA.

MacKenzie, J.J. and El-Ashry, T. (1988). *Ill Winds: Airborne Pollutant's Toll on Trees and Crops*. World Resource Institute, Washington, DC.

Mansdorf, S.Z. (1993). *Complete Manual of Industrial Safety*. Prentice Hall, Englewood Cliffs, NJ.

Masters, G.M. (1991). *Introduction to Environmental Engineering and Science*. Prentice-Hall, Englewood Cliffs, NJ.

Masters, G.M. (2007). *Introduction to Environmental Engineering and Science*, 3rd ed. Prentice Hall, Englewood Cliffs, NJ.

McElroy, F.E., Ed. (1980). *NSC Accident Prevention Manual for Industrial Operations: Engineering and Technology*, 8th ed. International Fire Chiefs Association, Merrifield, VA.

McHibben, B. (1995). *Hope, Human and Wild: True Stories of Living Lightly on the Earth*. Little, Brown & Company, Boston, MA.

Miller, G.T. (1997). *Environmental Science: Working with the Earth*, 5th ed. Wadsworth, Belmont, CA.

Miller, G.T. (2004). *Environmental Science*, 10 ed. Brooks/Cole, Belmont, CA.

Molina, M.J. and Rowland, F.S. (1974). Stratospheric sink for chlorofluoromethanes: chlorine atom-catalysed destruction of ozone. *Nature*, 249(5460):810–812.

Morrison, A. (1983). In third world villages, a simple handpump saves lives. *Civil Engineering*, 52:68–72.

Nathanson, J.A. (1997). *Basic Environmental Technology: Water Supply, Waste Management, and Pollution Control*. Prentice Hall, Upper Saddle River, NJ.

National Institute on Aging. (2002). *Practice Common Sense Safety Rules*. National Institutes of Health, U.S. Department of Health and Human Services, Washington, DC.

NFPA. (1989). *Cutting and Welding Processes*, NFPA 51B-1989. National Fire Protection Association, Quincy, MA.

NFPA. (1991). *Fire Protection Handbook*, 16th ed. National Fire Protection Association, Quincy, MA.

NHTSA. (1992). *Sudden Impact*. National Highway Traffic Safety Administration, Washington DC.

NIOSH. (1987). *Guide to Industrial Respiratory Protection*, NIOSH Publication No. 87-116. National Institute for Occupational Safety and Health, Cincinnati, OH.

NIOSH. (2005). *NIOSH Alert: Preventing Worker Injuries and Deaths from Traffic-Related Crashes*, NIOSH Publication No. 98-142. National Institute for Occupational Safety and Health, Cincinnati, OH.

Norse, E.A. (1985). The value of animal and plant species for agriculture, medicine, and industry, in *Animal Extinctions*, Hoage, R.J., Ed. Smithsonian Institution Press, Washington, DC, pp. 59–70.

NSC. (1985). *Accident Facts 1984*. National Safety Council, Chicago, IL.

NSC. (1986). *Accident Facts 1985*. National Safety Council, Chicago, IL.

NSC. (1987). *Guards: Safeguarding Concepts Illustrated*, 5th ed. National Safety Council, Chicago, IL.

NSC. (1992). *Accident Prevention Manual for Business and Industry: Engineering and Technology*. National Safety Council, Chicago, IL.

NYDEC. (2009). *Draft Supplemental Generic Environmental Impact Statement on the Oil, Gas and Solution Mining Regulatory Program*. New York Department of Environmental Conservation, Depew, NY (http://www.dec.ny.gov/energy/58440.html).

Ophuls, W. (1977). *Ecology and the Politics of Scarcity*. W.H. Freeman, New York.

Orfinger, B. (2002). *Saving More Lives: Red Cross Adds AED Training to CPR Course*. American Red Cross, Washington, DC.

OSHA. (1989). OSHA studies workplace deaths involving welding. *OSHA News*, February 8.

OSHA. (1992). *Concepts and Techniques of Machine Safeguarding*, OSHA 3067. Occupational Safety and Health Administration, Washington, DC.

OSHA. (2003). *Network of Employers for Traffic Safety Align with OSHA to Reduce Job-Related Traffic Injuries and Fatalities*, OSHA Trade Release. Occupational Safety and Health Administration, Washington, DC.

OTA. (1989). *Catching Our Breath: Next Steps for Reducing Urban Ozone*, OTA-O-412. Office of Technology Assessment, Washington, DC.

Pater, R. (1985). Fallsafe: reducing injuries from slips and falls. *Professional Safety*, 30(10):15–18.

Peavy, H.S., Rowe, D.R., and Tchobanoglous, G. (1985). *Environmental Engineering*. McGraw-Hill, New York.

Penton. (1986). *Right Off the Docket: Case Summaries and Court Rulings Involving Disputed Workers' Compensation and Occupational Disease Claims*. Penton Media, Cleveland, OH.

Pepper, I.L., Gerba, C.P., and Brusseau, M.L. (1996). *Pollution Science*. Academic Press, San Diego, CA.

Postel, S. (1987). Stabilizing chemical cycles, in *State of the World: A Worldwatch Institute Report on Progress Toward a Sustainable Society*, Brown, L.R., Ed. Norton, New York.

Rigzone. (2010). *Land Rig Utilization Trends*. Rigzone, Houston, TX (http://www.rigzone.com/data/analysis/2010_Q1_landRigReview_UtilTrends.pdf).

Rosen, S.I. (1983). *The Slip and Fall Handbook: Case Evaluation, Preparation, and Settlement*. Hanrow Press, Columbia, MD.

Royster, J.D. and Royster, L.H. (1990). *Hearing Conservation Programs: Practical Guidelines for Success*. Lewis Publishers, Chelsea, MI.

Ryan, M. (1998). How a little headwork saves a lot of children. *Parade Magazine*, May 24, pp. 4–5.

Santoro, R.L., Howarth, R.W., and Ingraffea, A.R. (2011). *Indirect Emissions of Carbon Dioxide from Marcellus Shale Gas Development*, Technical Report. Agriculture, Energy, and Environment Program, Cornell University, New York.

Sayers, D.L. and Walsh, J.L. (1998). *Thrones, Dominations*. St. Martin's Press, New York.

Smith, A.J. (1980). *Managing Hazardous Substances Accidents*. McGraw-Hill, New York.

Sonnenstuhl, W. and Trice, H. (1986). The social construction of alcohol problems in a union's peer counseling program. *Journal of Drug Issues*, 17(3):223–254.

Spellman, F.R. (1996a). *Stream Ecology and Self-Purification: An Introduction for Wastewater and Water Specialists*. Technomic, Lancaster, PA.

Spellman, F.R. (1996b). *Safe Work Practices for Wastewater Treatment Plants*. Technomic, Lancaster, PA.

Spellman, F.R. (1997). *A Guide to Compliance for Process Safety Management Planning (PSM/RMP)*. Technomic, Lancaster, PA.

Spellman, F.R. (1998). *Surviving an OSHA Audit*. Technomic, Lancaster, PA.

Spellman, F.R. (1999). *Confined Space Entry*. Technomic, Lancaster, PA.

Spellman, F.R. and Whiting, N. (2006). *Environmental Science and Technology: Concepts and Applications*. CRC Press, Boca Raton, FL.

STRONGER. (2008a). *History of STRONGER—Helping to Make an Experiment Work*. State Review of Oil and Natural Gas Regulations, Oklahoma City, OK (www.strongerinc.org/about/history.asp).

STRONGER. (2008b). *List of State Reviews*. State Review of Oil and Natural Gas Regulations, Oklahoma City, OK (www.strongerinc.org/reviews/reviews.asp).

Svoboda, E. (2010). The hard facts about fracking. *Popular Mechanics*, December 13 (www.popularmechanics.com/science/energy/coal-oil-gas/the-hard-facts-about-fracking).

Tower, E. (1995). *Environmental and Natural and Natural Resource Economics*. Eno River Press, New York.

Turner, D.B. (1994). *Workbook of Atmospheric Dispersion Estimates*, 2nd ed. CRC Press, Boca Raton, FL.

Urone, P. (1976). The primary air pollutants—gaseous: their occurrence, sources, and effects, in *Air Pollution*, Vol. 1, Stern, A.C., Ed. Academic Press, New York, pp. 23–75.

URS. (2008). *Sampling and Analysis Plan for Exploration and Production: Pit Solids and Fluids in Colorado Energy Basins*, Revision B. URS Corporation, Denver, CO.

USACE. (1987). *Safety and Health Requirements Manual*, rev. ed., EM 385-1-1. U.S. Army Corps of Engineers, Washington, DC.

USCC. (2009). *Analysis of Workers' Compensation Laws*. U.S. Chamber of Commerce, Washington, DC.

USDHHS. (2000). *Worker Deaths by Falls: A Summary of Surveillance Findings and Investigative Case Reports*. U.S. Department of Health and Human Services, Washington, DC.

USEPA. (1988). *Regulatory Determination for Oil and Gas and Geothermal Exploration, Development and Production Wastes*. U.S. Environmental Protection Agency, Washington, DC.

USEPA. (1998). *RCRA: Superfund and EPCRA Hotline Training Module*, EPA 540-R-98-022. U.S. Environmental Protection Agency, Washington, DC.

USEPA. (1999). *USEPA's Program to Regulate the Placement of Waste Water and Other Fluids Underground*, EPA 810-F-99-019. U.S. Environmental Protection Agency, Washington, DC.

USEPA. (2002). *Exemption of Oil and Gas Exploration and Production Wastes from Federal Hazardous Waste Regulations*. U.S. Environmental Protection Agency, Washington, DC (epa.gov/osw/nonhaz/industrial/special/oil/oil-gas.pdf).

USEPA. (2003). *Underground Injection Control Program*. U.S. Environmental Protection Agency, Washington, DC (http://water.epa.gov/type/groundwater/uic/index.cfm).

USEPA. (2004). *Understanding the Safe Drinking Water Act*, EPA 816-F-04-030. U.S. Environmental Protection Agency, Washington, DC (http://water.epa.gov/lawsregs/guidance/sdwa/basicinformation.cfm).

USEPA. (2005). *Basic Air Pollution Meteorology*. U.S. Environmental Protection Agency, Washington, DC (www.epa.gov/apti).

USEPA. (2007a). *National Ambient Air Quality Standards (NAAQS)*. U.S. Environmental Protection Agency, Washington, DC (www.epa.gov/air/criteria.html).

USEPA. (2007b). *Final Emission Standards of Performance for Stationary Spark Ignition Internal Combustion Engines; and Final Air Toxics Standards for Reciprocating Internal Combustion Engines*. U.S. Environmental Protection Agency, Washington, DC (www.epa.gov/ttn/atw/nsps/sinsps/sinspspg.html).

USEPA. (2008a). *Summary of the Clean Water Act*. U.S. Environmental Protection Agency, Washington, DC (www.epa.gov/lawsregs/laws/cwa.html).

USEPA. (2008b). *Water Quality and Technology-Based Permitting*. U.S. Environmental Protection Agency, Washington, DC (http://cfpub.epa.gov/npdes/generalissues/watertechnology.cfm).

USEPA. (2008c). *Effluent Limitations Guidelines and Standards*. U.S. Environmental Protection Agency, Washington, DC (http://cfpub.epa.gov/npdes/techbasedpermitting/effguide.cfm).

USEPA. (2008d). *Final Rule: Amendments to the Storm Water Regulations for Discharges Associated with Oil and Gas Construction Activities*. U.S. Environmental Protection Agency, Washington, DC (www.epa.gov/npdes/regulations/final_oil_gas_factsheet.pdf).

USEPA. (2008e). *The Green Book Nonattainment Areas for Criteria Pollutants*. U.S. Environmental Protection Agency, Washington, DC (www.epa.gov/oar/oaqps/greenbk/).

USEPA. (2008f). *Air Trends: Basic Information*. U.S. Environmental Protection Agency, Washington, DC (www.epa.gov/airtrends/sixpoll.html).

USEPA. (2008g). *Region 2. Solid Waste; RCRA Subtitle D*. U.S. Environmental Protection Agency, Washington, DC (www.epa.Gov/region2/waste/dsummary.htm).

USEPA. (2008h). *Region 9. Superfund*. U.S. Environmental Protection Agency, Washington, DC (http://www.epa.gov/region9/superfund/).

USEPA. (2008i). *Water. State, Tribal & Territorial Standards*. U.S. Environmental Protection Agency, Washington, DC (www.epa.gov/waterscience/standards/wqslibrary/).

USEPA. (2008j). *Summary of the Oil Pollution Act*. U.S. Environmental Protection Agency, Washington, DC (www.epa.gov/lawsregs/laws/opa.html).

USEPA. (2010). *Ozone Science: The Facts Behind the Phase-Out*. U.S. Environmental Protection Agency, Washington, DC (www.epa.gov/ozone/science/sc_fact.html).

USFS. (2011). *Endangered Species Program: Laws and Policies*. U.S. Fish and Wildlife Service, Arlington, VA (http://www.fws.gov/endangered/laws-policies/regulations-and-policies.html).

Walker, M. (1963). *The Nature of Scientific Thought*. Prentice Hall, Englewood Cliffs, NJ.

WEF. (1994). *Confined Space Entry*. Water Environment Federation, Alexandria, VA.

WRI. (1988). *World Resources 1988–1989: An Assessment of the Resource Base that Supports the Global Economy*. World Resources Institute, Washington, DC.

Zurer, P.S. (1988). Studies on ozone destruction expand beyond Antarctic. *C & E News*, 66:16–25.

Glossary

Abandonment pressure: The lowest gas pressure before a gas well must be abandoned.

Accelerator: An additive that increases the rate of a process such as cement setting.

Acid gas: A corrosive gas such as hydrogen sulfide or carbon dioxide that forms an acid with water.

Acid job: Refers to when acid is poured or pumped down a well to dissolve limestone and increase fluid flow.

Adsorption: Adhesion of gas molecules, ions, or molecules in solution to the surface of solid bodies.

Air drilling (pneumatic drilling): Rotary drilling with air pumped down the drill string instead of circulating drilling mud.

Alluvial aquifer: A water-bearing deposit of unconsolidated material (e.g., sand, gravel) left behind by a river or other flowing water.

Amphoteric: Having both basic and acidic properties.

Anaerobic bacteria: Bacteria that thrive in oxygen-poor environments.

Anisotropic: Having some physical property that varies with direction from a given location.

Annulus: The space between the casing (the material, typically steel, that is used to keep the well stable) in a well and the wall of the hole, or between two concentric strings of casing, or between casing and tubing.

Anticline: A fold of layered, sedimentary rocks whose core contains stratigraphically older rocks; the shape of the fold is generally convex upward.

Antifoam: An additive used to reduce foam.

Argillaceous: Shaly.

Associated gas: Natural gas that is in contact with crude oil in the reservoir.

Attenuate: To reduce the amplitude of sound pressure (noise).

Audible range: The frequency range over which normal ears hear, approximately 20 to 20,000 Hz.

Audiogram: A chart, graph, or table resulting from an audiometric test showing an individual's hearing threshold levels as a function of frequency.

Audiologist: A professional, specializing in the study and rehabilitation of hearing, who is certified by the American Speech–Language–Hearing Association or licensed by a state board of examiners.

Aulacogen: A long, narrow rift in a continent, often filled with thick sediments.

Aureole: A ring surrounding a volcanic intrusion where the surrounding rock has been altered.
Authigenic: Refers to a mineral that was formed by a chemical reaction in the subsurface.
Azimuth: The direction of a horizontal line as measured on an imaginary horizontal circle
Backflush: To pump an injected fluid back out of a well.
Background noise: Noise coming from sources other than the particular noise sources being monitored.
Backoff operation: Method used to remove stuck pipe from a well.
Barrels of oil equivalent (BOE): The amount of natural gas that has the same heat content as an average barrel of oil. It is about 6000 cubic feet of gas.
Baseline audiogram: The audiogram against which future audiograms are compared.
Bedrock aquifer: An aquifer located in the solid rock underlying unconsolidated surface materials (i.e., sediment). Solid rock can bear water when it is fractured.
Bentonite: A clay mineral used to make common drilling mud.
Billion cubic feet (Bcf): A unit typically used to define gas production volumes in the coalbed methane industry; 1 Bcf is roughly equivalent to the volume of gas required to heat approximately 12,000 households for 1 year (based on 2001 U.S. Department of Energy average household energy consumption statistics).
Biogenic: A direct product of the physiological activities of organisms.
Biotite: Black mica.
Bitumen: Solid hydrocarbons such as tar in sedimentary rocks.
Bituminous: From the base word *bitumen*, a general term for various solid and semisolid hydrocarbons that are able to join together and are soluble in carbon bisulfide (e.g., asphalts).
Blowout: An uncontrolled flow of fluid from a well.
Bottom water: A mixture of freshwater and brine.
Breaker: A fracturing fluid additive that breaks down the viscosity of the fluid.
Breccia: A coarse-grained clastic rock composed of angular broken rock fragments held together by a mineral cement or a fine-grained matrix.
Brecciated: Consisting of angular fragments cemented together.
British thermal unit (Btu): A unit of measure used to define energy.
Butt cleat: A short, poorly defined vertical cleavage plane in a coal seam, usually at right angles to the long face cleat; the coal cleat set that abuts into face cleats.
Buttress sand: Sand deposited on top of an unconformity.
Caprock: (1) Impermeable rock layer that forms the seal on top of an oil or gas reservoir; or (2) insoluble rock on the top of a salt plug.

Capture zone: The portion of an aquifer that contributes water to a particular pumping well.

Carboxymethyl hydroxypropyl guar (CMHPG): A form of guar gel.

Casing: Steel pipe that is relatively slim walled and large diameter (5.5 to 13.37 inches). Joints of casing are screwed together to form a casing string, which is run into a well and cemented to the sides of the well.

Casinghead gas: Natural gas that bubbles out of oil on the surface of the well.

Cathead: A hub on a shaft (catshaft) on the drawworks of a drilling rig that is used to pull a line (catline) to lift or pull equipment.

Cavitation cycling: Also known as *cavity completion*, an alternative completion technique to hydraulic fracturing, in which a cavity is generated by alternately pumping in nitrogen and blowing down pressure.

Cement: (1) Minerals that naturally grow between clastic grains and solidify a sedimentary rock; or (2) Portland cement used to bind the casing strings to the well walls.

Cement job: Refers to cementing casing into a well.

Centralizer: An attachment to the outside of a casing string that uses steel bands to keep the string central in the well.

Charcoal test: A test used to measure the amount of condensate in natural gas. Activated charcoal is used to absorb the condensate from a volume of natural gas.

Christmas tree: The fittings, valves, and gauges that are bolted to the wellhead of a flowing well to control the flow from the well.

Clean sands: Well-sorted sands.

Cleats: Natural fractures in coal that often occur in systematic sets through which gas and water can flow.

Compounder: A system of pulleys, belts, shafts, chains, and gears that transmit power from the prime movers to the drilling rig.

Connate: Saline, subsurface water.

Coquina: Sedimentary rock composed of broken shells.

Craton: A part of the Earth's crust that has attained stability and has been relatively undeformed for a long time; the term is restricted to continents and includes both shield and platform.

Criterion sound level: A sound level of 90 decibels.

Crosslinked gel: A gel to which a crosslinker has been added.

Crosslinker: An additive that, when added to a linear gel, creates a complex, high-viscosity, pseudoplastic fracturing fluid.

Crude stream: Crude oil from a single field or a mixture from fields that is offered for sale by an exporting country.

Cyclotherm: Alternating marine and nonmarine sedimentary rocks.

Darcy: A measure of the permeability of rock or sediment.

Decibel (dB): Unit of measurement of sound level.

Demulsifier: A chemical used to break an emulsion.

Desorption: Liberation of tightly held methane gas molecules previously bound to the solid surface of the coal.

Detrital: A sediment grain that has been transported and deposited to a whole particle such as a sand grain.

Dip-slip fault: A fault with predominately vertical displacement. It can be either a normal or reverse dip-slip fault.

Doghouse: The room or vehicle that houses the seismic recording equipment.

Dolomite: A mineral composed of $CaMg(CO_3)_2$ and formed by the natural alteration of calcite. A rock composed of dolomite is called *dolostone* and can be a reservoir rock.

Double hearing protection: A combination of both earplug and earmuff types of hearing protection devices is required for employees who have demonstrated temporary threshold shift during audiometric examination and for those who have been advised by a medical doctor to wear double protection in work areas that exceed 104 dBA.

Drawworks: A drum in a steel frame used on the floor of a drilling rig to raise and lower equipment in a well. It is driven by the prime movers. Hoisting line is wound around the reel.

Drilling mud: A viscous mixture of clay (usually bentonite) and additives with water, oil, an emulsion of water with droplets of oil, or a synthetic fluid.

Duster: A well that did not encounter commercial amounts of petroleum.

Edge water: Water located in the reservoir to the side of the oil.

Emulsion: Droplets of one liquid suspended in a different liquid, such as water in oil.

Epiclastic: Formed from the fragments or particles broken away (by weathering and erosion) from preexisting rocks to form an altogether new rock in a new place.

Evapotranspiration: The portion of precipitation returned to the air through evaporation and transpiration.

Explosive fracturing: To explode nitroglycerin in a torpedo at reservoir depth in a well to fracture the reservoir and stimulate production.

Face cleat: The major joint or cleavage system in a coal seam.

Fairway (or trend): The area along which the play has been proven and more fields could be found.

Flowback: The process of causing fluid to flow back to the well of a fracture after a hydraulic fracturing event is complete.

Formation: A mappable rock layer; it has a sharp top and bottom.

Fracture conductivity: The capability of the fracture to conduct fluids under a given hydraulic head difference.

Frequency: Rate at which pressure oscillations are produced; measured in hertz (Hz).

Gas cap: The uppermost portion of a saturated oil reservoir.

Gathering system: A system of flowlines that conducts produced fluids from wells to a central processing unit.

Geophone: A seismic detector, placed on or in the ground, that responds to ground motion at its point of location.

Geothermal gradient: Increase in temperature with depth in the Earth.

Graben: An elongated depression bounded by nearly parallel faults.

Graywacke: A poorly sorted, dark-colored sandstone.

Guar: Organic powder thickener, typically used to make viscous fracturing fluids; it is completely soluble in hot and cold water and insoluble in oils, grease, and hydrocarbons.

HCl: Molecular formula for hydrochloric acid, which can be used in diluted form in the hydraulic fracturing process to fracture limestone formations and to clean up perforations in coalbed methane fracturing treatments.

Hearing conservation record: Employee's audiometric record that includes name, age, job classification, time-weighted average (TWA) exposure, date of audiogram, and name of audiometric technician. It is to be retained for the duration of employment for OSHA and kept indefinitely for workers' compensation.

Heavy oil: Viscous high-density oil with gravity less than 25°API.

Hertz (Hz): Unit of measurement of frequency, numerically equal to cycles per second.

Hydraulic conductivity: See *permeability*.

Hydraulic fracturing: A well-stimulation method in which liquid under high pressure is pumped down a well to fracture the reservoir rock adjacent to the wellbore. Propping agents are used to keep the fractures open.

Hydroxyethylcellulose (HEC): A form of guar gel.

Injectate: In relation to the coalbed methane industry, this is the fracturing fluid injected into a coalbed methane well.

Interfinger: A boundary between two rock types in which both form distinctive wedges protruding into each other.

Isopach: A line on a map connecting points of equal true thickness of a designated stratigraphic unit or group of stratigraphic units.

Isotopic: Rocks formed in the same environment (i.e., in the same sedimentary basin or geologic province).

Isotropic: A medium, such as unconsolidated sediments or a rock formation, whose properties are the same in all directions.

Junk: A tool or broken pipe that has fallen to the bottom of a well.

KCl: Molecular formula for potassium chloride.

Kelly: A strong four- or six-sided steel pipe that is located at the top of the drill string; it runs through the kelly bushing.

Kerogen: Insoluble organic matter in sedimentary rocks.

Lacustrine: Pertaining to, produced by, or formed in a lake or lakes.

Laminar flow: Water flow in which the stream lines remain distinct and the flow direction at every point remains unchanged with time; nonturbulent flow.

Leakoff: The magnitude of pressure exerted on a formation that causes fluid to be forced into the formation. In common usage, leakoff is often considered the movement of fluid out of primary fractures and into a geologic formation, either through small existing permeable paths (connected pores and natural fracture networks) or through small pathways created or enlarged in the rock through the fracturing process.

Lenticular: Pertaining to a discontinuous lens-shaped (saucer-shaped) stratigraphic body.

Lift gas: Inert gas, usually natural gas, that is used for gas lift.

Linear gel: A simple guar-based fracturing fluid usually formulated using guar and water with additives or guar with diesel fuel.

Lithology: Description of rocks based on mineralogic composition and texture.

Make up: To screw together pipe.

Medical pathology: A disorder or disease; for example, a condition or disease affecting the ear that is treated by a physician specialist.

Milligrams per liter (mg/L): Typically used to define the concentration of a compound dissolved in a fluid.

Millidarcy (md): The customary unit of measurement of fluid permeability; equivalent to 0.001 darcy.

Million cubic feet (MMcf): A unit typically used to define gas production volumes in the coalbed methane industry; 1 MMcf is roughly equivalent to the volume of gas required to heat approximately 12 households for one year (based on 2001 U.S. Department of Energy average household energy consumption statistics).

Mined-through studies: Projects in which coalbeds have been actually mined through (i.e., the coal has been removed) so that remaining coal and surrounding rock can be inspected, after the coalbeds have been hydraulically fractured. These studies provide unique subsurface access to investigate coalbeds and surrounding rock after hydraulic fracturing.

Moduli (plural of modulus): Also referred to as *bulk modulus*, the ratio of stress to strain, abbreviated as k. It is an elastic constant equal to the applied stress divided by the ratio of the change in volume to the original volume of a body.

Natural gas: A gas composed of a mixture of hydrocarbon molecules that have one, two, three, and four carbon atoms.

NIOSH: National Institute of Occupational Safety and Health.

Noise dose: As defined by OSHA, the ratio, expressed as a percentage, of (1) the time integral, over a stated time or event, of the 0.6 power of the measured SLOW exponential time-averaged, squared, A-weighted sound pressure; or (2) the product of the criterion duration (8 hours) and the 0.6 power of the squared sound pressure corresponding to the criterion sound level (90 dB).

Noise dosimeter: An instrument that integrates a function of sound pressure over a period of time to directly indicate a noise dose.

Noise hazard area: Any area where noise levels are equal to or exceed 85 dBA. OSHA requires employers to designate such work areas as such, post warning signs, and warn employees when work practices exceed 90 dBA. Hearing protection must be worn whenever a level of 90 dBA is reached or exceeded.

Noise hazard work practice: Performing or observing work where 90 dBA is equaled or exceeded; however, some work practices will be specified as a rule of thumb. An example is when shouting must be employed to be heard when attempting to hold a normal conversation with someone who is 1 foot away; one can assume that a 90-dBA noise level or greater exists and that hearing protection is required. Typical examples of work practices where hearing protection is required are jackhammering, heavy grinding, heavy equipment operations, and similar activities.

Noise level measurement: Total sound level within an area; includes workplace measurements indicating the combined sound levels of tool noise (e.g., from ventilation systems, cooling compressors, circulation pumps).

Noise reduction ratio: The number of decibels of sound reduction actually achieved by a particular hearing protection device.

Oilfield brine: Very saline water that is produced with oil.

Otolaryngologist: A physician specializing in the diagnosis and treatment of disorders of the ear, nose, and throat.

Otoscopic examination: Inspection of the external ear canal and tympanic membrane.

Overthrust: A large-scale, low-angle thrust fault, with total displacement (lateral or vertical) generally measured in kilometers.

Pad: An initial volume of fluid that is used to initiate and propagate a fracture before a proppant is placed.

Paleochannels: Old or ancient river channels preserved in the subsurface as lenticular sandstones.

Paraffin: A member of the hydrocarbon series of molecules. They are straight chains with single bonds. All hydrocarbon molecules in natural gas and some in crude oil are paraffins.

Parts per million (ppm): Number of weight or volume units of a constituent present with each 1 million units of a solution or mixture. Formerly used to express the results of most water and wastewater analyses, parts per million is being replaced by milligrams per liter (mg/L). For drinking water analyses, concentration in parts per million and milligrams per liter are equivalent. Comparatively, a single PPM can be compared to a shot glass full of water inside a swimming pool.

Permanent threshold shift (PTS): Hearing loss with less than normal recovery.

Permeability: The capacity of a porous rock, sediment, or soil to transmit a fluid; it is a measure of the relative ease of fluid flow under equal pressure and from equal elevations.

Personal protective device: Examples include earplugs or earmuffs worn as protection against hazardous noise.

Physiographic: Refers to a region where all parts are similar in geologic structure and climate and which has had a unified geomorphic history; its relief features differ significantly from those of adjacent regions.

Play: A productive coalbed methane formation or a productive oil or gas deposit; a particular combination of trap, reservoir rock, and seal that has been shown by previously discovered fields to contain natural gas and oil.

Potentiometric: The total head of groundwater, defined by the level to which water will rise in a well.

Pounds per square inch (psi): A unit of pressure.

Presbycusis: Hearing loss due to age.

Primacy: The right to self-establish, self-enforce, and self-regulate environmental standards; this enforcement responsibility is granted by the USEPA to states and Indian tribes.

Primary porosity: The porosity preserved from some time between sediment deposition and the final rock-forming process (e.g., spaces between grains of sediment).

Proppant: Granules of sand, ceramic, or other minerals that are wedged within the fracture and act to prop it open after the fluid pressure from fracture injection has dissipated.

Prospect: Location where the geological and economical conditions are favorable for drilling an exploratory well.

Quartz: A common mineral composed of SiO_2. Sandstones are usually composed of quartz sand grains.

Rank: The degree of metamorphism in coal; the basis of coal classification into a natural series from lignite to anthracite.

Recovery factor: The percentage of oil and gas in place that will be produced from a reservoir.

Representative exposure: Measurements of an employee's noise dose or 8-hour time-weighted average (TWA) sound level that the employer deems to be representative of the exposures of other employees in the workplace.

Sample log: A record of the physical properties of rocks in a well. It includes composition, texture, color, presence of pore spaces, and oil staining.

Scale: Salts that have precipitated out of water. Calcium carbonate, barium sulfate, and calcium sulfate are common in oil fields.

Screen-out: Refers to a fracturing job where proppant placement has failed.

Secondary porosity: The porosity created through alteration of rock, commonly by processes such as dissolution and fracturing.

Semianthracite: Term used to identify coal rank; specifically refers to coal that possesses a fixed-carbon content of 86 to 92%.

Sensorineural: Type of hearing loss characterized as having been induced by industrial noise exposure. This hearing loss is permanent.

Shale: Very common sedimentary rock composed of clay-sized particles. Black shales are source rocks for petroleum.

Siltstone: Sedimentary rock composed primarily of silt-sized particles.

Solution gas: The dissolved natural gas that bubbles out of crude oil on the surface when the pressure drops during production.

Sound level: As defined by OSHA, ten times the common logarithm of the ratio of the square of the measured A-weighted sound pressure to the square of the standard reference pressure of 20 micropascals; measured in decibels (dB).

Sound level meter: An instrument that measures sound levels.

Stratigraphy: The study of rock strata, including all characteristics and attributes of rocks and their interpretation in terms of mode of origin and geologic history.

Subbituminous: A black coal, intermediate in rank between lignite and bituminous.

Subgraywacke: Sedimentary rock (sandstone) that contains less feldspar and more and better-rounded quartz grains than graywacke. Intermediate in composition between graywacke and orthoquartzite, it is lighter colored and better sorted and has less matrix than graywacke.

Surficial: Pertaining to or lying in or on a surface; specific to the surface of the Earth.

Syncline: A fold of layered, sedimentary rocks whose core contains stratigraphically younger rocks; the shape of the fold is generally concave upward.

Tank battery: Two or more stock tanks connected in line.

Temporary threshold shift (TTS): Temporary loss of normal hearing level brought on by brief exposure to high-level sound. TTS is greatest immediately after exposure to excessive noise and progressively diminishes with increasing rest time.

Thermogenic: A direct product of high temperatures (e.g., thermogenic methane).

Time-weighted average (TWA) sound level: The sound level that, if constant over an 8-hour exposure, would result in the same noise dose as is measured.

Toughness: The point at which enough stress intensity has been applied to a rock formation so that a fracture initiates and propagates.

Transmissivity: A measure of the amount of water that can be transmitted horizontally through a unit width by the full saturated thickness of the aquifer under a hydraulic gradient of one.

Trillion cubic feet (Tcf): A unit typically used to define gas production volumes in the coalbed methane industry; 1 Tcf is roughly equivalent to the volume of gas required to heat approximately 12 million households for one year (based on 2001 U.S. Department of Energy average household energy consumption statistics).

Unsaturated pool: An oil reservoir without a free gas cap.

Upwarp: The uplift of a region; usually a result of the release of isostatic pressure (e.g., the melting of an ice sheet).

Viscosity: The property of a substance to offer internal resistance to flow; internal friction.

Volcaniclastic: Composed of fragments or particles and related to volcanic processes either by forming as the result of explosive processes or due to the weathering and erosion of volcanic rocks.

Water table: The subsurface level below which the pores in the soil or rock are filled with water.

Zone: A rock layer identified by a characteristic microfossil species.

Appendix A. Chemicals Used in Hydraulic Fracturing

Chemical Component	Chemical Abstract Service Number	No. of Products Containing Chemical
1-(1-Naphthylmethyl)quinolinium chloride	65322-65-8	1
1,2,3-Propanetricarboxylic acid, 2-hydroxy-, trisodium salt, dihydrate	6132-04-3	1
1,2,3-Trimethylbenzene	526-73-8	1
1,2,4-Trimethylbenzene	95-63-6	21
1,2-Benzisothiazol-3	2634-33-5	1
1,2-Dibromo-2,4-dicyanobutane	35691-65-7	1
1,2-Ethanediaminium, *N,N′-bis*[2-[*bis*(2-hydroxyethyl) methylammonio]ethyl]-*N,N′-bis*(2-hydroxyethyl)-*N,N′*-dimethyl-, tetrachloride	138879-94-4	2
1,3,5-Trimethylbenzene	108-67-8	3
1,6-Hexanediamine dihydrochloride	6055-52-3	1
1,8-Diamino-3,6-dioxaoctane	929-59-9	1
1-Hexanol	111-27-3	1
1-Methoxy-2-propanol	107-98-2	3
2,2′-Azobis (2-amidopropane) dihydrochloride	2997-92-4	1
2,2-Dibromo-3-nitrilopropionamide	10222-01-2	27
2-Acrylamido-2-methylpropanesulphonic acid sodium salt polymer	—[a]	1
2-Bromo-2-nitropropane-1,3-diol	52-51-7	4
2-Butanone oxime	96-29-7	1
2-Hydroxypropionic acid	79-33-4	2
2-Mercaptoethanol (thioglycol)	60-24-2	13
2-Methyl-4-isothiazolin-3-one	2682-20-4	4
2-Monobromo-3-nitrilopropionamide	1113-55-9	1
2-Phosphonobutane-1,2,4-tricarboxylic acid	37971-36-1	2
2-Phosphonobutane-1,2,4-tricarboxylic acid, potassium salt	93858-78-7	1
2-Substituted aromatic amine salt	—[a]	1
4,4′-Diaminodiphenyl sulfone	80-08-0	3
5-Chloro-2-methyl-4-isothiazolin-3-one	26172-55-4	5
Acetaldehyde	75-07-0	1
Acetic acid	64-19-7	56
Acetic anhydride	108-24-7	7
Acetone	67-64-1	3
Acetophenone	98-86-2	1

Chemical Component	Chemical Abstract Service Number	No. of Products Containing Chemical
Acetylenic alcohol	—[a]	1
Acetyltriethyl citrate	77-89-4	1
Acrylamide	79-06-1	2
Acrylamide copolymer	—[a]	1
Acrylamide copolymer	38193-60-1	1
Acrylate copolymer	—[a]	1
Acrylic acid, 2-hydroxyethyl ester	818-61-1	1
Acrylic acid/2-acrylamido-methylpropylsulfonic acid copolymer	37350-42-8	1
Acrylic copolymer	403730-32-5	1
Acrylic polymers	—[a]	1
Acrylic polymers	26006-22-4	2
Acyclic hydrocarbon blend	—[a]	1
Adipic acid	124-04-9	6
Alcohol alkoxylate	—[a]	5
Alcohol ethoxylates	—[a]	2
Alcohols	—[a]	9
Alcohols, C11–C15-secondary, ethoxylated	68131-40-8	1
Alcohols, C12–C14-secondary	126950-60-5	4
Alcohols, C12–C14-secondary, ethoxylated	84133-50-6	19
Alcohols, C12–C15, ethoxylated	68131-39-5	2
Alcohols, C12–C16, ethoxylated	103331-86-8	1
Alcohols, C12–C16, ethoxylated	68551-12-2	3
Alcohols, C14–C15, ethoxylated	68951-67-7	5
Alcohols, C9–C11-iso-, C10-rich, ethoxylated	78330-20-8	4
Alcohols, C9–C22	—[a]	1
Aldehyde	—[a]	4
Aldol	107-89-1	1
α-Alumina	—[a]	5
Aliphatic acid	—[a]	1
Aliphatic alcohol polyglycol ether	68015-67-8	1
Aliphatic amine derivative	120086-58-0	2
Alkaline bromide salts	—[a]	2
Alkanes, C10–C14	93924-07-3	2
Alkanes, C13–C16-*iso*	68551-20-2	2
Alkanolamine	150-25-4	3
Alkanolamine chelate of zirconium alkoxide (zirconium complex)	197980-53-3	4
Alkanolamine/aldehyde condensate	—[a]	1
Alkenes	—[a]	1
Alkenes, C > 10	64743-02-8	3
Alkenes, C > 8	68411-00-7	2

Chemical Component	Chemical Abstract Service Number	No. of Products Containing Chemical
Alkoxylated alcohols	—[a]	1
Alkoxylated amines	—[a]	6
Alkoxylated phenol formaldehyde resin	63428-92-2	1
Alkyaryl sulfonate	—[a]	1
Alkyl (C12–C16) dimethyl benzyl ammonium chloride	68424-85-1	7
Alkyl (C6–C12) alcohol, ethoxylated	68439-45-2	2
Alkyl (C9–C11) alcohol, ethoxylated	68439-46-3	1
Alkyl alkoxylate	—[a]	9
Alkyl amine	—[a]	2
Alkyl amine blend in a metal salt solution	—[a]	1
Alkyl aryl amine sulfonate	255043-08-04	1
Alkyl benzenesulfonic acid	68584-22-5	2
Alkyl esters	—[a]	2
Alkyl hexanol	—[a]	1
Alkyl orthophosphate ester	—[a]	1
Alkyl phosphate ester	—[a]	3
Alkyl quaternary ammonium chlorides	—[a]	4
Alkylaryl sulfonate	—[a]	1
Alkylaryl sulfonic acid	27176-93-9	1
Alkylated quaternary chloride	—[a]	5
Alkylbenzenesulfonic acid	—[a]	1
Alkylethoammonium sulfates	—[a]	1
Alkylphenol ethoxylates	—[a]	1
Almandite and pyrope garnet	1302-62-1	1
Aluminium isopropoxide	555-31-7	1
Aluminum	7429-90-5	2
Aluminum chloride	—[a]	3
Aluminum chloride	1327-41-9	2
Aluminum oxide (α-alumina)	1344-28-1	24
Aluminum oxide silicate	12068-56-3	1
Aluminum silicate (mullite)	1302-76-7	38
Aluminum sulfate hydrate	10043-01-3	1
Amides, tallow, *n*-[3-(dimethylamino)propyl], *n*-oxides	68647-77-8	4
Amidoamine	—[a]	1
Amine	—[a]	7
Amine bisulfite	13427-63-9	1
Amine oxides	—[a]	1
Amine phosphonate	—[a]	3
Amine salt	—[a]	2
Amines, C14–C18; C16–C18-unsaturated, alkyl, ethoxylated	68155-39-5	1

Chemical Component	Chemical Abstract Service Number	No. of Products Containing Chemical
Amines, coco alkyl, acetate	61790-57-6	3
Amines, polyethylenepoly-, ethoxylated, phosphonomethylated	68966-36-9	1
Amines, tallow alkyl, ethoxylated	61791-26-2	2
Amino compounds	—[a]	1
Amino methylene phosphonic acid salt	—[a]	1
Amino trimethylene phosphonic acid	6419-19-8	2
Ammonia	7664-41-7	7
Ammonium acetate	631-61-8	4
Ammonium alcohol ether sulfate	68037-05-8	1
Ammonium bicarbonate	1066-33-7	1
Ammonium bifluoride (ammonium hydrogen difluoride)	1341-49-7	10
Ammonium bisulfate	7783-20-2	3
Ammonium bisulfite	10192-30-0	15
Ammonium C6–C10 alcohol ethoxysulfate	68187-17-7	4
Ammonium C8–C10 alkyl ether sulfate	68891-29-2	4
Ammonium chloride	12125-02-9	29
Ammonium fluoride	12125-01-8	9
Ammonium hydroxide	1336-21-6	4
Ammonium nitrate	6484-52-2	2
Ammonium persulfate (diammonium peroxidisulfate)	7727-54-0	37
Ammonium salt	—[a]	1
Ammonium salt of ethoxylated alcohol sulfate	—[a]	1
Amorphous silica	99439-28-8	1
Amphoteric alkyl amine	61789-39-7	1
Anionic copolymer	—[a]	3
Anionic polyacrylamide	—[a]	1
Anionic polyacrylamide	25085-02-3	6
Anionic polyacrylamide copolymer	—[a]	3
Anionic polymer	—[a]	2
Anionic polymer in solution	—[a]	1
Anionic polymer, sodium salt	9003-04-7	1
Anionic water-soluble polymer	—[a]	2
Antifoulant	—[a]	1
Antimonate salt	—[a]	1
Antimony pentoxide	1314-60-9	2
Antimony potassium oxide	29638-69-5	4
Antimony trichloride	10025-91-9	2
a-Organic surfactants	61790-29-8	1
Aromatic alcohol glycol ether	—[a]	2
Aromatic aldehyde	—[a]	2

Chemical Component	Chemical Abstract Service Number	No. of Products Containing Chemical
Aromatic ketones	224635-63-6	2
Aromatic polyglycol ether	—[a]	1
Barium sulfate	7727-43-7	3
Bauxite	1318-16-7	16
Bentonite	1302-78-9	2
Benzene	71-43-2	3
Benzene, C10–C16, alkyl derivatives	68648-87-3	1
Benzenecarboperoxoic acid, 1,1-dimethylethyl ester	614-45-9	1
Benzenemethanaminium	3844-45-9	1
Benzenesulfonic acid, C10–C16-alkyl derivatives, potassium salts	68584-27-0	1
Benzoic acid	65-85-0	11
Benzyl chloride	100-44-7	8
Biocide component	—[a]	3
bis(1-Methylethyl)naphthalenesulfonic acid, cyclohexylamine salt	68425-61-6	1
bis(Hexamethylenetriamine) penta(methylene phosphonic acid)	35657-77-3	1
Bisphenol A/epichlorohydrin resin	25068-38-6	5
Bisphenol A/novolac epoxy resin	28906-96-9	1
Borate	12280-03-4	2
Borate salts	—[a]	5
Boric acid	10043-35-3	18
Boric acid, potassium salt	20786-60-1	1
Boric acid, sodium salt	1333-73-9	2
Boric oxide	1303-86-2	1
β-Tricalcium phosphate	7758-87-4	1
Butanedioic acid	2373-38-8	4
Butanol	71-36-3	3
Butyl glycidyl ether	2426-08-6	5
Butyl lactate	138-22-7	4
C10–C16 ethoxylated alcohol	68002-97-1	4
C11–C14 *n*-alkanes, mixed	—[a]	1
C12–C14 alcohol, ethoxylated	68439-50-9	3
Calcium carbonate	471-34-1	1
Calcium carbonate (limestone)	1317-65-3	9
Calcium chloride	10043-52-4	17
Calcium chloride, dihydrate	10035-04-8	1
Calcium fluoride	7789-75-5	2
Calcium hydroxide	1305-62-0	9
Calcium hypochlorite	7778-54-3	1
Calcium oxide	1305-78-8	6

Chemical Component	Chemical Abstract Service Number	No. of Products Containing Chemical
Calcium peroxide	1305-79-9	5
Carbohydrates	—[a]	3
Carbon dioxide	124-38-9	4
Carboxymethyl guar gum, sodium salt	39346-76-4	7
Carboxymethyl hydroxypropyl guar	68130-15-4	11
Cellophane	9005-81-6	2
Cellulase	9012-54-8	7
Cellulase enzyme	—[a]	1
Cellulose	9004-34-6	1
Cellulose derivative	—[a]	2
Chloromethylnaphthalene quinoline quaternary amine	15619-48-4	3
Chlorous ion solution	—[a]	2
Choline chloride	67-48-1	3
Chromates	—[a]	1
Chromium (iii) acetate	1066-30-4	1
Cinnamaldehyde (3-phenyl-2-propenal)	104-55-2	5
Citric acid (2-hydroxy-1,2,3 propanetricarboxylic acid)	77-92-9	29
Citrus terpenes	94266-47-4	11
Coal, granular	50815-10-6	1
Cobalt acetate	71-48-7	1
Cocaidopropyl betaine	61789-40-0	2
Cocamidopropylamine oxide	68155-09-9	1
Coco *bis*(2-hydroxyethyl) amine oxide	61791-47-7	1
Cocoamidopropyl betaine	70851-07-9	1
Cocomidopropyl dimethylamine	68140-01-2	1
Coconut fatty acid diethanolamide	68603-42-9	1
Collagen (gelatin)	9000-70-8	6
Complex alkylaryl polyo-ester	—[a]	1
Complex aluminum salt	—[a]	2
Complex organometallic salt	—[a]	2
Complex substituted keto-amine	143106-84-7	1
Complex substituted keto-amine hydrochloride	—[a]	1
Copolymer of acrylamide and sodium acrylate	25987-30-8	1
Copper	7440-50-8	1
Copper iodide	7681-65-4	1
Copper sulfate	7758-98-7	3
Corundum (aluminum oxide)	1302-74-5	48
Crotonaldehyde	123-73-9	1
Crystalline silica—cristobalite	14464-46-1	44
Crystalline silica—quartz (SiO_2)	14808-60-7	207
Crystalline silica, tridymite	15468-32-3	2

Chemical Component	Chemical Abstract Service Number	No. of Products Containing Chemical
Cumene	98-82-8	6
Cupric chloride	7447-39-4	10
Cupric chloride dihydrate	10125-13-0	7
Cuprous chloride	7758-89-6	1
Cured acrylic resin	—[a]	7
Cured resin	—[a]	4
Cured silicone rubber-polydimethylsiloxane	63148-62-9	1
Cured urethane resin	—[a]	3
Cyclic alkanes	—[a]	1
Cyclohexane	110-82-7	1
Cyclohexanone	108-94-1	1
Decanol	112-30-1	2
Decyl-dimethyl amine oxide	2605-79-0	4
Dextrose monohydrate	50-99-7	1
D-Glucitol	50-70-4	1
Di(2-ethylhexyl) phthalate	117-81-7	3
Di(ethylene glycol) ethyl ether acetate	112-15-2	4
Diatomaceous earth	61790-53-2	3
Diatomaceous earth, calcined	91053-39-3	7
Dibromoacetonitrile	3252-43-5	1
Dibutylaminoethanol (2-dibutylaminoethanol)	102-81-8	4
Di-calcium silicate	10034-77-2	1
Dicarboxylic acid	—[a]	1
Didecyl dimethyl ammonium chloride	7173-51-5	1
Diesel	—[a]	1
Diesel	68334-30-5	3
Diesel	68476-30-2	4
Diesel	68476-34-6	43
Diethanolamine (2,2-iminodiethanol)	111-42-2	14
Diethylbenzene	25340-17-4	1
Diethylene glycol	111-46-6	8
Diethylene glycol monomethyl ether	111-77-3	4
Diethylene triaminepenta (methylene phosphonic acid)	15827-60-8	1
Diethylenetriamine	111-40-0	2
Diethylenetriamine, tall oil fatty acids reaction product	61790-69-0	1
Diisopropylnaphthalenesulfonic acid	28757-00-8	2
Dimethyl formamide	68-12-2	5
Dimethyl glutarate	1119-40-0	1
Dimethyl silicone	—[a]	2
Dioctyl sodium sulfosuccinate	577-11-7	1

Chemical Component	Chemical Abstract Service Number	No. of Products Containing Chemical
Dipropylene glycol	25265-71-8	1
Dipropylene glycol monomethyl ether (2-methoxymethylethoxy propanol)	34590-94-8	12
Di-secondary-butylphenol	53964-94-6	3
Disodium EDTA	139-33-3	1
Disodium ethylenediaminediacetate	38011-25-5	1
Disodium ethylenediaminetetraacetate dihydrate	6381-92-6	1
Disodium octaborate tetrahydrate	12008-41-2	1
Dispersing agent	—[a]	1
D-Limonene	5989-27-5	11
Dodecyl alcohol ammonium sulfate	32612-48-9	2
Dodecylbenzene sulfonic acid	27176-87-0	14
Dodecylbenzene sulfonic acid salts	42615-29-2	2
Dodecylbenzene sulfonic acid salts	68648-81-7	7
Dodecylbenzene sulfonic acid salts	90218-35-2	1
Dodecylbenzenesulfonate isopropanolamine	42504-46-1	1
Dodecylbenzenesulfonic acid, monoethanolamine salt	26836-07-7	1
Dodecylbenzenesulphonic acid, morpholine salt	12068-08-5	1
EDTA/copper chelate	—[a]	2
EO-C7–C9-iso-, C8-rich alcohols	78330-19-5	5
Epichlorohydrin	25085-99-8	5
Epoxy resin	—[a]	5
Erucic amidopropyl dimethyl betaine	149879-98-1	3
Erythorbic acid	89-65-6	2
Essential oils	—[a]	6
Ethanaminium, *N,N,N*-trimethyl-2-[(1-oxo-2-propenyl)oxy]-, chloride, polymer with 2-propenamide	69418-26-4	4
Ethanol (ethyl alcohol)	64-17-5	36
Ethanol, 2-(hydroxymethylamino)-	34375-28-5	1
Ethanol, 2,2′-(octadecylamino)*bis*-	10213-78-2	1
Ethanoldiglycine disodium salt	135-37-5	1
Ether salt	25446-78-0	2
Ethoxylated 4-nonylphenol (nonylphenol ethoxylate)	26027-38-3	9
Ethoxylated alcohol	104780-82-7	1
Ethoxylated alcohol	78330-21-9	2
Ethoxylated alcohols	—[a]	3
Ethoxylated alkyl amines	—[a]	1
Ethoxylated amine	—[a]	1
Ethoxylated amines	61791-44-4	1
Ethoxylated fatty acid ester	—[a]	1
Ethoxylated nonionic surfactant	—[a]	1

Chemical Component	Chemical Abstract Service Number	No. of Products Containing Chemical
Ethoxylated nonylphenol	—[a]	8
Ethoxylated nonylphenol	68412-54-4	10
Ethoxylated nonylphenol	9016-45-9	38
Ethoxylated octylphenol	68987-90-6	1
Ethoxylated octylphenol	9002-93-1	1
Ethoxylated octylphenol	9036-19-5	3
Ethoxylated oleyl amine	13127-82-7	2
Ethoxylated oleyl amine	26635-93-8	1
Ethoxylated sorbitol esters	—[a]	1
Ethoxylated tridecyl alcohol phosphate	9046-01-9	2
Ethoxylated undecyl alcohol	127036-24-2	2
Ethyl acetate	141-78-6	4
Ethyl acetoacetate	141-97-9	1
Ethyl octynol (1-octyn-3-ol,4-ethyl-)	5877-42-9	5
Ethylbenzene	100-41-4	28
Ethylene glycol (1,2-ethanediol)	107-21-1	119
Ethylene glycol monobutyl ether (2-butoxyethanol)	111-76-2	126
Ethylene oxide	75-21-8	1
Ethylene oxide-nonylphenol polymer	—[a]	1
Ethylenediaminetetraacetic acid	60-00-4	1
Ethylene-vinyl acetate copolymer	24937-78-8	1
Ethylhexanol (2-ethylhexanol)	104-76-7	18
Fatty acid ester	—[a]	1
Fatty acid, tall oil, hexa esters with sorbitol, ethoxylated	61790-90-7	1
Fatty acids	—[a]	1
Fatty alcohol alkoxylate	—[a]	1
Fatty alkyl amine salt	—[a]	1
Fatty amine carboxylates	—[a]	1
Fatty quaternary ammonium chloride	61789-68-2	1
Ferric chloride	7705-08-0	3
Ferric sulfate	10028-22-5	7
Ferrous sulfate, heptahydrate	7782-63-0	4
Fluoroaliphatic polymeric esters	—[a]	1
Formaldehyde	50-00-0	12
Formaldehyde polymer	—[a]	2
Formaldehyde, polymer with 4-(1,1-dimethyl)phenol, methyloxirane, and oxirane	30704-64-4	3
Formaldehyde, polymer with 4-nonylphenol and oxirane	30846-35-6	1
Formaldehyde, polymer with ammonia and phenol	35297-54-2	2
Formamide	75-12-7	5

Chemical Component	Chemical Abstract Service Number	No. of Products Containing Chemical
Formic acid	64-18-6	24
Fumaric acid	110-17-8	8
Furfural	98-01-1	1
Furfuryl alcohol	98-00-0	3
Glass fiber	65997-17-3	3
Gluconic acid	526-95-4	1
Glutaraldehyde	111-30-8	20
Glycerol (1,2,3-propanetriol, glycerine)	56-81-5	16
Glycol ethers	—[a]	9
Glycol ethers	9004-77-7	4
Glyoxal	107-22-2	3
Glyoxylic acid	298-12-4	1
Guar gum	9000-30-0	41
Guar gum derivative	—[a]	12
Haloalkyl heteropolycycle salt	—[a]	6
Heavy aromatic distillate	68132-00-3	1
Heavy aromatic petroleum naphtha	64742-94-5	45
Heavy catalytic reformed petroleum naphtha	64741-68-0	10
Hematite	—[a]	5
Hemicellulase	9025-56-3	2
Hexahydro-1,3,5-*tris*(2-hydroxyethyl)-*s*-triazine (triazine)	4719-04-4	4
Hexamethylenetetramine	100-97-0	37
Hexanediamine	124-09-4	1
Hexanes	—[a]	1
Hexylene glycol	107-41-5	5
Hydrated aluminum silicate	1332-58-7	4
Hydrocarbon mixtures	8002-05-9	1
Hydrocarbons	—[a]	3
Hydrodesulfurized kerosine (petroleum)	64742-81-0	3
Hydrodesulfurized light catalytic cracked distillate (petroleum)	68333-25-5	1
Hydrodesulfurized middle distillate (petroleum)	64742-80-9	1
Hydrogen chloride (hydrochloric acid)	7647-01-0	42
Hydrogen fluoride (hydrofluoric acid)	7664-39-3	2
Hydrogen peroxide	7722-84-1	4
Hydrogen sulfide	7783-06-4	1
Hydrotreated and hydrocracked base oil	—[a]	2
Hydrotreated heavy naphthenic distillate	64742-52-5	3
Hydrotreated heavy paraffinic petroleum distillates	64742-54-7	1
Hydrotreated heavy petroleum naphtha	64742-48-9	7
Hydrotreated light petroleum distillates	64742-47-8	89

Chemical Component	Chemical Abstract Service Number	No. of Products Containing Chemical
Hydrotreated middle petroleum distillates	64742-46-7	3
Hydroxyacetic acid (glycolic acid)	79-14-1	6
Hydroxyethylcellulose	9004-62-0	1
Hydroxyethylethylenediaminetriacetic acid, trisodium salt	139-89-9	1
Hydroxylamine hydrochloride	5470-11-1	1
Hydroxypropyl guar gum	39421-75-5	2
Hydroxysultaine	—[a]	1
Inner salt of alkyl amines	—[a]	2
Inorganic borate	—[a]	3
Inorganic particulate	—[a]	1
Inorganic salt	—[a]	1
Inorganic salt	533-96-0	1
Inorganic salt	7446-70-0	1
Instant coffee purchased off the shelf	—[a]	1
Inulin, carboxymethyl ether, sodium salt	430439-54-6	1
Iron oxide	1332-37-2	2
Iron oxide (ferric oxide)	1309-37-1	18
Isoamyl alcohol	123-51-3	1
Iso-alkanes/*n*-alkanes	—[a]	10
Isobutanol (isobutyl alcohol)	78-83-1	4
Isomeric aromatic ammonium salt	—[a]	1
Isooctanol	26952-21-6	1
Isooctyl alcohol	68526-88-0	1
Isooctyl alcohol bottoms	68526-88-5	1
Isopropanol (isopropyl alcohol, propan-2-ol)	67-63-0	274
Isopropylamine	75-31-0	1
Isotridecanol, ethoxylated	9043-30-5	1
Kerosene	8008-20-6	13
Lactic acid	10326-41-7	1
Lactic acid	50-21-5	1
L-Dilactide	4511-42-6	1
Lead	7439-92-1	1
Light aromatic solvent naphtha	64742-95-6	11
Light catalytic cracked petroleum distillates	64741-59-9	1
Light naphtha distillate, hydrotreated	64742-53-6	1
Low toxicity base oils	—[a]	1
Maghemite	—[a]	2
Magnesium carbonate	546-93-0	1
Magnesium chloride	7786-30-3	4
Magnesium hydroxide	1309-42-8	4
Magnesium iron silicate	1317-71-1	3

Chemical Component	Chemical Abstract Service Number	No. of Products Containing Chemical
Magnesium nitrate	10377-60-3	5
Magnesium oxide	1309-48-4	18
Magnesium peroxide	1335-26-8	2
Magnesium peroxide	14452-57-4	4
Magnesium phosphide	12057-74-8	1
Magnesium silicate	1343-88-0	3
Magnesium silicate hydrate (talc)	14807-96-6	2
Magnetite	—[a]	3
Medium aliphatic solvent petroleum naphtha	64742-88-7	10
Metal salt	—[a]	2
Metal salt solution	—[a]	1
Methanol (methyl alcohol)	67-56-1	342
Methyl isobutyl carbinol (methyl amyl alcohol)	108-11-2	3
Methyl salicylate	119-36-8	6
Methyl vinyl ketone	78-94-4	2
Methylcyclohexane	108-87-2	1
Mica	12001-26-2	3
Microcrystalline silica	1317-95-9	1
Mineral	—[a]	1
Mineral filler	—[a]	1
Mineral spirits (Stoddard solvent)	8052-41-3	2
Mixed titanium ortho ester complexes	—[a]	1
Modified alkane	—[a]	1
Modified cycloaliphatic amine adduct	—[a]	3
Modified lignosulfonate	—[a]	1
Monoethanolamine (ethanolamine)	141-43-5	17
Monoethanolamine borate	26038-87-9	1
Morpholine	110-91-8	2
Mullite	1302-93-8	55
N,N′-Dibutylthiourea	109-46-6	1
N,N-Dimethyl-1-octadecanamine-HCl	—[a]	1
N,N-Dimethyloctadecylamine	124-28-7	3
N,N-Dimethyloctadecylamine hydrochloride	1613-17-8	2
N,N′-Methylenebisacrylamide	110-26-9	1
n-Alkyl dimethyl benzyl ammonium chloride	139-08-2	1
Naphthalene	91-20-3	44
Naphthalene derivatives	—[a]	1
Naphthalenesulphonic acid, *bis*(1-methylethyl)-methyl derivatives	99811-86-6	1
Natural asphalt	12002-43-6	1
N-Cocoamidopropyl-*N,N*-dimethyl-*N*-2-hydroxypropylsulfobetaine	68139-30-0	1

Chemical Component	Chemical Abstract Service Number	No. of Products Containing Chemical
N-Dodecyl-2-pyrrolidone	2687-96-9	1
N-Heptane	142-82-5	1
Nickel sulfate hexahydrate	10101-97-0	2
Nitrilotriacetamide	4862-18-4	4
Nitrilotriacetic acid	139-13-9	6
Nitrilotriacetonitrile	7327-60-8	3
Nitrogen	7727-37-9	9
N-Methylpyrrolidone	872-50-4	1
Nonane, all isomers	—[a]	1
Nonhazardous salt	—[a]	1
Nonionic surfactant	—[a]	1
Nonylphenol ethoxylate	—[a]	2
Nonylphenol ethoxylate	9016-45-6	2
Nonylphenol ethoxylate	9018-45-9	1
Nonylphenol	25154-52-3	1
Nonylphenol, ethoxylated and sulfated	9081-17-8	1
N-Propyl zirconate	—[a]	1
N-Tallowalkyltrimethylenediamines	—[a]	1
Nuisance particulates	—[a]	2
Nylon fibers	25038-54-4	2
Octanol	111-87-5	2
Octyltrimethylammonium bromide	57-09-0	1
Olefinic sulfonate	—[a]	1
Olefins	—[a]	1
Organic acid salt	—[a]	3
Organic acids	—[a]	1
Organic phosphonate	—[a]	1
Organic phosphonate salts	—[a]	1
Organic phosphonic acid salts	—[a]	6
Organic salt	—[a]	1
Organic sulfur compound	—[a]	2
Organic titanate	—[a]	2
Organiophilic clay	—[a]	2
Organo-metallic ammonium complex	—[a]	1
Other inorganic compounds	—[a]	1
Oxirane, methyl-, polymer with oxirane, mono-C10–C16-alkyl ethers, phosphates	68649-29-6	1
Oxyalkylated alcohol	—[a]	6
Oxyalkylated alcohols	228414-35-5	1
Oxyalkylated alkyl alcohol	—[a]	1
Oxyalkylated alkylphenol	—[a]	1
Oxyalkylated fatty acid	—[a]	2

Chemical Component	Chemical Abstract Service Number	No. of Products Containing Chemical
Oxyalkylated phenol	—[a]	1
Oxyalkylated polyamine	—[a]	1
Oxylated alcohol	—[a]	1
Paraffin wax	8002-74-2	1
Paraffinic naphthenic solvent	—[a]	1
Paraffinic solvent	—[a]	5
Paraffins	—[a]	1
Perlite	93763-70-3	1
Petroleum distillates	26	
Petroleum distillates	64742-65-0	1
Petroleum distillates	64742-97-5	1
Petroleum distillates	68477-31-6	3
Petroleum gas oils	—[a]	1
Petroleum gas oils	64741-43-1	1
Phenol	108-95-2	5
Phenol-formaldehyde resin	9003-35-4	32
Phosphate ester	—[a]	6
Phosphate esters of alkyl phenyl ethoxylate	68412-53-3	1
Phosphine	—[a]	1
Phosphonic acid	—[a]	1
Phosphonic acid	129828-36-0	1
Phosphonic acid	13598-36-2	3
Phosphonic acid (dimethlamino(methylene)	29712-30-9	1
Phosphonic acid, [nitrilotris(methylene)]*tris*-, pentasodium salt	2235-43-0	1
Phosphoric acid	7664-38-2	7
Phosphoric acid ammonium salt	—[a]	1
Phosphoric acid, mixed decyl, octyl and ethyl esters	68412-60-2	3
Phosphorous acid	10294-56-1	1
Phthalic anhydride	85-44-9	2
Pine oil	8002-09-3	5
Plasticizer	—[a]	1
Poly(oxy-1,2-ethanediyl)	24938-91-8	1
Poly(oxy-1,2-ethanediyl), alpha-(4-nonylphenyl)-omega-hydroxy-, branched (nonylphenol ethoxylate)	127087-87-0	3
Poly(oxy-1,2-ethanediyl), alpha-hydro-omega-hydroxy	65545-80-4	1
Poly(oxy-1,2-ethanediyl), alpha-sulfo-omega-(hexyloxy)-, ammonium salt	63428-86-4	3
Poly(oxy-1,2-ethanediyl), alpha-(nonylphenyl)-omega-hydroxy-, phosphate	51811-79-1	1
Poly(oxy-1,2-ethanediyl), alpha-undecyl-omega-hydroxy	34398-01-1	6

Chemical Component	Chemical Abstract Service Number	No. of Products Containing Chemical
Poly(sodium-*p*-styrenesulfonate)	25704-18-1	1
Poly(vinyl alcohol)	25213-24-5	2
Polyacrylamides	9003-05-8	2
Polyacrylamides	—[a]	1
Polyacrylate	—[a]	1
Polyamine	—[a]	2
Polyanionic cellulose	—[a]	2
Polyepichlorohydrin, trimethylamine quaternized	51838-31-4	1
Polyetheramine	9046-10-0	3
Polyether-modified trisiloxane	27306-78-1	1
Polyethylene glycol	25322-68-3	20
Polyethylene glycol ester with tall oil fatty acid	9005-02-1	1
Polyethylene polyammonium salt	68603-67-8	2
Polyethylene-polypropylene glycol	9003-11-6	5
Polylactide resin	—[a]	3
Polyoxyalkylenes	—[a]	1
Polyoxyethylene castor oil	61791-12-6	1
Polyphosphoric acid, esters with triethanolamine, sodium salts	68131-72-6	1
Polypropylene glycol	25322-69-4	1
Polysaccharide	—[a]	20
Polyvinyl alcohol	—[a]	1
Polyvinyl alcohol	9002-89-5	2
Polyvinyl alcohol/polyvinylacetate copolymer	—[a]	1
Potassium acetate	127-08-2	1
Potassium carbonate	584-08-7	12
Potassium chloride	7447-40-7	29
Potassium formate	590-29-4	3
Potassium hydroxide	1310-58-3	25
Potassium iodide	7681-11-0	6
Potassium metaborate	13709-94-9	3
Potassium metaborate	16481-66-6	3
Potassium oxide	12136-45-7	1
Potassium pentaborate	—[a]	1
Potassium persulfate	7727-21-1	9
Propanol (propyl alcohol)	71-23-8	18
Propanol, [2(2-methoxy-methylethoxy)methylethoxyl]	20324-33-8	1
Propargyl alcohol (2-propyn-1-ol)	107-19-7	46
Propylene carbonate (1,3-dioxolan-2-one, methyl-)	108-32-7	2
Propylene glycol (1,2-propanediol)	57-55-6	18
Propylene oxide	75-56-9	1
Propylene pentamer	15220-87-8	1

Chemical Component	Chemical Abstract Service Number	No. of Products Containing Chemical
p-Xylene	106-42-3	1
Pyridinium, 1-(phenylmethyl)-, ethyl methyl derivatives, chlorides	68909-18-2	9
Pyrogenic silica	112945-52-5	3
Quaternary amine compounds	—[a]	3
Quaternary amine compounds	61789-18-2	1
Quaternary ammonium compounds	—[a]	9
Quaternary ammonium compounds	19277-88-4	1
Quaternary ammonium compounds	68989-00-4	1
Quaternary ammonium compounds	8030-78-2	1
Quaternary ammonium compounds, dicoco alkyldimethyl, chlorides	61789-77-3	2
Quaternary ammonium salts	—[a]	2
Quaternary compound	—[a]	1
Quaternary salt	—[a]	2
Quaternized alkyl nitrogenated compound	68391-11-7	2
Rafinnates (petroleum), sorption process	64741-85-1	2
Residues (petroleum), catalytic reformer fractionator	64741-67-9	10
Resin	8050-09-7	2
Rutile	1317-80-2	2
Salt of phosphate ester	—[a]	3
Salt of phosphono-methylated diamine	—[a]	1
Salts of oxyalkylated fatty amines	68551-33-7	1
Secondary alcohol	—[a]	7
Silica (silicon dioxide)	7631-86-9	47
Silica, amorphous	—[a]	3
Silica, amorphous precipitated	67762-90-7	1
Silicon carboxylate	681-84-5	1
Silicon dioxide (fused silica)	60676-86-0	7
Silicone emulsion	—[a]	1
Sodium (C14–C16) olefin sulfonate	68439-57-6	4
Sodium 2-ethylhexyl sulfate	126-92-1	1
Sodium acetate	127-09-3	6
Sodium acid pyrophosphate	7758-16-9	5
Sodium alkyl diphenyl oxide sulfonate	28519-02-0	1
Sodium aluminate	1302-42-7	1
Sodium aluminum phosphate	7785-88-8	1
Sodium bicarbonate (sodium hydrogen carbonate)	144-55-8	10
Sodium bisulfite	7631-90-5	6
Sodium bromate	7789-38-0	10
Sodium bromide	7647-15-6	1
Sodium carbonate	497-19-8	14

Chemical Component	Chemical Abstract Service Number	No. of Products Containing Chemical
Sodium chlorate	7775-09-9	1
Sodium chloride	7647-14-5	48
Sodium chlorite	7758-19-2	8
Sodium cocaminopropionate	68608-68-4	2
Sodium diacetate	126-96-5	2
Sodium erythorbate	6381-77-7	4
Sodium glycolate	2836-32-0	2
Sodium hydroxide (caustic soda)	1310-73-2	80
Sodium hypochlorite	7681-52-9	14
Sodium lauryl-ether sulfate	68891-38-3	3
Sodium metabisulfite	7681-57-4	1
Sodium metaborate	7775-19-1	2
Sodium metaborate tetrahydrate	35585-58-1	6
Sodium metasilicate, anhydrous	6834-92-0	2
Sodium nitrite	7632-00-0	1
Sodium oxide (Na_2O)	1313-59-3	1
Sodium perborate	1113-47-9	1
Sodium perborate	7632-04-4	1
Sodium perborate tetrahydrate	10486-00-7	4
Sodium persulfate	7775-27-1	6
Sodium phosphate	—[a]	2
Sodium polyphosphate	68915-31-1	1
Sodium salicylate	54-21-7	1
Sodium silicate	1344-09-8	2
Sodium sulfate	7757-82-6	7
Sodium tetraborate	1330-43-4	7
Sodium tetraborate decahydrate	1303-96-4	10
Sodium thiosulfate	7772-98-7	10
Sodium thiosulfate pentahydrate	10102-17-7	3
Sodium trichloroacetate	650-51-1	1
Sodium tripolyphosphate	7758-29-4	2
Sodium xylene sulfonate	1300-72-7	3
Sodium zirconium lactate	174206-15-6	1
Solvent refined heavy naphthenic petroleum distillates	64741-96-4	1
Sorbitan monooleate	1338-43-8	1
Stabilized aqueous chlorine dioxide	10049-04-4	1
Stannous chloride	7772-99-8	1
Stannous chloride dihydrate	10025-69-1	6
Starch	9005-25-8	5
Steam-cracked distillate, cyclodiene dimer, dicyclopentadiene polymer	68131-87-3	1
Steam-cracked petroleum distillates	64742-91-2	6

Chemical Component	Chemical Abstract Service Number	No. of Products Containing Chemical
Straight run middle petroleum distillates	64741-44-2	5
Substituted alcohol	—[a]	2
Substituted alkene	—[a]	1
Substituted alkylamine	—[a]	2
Sucrose	57-50-1	1
Sulfamic acid	5329-14-6	6
Sulfate	—[a]	1
Sulfonate acids	—[a]	1
Sulfonate surfactants	—[a]	1
Sulfonic acid salts	—[a]	1
Sulfonic acids, petroleum	61789-85-3	1
Sulfur compound	—[a]	1
Sulfuric acid	7664-93-9	9
Sulfuric acid, monodecyl ester, sodium salt	142-87-0	2
Sulfuric acid, monooctyl ester, sodium salt	142-31-4	2
Surfactants	—[a]	13
Sweetened middle distillate	64741-86-2	1
Synthetic organic polymer	9051-89-2	2
Tall oil (fatty acids)	61790-12-3	4
Tall oil, compound with diethanolamine	68092-28-4	1
Tallow soap	—[a]	2
Tar bases, quinoline derivatives, benzyl chloride-quaternized	72480-70-7	5
Tergitol	68439-51-0	1
Terpene hydrocarbon byproducts	68956-56-9	3
Terpenes	—[a]	1
Terpenes and terpenoids, sweet orange-oil	68647-72-3	2
Terpineol	8000-41-7	1
tert-Butyl hydroperoxide	75-91-2	6
Tetracalcium-alumino-ferrite	12068-35-8	1
Tetraethylene glycol	112-60-7	1
Tetraethylenepentamine	112-57-2	2
Tetrahydro-3,5-dimethyl-2H-1,3,5-thiadiazine-2-thione (Dazomet)	533-74-4	13
Tetrakis (hydroxymethyl) phosphonium sulfate	55566-30-8	12
Tetramethyl ammonium chloride	75-57-0	14
Tetrasodium 1-hydroxyethylidene-1,1-diphosphonic acid	3794-83-0	1
Tetrasodium ethylenediaminetetraacetate	64-02-8	10
Thiocyanate sodium	540-72-7	1
Thioglycolic acid	68-11-1	6

Chemical Component	Chemical Abstract Service Number	No. of Products Containing Chemical
Thiourea	62-56-6	9
Thiourea polymer	68527-49-1	3
Titanium complex	—[a]	1
Titanium oxide	13463-67-7	19
Titanium, isopropoxy (triethanolaminate)	74665-17-1	2
Toluene	108-88-3	29
Treated ammonium chloride (with anticaking agent a or b)	12125-02-9	1
Tributyl tetradecyl phosphonium chloride	81741-28-8	5
Tricalcium silicate	12168-85-3	1
Tridecyl alcohol	112-70-9	1
Triethanolamine (2,2,2-nitrilotriethanol)	102-71-6	21
Triethanolamine polyphosphate ester	68131-71-5	3
Triethanolamine titanate	36673-16-2	1
Triethanolamine zirconate	101033-44-7	6
Triethanolamine zirconium chelate	—[a]	1
Triethyl citrate	77-93-0	1
Triethyl phosphate	78-40-0	1
Triethylene glycol	112-27-6	3
Triisopropanolamine	122-20-3	5
Trimethylammonium chloride	593-81-7	1
Trimethylbenzene	25551-13-7	5
Trimethyloctadecylammonium (1-octadecanaminium, *N,N,N*-trimethyl-, chloride)	112-03-8	6
Tris(hydroxymethyl)aminomethane	77-86-1	1
Trisodium ethylenediaminetetraacetate	150-38-9	1
Trisodium ethylenediaminetriacetate	19019-43-3	1
Trisodium nitrilotriacetate	18662-53-8	8
Trisodium nitrilotriacetate (nitrilotriacetic acid, trisodium salt monohydrate)	5064-31-3	9
Trisodium orthophosphate	7601-54-9	1
Trisodium phosphate dodecahydrate	10101-89-0	1
Ulexite	1319-33-1	1
Urea	57-13-6	3
Wall material	—[a]	1
Walnut hulls	—[a]	2
White mineral oil	8042-47-5	8
Xanthan gum	11138-66-2	6
Xylene	1330-20-7	44
Zinc chloride	7646-85-7	1
Zinc oxide	1314-13-2	2

Chemical Component	Chemical Abstract Service Number	No. of Products Containing Chemical
Zirconium complex	—[a]	10
Zirconium dichloride oxide	7699-43-6	1
Zirconium oxide sulfate	62010-10-0	2
Zirconium sodium hydroxy lactate complex (sodium zirconium lactate)	113184-20-6	2

Source: U.S. House of Representatives, *Chemicals Used in Hydraulic Fracturing*, U.S. Government Printing Office, Washington, DC, 2011.

[a] These components appeared on at least one MSDS without an identifying CAS number. The MSDSs in these cases marked the CAS as proprietary, noted that the CAS was not available, or left the CAS field blank. These components may be duplicative of other components on this list, but it was not possible to identify such duplicates without the identifying CAS number.

Appendix B. Case Study Locations for Hydraulic Fracturing Study[*]

To determine the potential impacts of hydraulic fracturing on drinking water resources under a variety of circumstances, the USEPA has selected seven case studies located in various formation locations across the country that the Agency believes will provide the most useful information. The study includes a review of published literature, analysis of existing data, scenario evaluation and modeling, industry studies, and case studies. Two perspective case studies, where the USEPA will monitor key aspects of the hydraulic fracturing process at future hydraulic fracturing sites, are located in

- Haynesville Shale, DeSoto Parish, Louisiana
- Marcellus Shale, Washington County, Pennsylvania

The USEPA's partners in conducting the prospective case studies are Chesapeake Energy Corporation in DeSoto Parish, Louisiana, and Range Resources Corporation in Washington County, Pennsylvania. The USEPA will release initial study results in a 2012 report and an additional report at the end of 2014.

Criteria for Case Study Location Selection

The sites were identified, prioritized, and selected based on a rigorous set of criteria and represent a wide range of conditions and impacts that my result from hydraulic fracturing activities. These criteria included the following:

- Proximity of population and drinking water supplies
- Evidence of impaired water quality (retrospective)
- Health and environmental concerns (retrospective only)
- Knowledge gaps that could be filled by the case study

Sites were prioritized based on the following criteria:

* Adapted from USEPA, *Case Study Locations for Hydraulic Fracturing Study*, U.S. Environmental Protection Agency, Washington, DC, 2011 (http://www.epa.gov/hfstudy/casestudies.html).

- Geographic and geologic diversity
- Population at risk
- Site status (planned, active or completed)
- Unique geological or hydrological features
- Characteristics of water resources
- Land use

Five retrospective case studies that will investigate reported drinking water contamination due to hydraulic fracturing operations at existing sites are identified in Table B.1.

TABLE B.1

Retrospective Hydraulic Fracturing Case Studies

Locations	Key Issues To Be Investigated	Potential Outcomes	Companies Involved
Dunn County, North Dakota (Bakken Shale)	Production well failure during hydraulic fracturing Suspected drinking water aquifer contamination	Identify sources of well failure Determine if drinking water resources are contaminated and to what extent	Denbury Resources, Inc.
Wise County, Texas (Barnett Shale)	Possible drinking water well contamination Spills and runoff leading to suspected drinking water well contamination	Determine if private water wells are contaminated Obtain information about the likelihood of transport of contaminants via spills, leaks, and runoff	Aruba Petroleum, Inc. Primexx Energy Partners, Ltd. XR-5, LLC White
Bradford and Susquehanna Counties, Pennsylvania (Marcellus Shale)	Groundwater and drinking water well contamination Suspected surface water contamination from a spill of fracturing fluids Methane contamination of multiple drinking water wells	Determine if drinking water wells are contaminated Determine source of methane in private wells Transferable results due to common types of impacts	Chesapeake Energy Corp.
Washington County, Pennsylvania (Marcellus Shale)	Changes in drinking water quality, suspected contamination Stray gas in wells, surface spills	Determine if drinking water wells are contaminated Determine if surface spills affect surface and ground water If contamination exists, determine potential source of contaminants in drinking water	Range Resources Corporation Atlas Energy, LP
Las Animas County, Colorado (Raton Basin)	Potential drinking water well contamination (methane and other contaminants) in an area with intense concentration of gas wells in shallow surficial aquifer (coalbed methane)	Determine source of methane Identify presence/source of contamination in drinking water wells	Pioneer Natural Resources Co. Petroglyph Energy, Inc.

Index

2-butoxyethanol (2-BE), 143–144
29 CFR 1910, 166, 253–255, 262–402
  Subpart D, Walking–Working Surfaces, 262–264
  Subpart E, Means of Egress, 264–270
  Subpart F, Powered Platforms, Manlifts, and Vehicle-Mounted Work Platforms, 271–278
  Subpart G, Occupational and Environmental Control, 278–281
  Subpart H, Hazardous Materials, 281–284
  Subpart I, Personal Protective Equipment, 285–304
  Subpart J, General Environmental Controls, 305–341
  Subpart K, Medical and First Aid, 341–342
  Subpart L, Fire Protection, 342–346
  Subpart N, Material Handilng and Storage, 346–361
  Subpart O, Machinery and Machine Guarding, 361–375
  Subpart P, Hand and Portable Powered Tools, 375–376
  Subpart Q, Welding, Cutting, and Brazing, 376–389
  Subpart S, Electrical, 389–396
  Subpart Z, Toxic and Hazardous Substances, 397–402
40 CFR 141, 193
40 CFR 143, 193
40 CFR 435, 185

A

abrasive wheel, 375
acceptable confined space entry conditions, 306
access roads, 154–156
accident-prevention tags, 305
accident reporting, 397
acid rain, 172, 211, 222, 223, 224, 233–234
acids, in fracturing fluids, 132, 135
acoustic barriers, 161
acrylamide, 144
action level (AL), 189
adiabatic rate, 214–215
adjustable guards, 366
advanced oxidation process (AOP), 125
aerosol, 292
aftershocks, 5
air parcels, 213–215
air permits, 244
air pollutants, 147–148, 171, 172, 209–210, 221–228
  dispersion of, 219–221, 240
  primary vs. secondary, 222, 226
  toxic, 233, 243
Air Pollution Control Act of 1955, 229
air-purifying respirators, 289
air quality, 9; *see also* air pollutants
  impact analysis, 238–239
  models, 219–221, 236–237, 238, 239, 240
  natural gas production, and, 3
  regulations, 208–228
  weather, and, 211
air stability, 212–213
air-supplying respirators, 289, 300–301
aisles and passageways, 263
albite, 38
algae, 39, 44, 45, 47, 51
  lacustrine, 42, 44, 45
alternative energy, 25–27
alumina, 49
ambient air, 209, 215
amine processes, 30
ammonia, 29
anthracite, 54
Antrim Shale, 71, 83–85, 108
  gas content, 84–85
  original gas-in-place estimate, 85
  technically recoverable resources, 85
Application for Permit to Drill (APD), 249

aquifer, 16, 93, 94, 95, 96, 97, 99, 106–107, 108, 109, 126, 161–162, 180, 190, 191, 192
defined, 9
sole or principal source, 191
arenaceous shale, 38
argillaceous shale, 38
artificial lift, 100
asbestos, 292
ASTM Method D-3904-80, 41
atmospheric dispersion, 210–219
atmospheric testing, 318, 327–328
attendant, confined space entry, 306, 318–319, 323, 324
audit, OSHA, 166, 167, 253, 256, 257, 262, 264, 288, 311, 321, 325, 328–331
authorized entrant, confined space entry, 306
Automated External Defibrillator (AED), 342

B

background radiation, 140
bacteria, 39, 42, 47, 69, 138, 190, 198, 223
banana oil, 293
Barnett Shale, 61, 64, 71, 74–75, 76, 77, 84, 110, 157
formation water, 113
gas content, 74
horizontal vs. vertical wells in, 158
local ordinances, and, 161
original gas-in-place estimate, 75
ozone nonattainment area, 241
permanent pipelines in, 160
regulatory agency, 175
technically recoverable resources, 75
traffic volume in, 160
water needs per well, 121
barriers, confined space, 315
basin, defined, 9
Batesville Sandstone, 75
Bcf (billion cubic feet), 9, 63
Bedford Shale, 83
bentonite clay, 104
Bentsen Amendment, 246
benzene, 30, 133, 144, 147, 196, 242
best available control measures (BACMs), 231
best available technology (BAT), 185
best conventional technology (BCT), 185
best management practices (BMPs), 154, 156, 161, 189
best practicable technology (BPT), 185
best professional judgment (BPJ), 184
Bevill Amendment, 246
bicarbonates, 47
biochemical sedimentary rocks, 52, 53–59
biocides, 117
biogenic gas, 10, 210
biomass, 26
bitumen, 45
bitumen-impregnated rock, 44
bituminite, 42, 43, 44, 45, 51
bituminous coal, 53, 54
blanking and blinding, 306
blasting abrasive, 293
block and tackle, 101
bloodborne pathogens, 265, 397
blow-off valves, 99
blowout preventers, 103
bonding, 392, 394
bored wells, 96
borehole, 97, 98, 117, 161, 162, 243
breathing resistance, 293
breccias, 38
Briggs' plume rise formula, 218
BTEX compounds, 147
Btu (British thermal unit), 10, 22, 41, 63
buoyancy, 213–214
Bureau of Land Management (BLM), 249
butane, 62
butane (C_4H_{10}), 27, 30, 62

C

calcareous shale, 38
calcite, 38, 52
cannel coal, 44
capacitance grounds, 393
capacitance presence-sensing device, 367
carbonaceous rock, 36, 38
carbonate processes, 30
carbonates, 47
carbon dioxide (CO_2), 23, 29, 209, 221, 222, 227, 345

carbon monoxide (CO), 209, 221, 222, 224, 230, 231, 232, 237, 318
carcinogens, 133, 142, 144–146, 147, 194, 196
casing, 10
casing head, 102
casing string, 106, 107, 161, 162
cement log, 107, 109, 163
cement, manufacture of, 38, 46
chains, 351, 352
chalk, 53
Chappel Limestone, 74
check valves, 99
chemical additives, 117–121, 131–149
chemical hazard, defined, 293
chemical inventories, MSDSs and, 251
chemical precipitation, 36, 47
Chemical Safety Board, 233
chemical sedimentary rocks, 52–54
chemical spills, 252, 267
chert, 52
chlorinated solvents, 235
chlorination, 196
chlorofluorocarbons (CFCs), 209, 235
chloroform, 196
circuit breakers, 391
city plume, 218
Class A/B/C fire extinguishers, 345
Class II injection wells, 107
clastic sedimentary rocks, 36, 37–38
Clean Air Act (CAA), 144, 166, 174, 208–244, 246; *see also* air quality, emissions
- amendments, 220, 229–241
- hydraulic fracturing, and, 241–244
- titles, 230–236

cleaning respirators, 293
Clean Water Act, 166, 174, 179, 180, 181–187, 190, 207, 246
- amendments, 182

clean water reform, 179–189
coal, 52, 53–54
- bituminous, 53
- heating value of, 40
- types of, 53–54
- vs. oil shale, 39

coal bed methane gas (CBMG), 10, 62, 63
coal bed natural gas (CBNG), 19, 84, 85
coal mining, surface, 17
coal vs. natural gas, 23
coliform bacteria, 198
color-coded accident-prevention tags, 305
commercial chemical products, 283
community impacts, 159–161
Community Right-to-Know Act, 250
compaction, 37–38
completion, gas well, 10
Comprehensive Environmental Response, Compensation, and Liability Act (CERCLA), 250–251, 252, 284
computer models, 112–113, 124
concretions, 55
condensate, wet gas, 3
confined space, defined, 307, 311
confined space entry, 259–262
- by welders, 385
- equipment, 313–317
- fire watch requirements, 378–379
- hot work permit procedure, 377–379, 383
- lockout/tagout, and, 332
- non-permit, 310, 311
- permit-required, 260, 261, 290, 305–331
  - defined, 310
- pre-entry requirements, 317–319
- rescue, 325–327
- respiratory protection, 289–305
- training, 320–323
- written program for, 312–313

Confined Space Entry Program (CSEP), 306
conglomerates, sedimentary rock, 36, 37, 38
Consumer Confidence Report (CCR), 189
contamination, 161, 170, 171, 192, 193
- as pollution, 171
- bloodborne pathogen, 265
- CERCLA, and, 250
- due to fracking, 3–4, 106, 108, 138, 206
- background level of, 219
- fecal, 198
- microbial, 201
- radioactive, 198–199
- water well sites, and, 95, 96

control valves, water well, 100
conventional drive, 100
conventional gas resources, 21
coquina, 52, 53
core, of Earth, 7–8
corridor, defined, 10
corrosivity, 283
Cotton Valley Group, 78
Council on Regulatory Needs, 246
criteria pollutants, 242
cross-bedding, 54
cross-linked gels, 133–134
crown block, 101
crude oil, 40
crust, of Earth, 7–8, 16
Cryptosporidium, 198, 201

D

darcies, 12–13
dawsonite, 40, 47, 48, 49
deep natural gas, 62
deep water wells, 95, 96–97
deposition, atmospheric, 210–219
deposition, dry/wet, 211, 240
deposition, sediment, 37
derrick, 101
detrital sedimentary rocks, 36
detritus, 35
Devonian–Mississippian oil shales, 45, 46, 50–52
diagenesis, 37–38
diesel fuel, 133, 144, 147, 210
dikes, storage, 282, 346
directional drilling, 64, 100, 105–106, 257
  defined, 10
disinfectants, 189
dispersion, in air, 9, 210–221, 233, 240
  models, 219–221, 239
disposal, of wastewater, 3
disposal well, 10
dolomite, 38, 52
dolostone, 36
double block and bleed, 307
drill bit, 103, 104
drilled wells, 97, 101
drilling, air-based, 109, 162
drilling, closed-loop, 109
drilling fluids, 101, 102, 103, 104, 106, 107, 109–110
drilling line, 243
drilling mud, 103, 245
  functions of, 104
drilling technology, 100–110
drill line, 101
drill rig, 10
drill string, 103, 104, 105
driven wells, 96
drop pipe, 99
dry adiabatic lapse rate, 214
dry deposition, 211, 240
dug wells, 96
dust, 293

E

Earth, internal structure of, 7–8
earthquakes, 3–8, 174
  cause of, 5
  intensity, 6, 7
  magnitude, 6
  volcanoes, and, 5
effluent limitation, 182, 184
  guidelines, 184, 185
egress, means of, 315–316
ejection methods, as safety control, 369
elastic rebound, 5
electrical safety, 389–396
electromechanical presence-sensing device, 367
emergency, confined space entry, 307
Emergency Planning and Community Right-to-Know Act (EPCRA), 251–262
emergency response plans, 265–268
  components of, 268–269
emissions, 25, 147, 174, 240; *see also* Clean Air Act, dispersion
  air quality, and, 211, 218, 220, 229
  anthropogenic, 226–227
  banking, 229
  carbon dioxide, 25
  carbon monoxide, 224
  coal, 53, 223, 224
  greenhouse gas, 23
  hydrocarbon, 224

motor vehicle, 231–236
natural gas production and, 3, 10, 30, 32, 174–175, 241, 242, 244
nitric oxide, 223, 224
particulate, 228
standards, 242
stationary internal combustion engine, 242–243
stationary reciprocating internal combustion engine, 243–244
volatile organic chemical, 237
endangered species, 10
Endangered Species Act (ESA), 247–249
energy isolating device, 333
Energy Policy Act, 185
engineering controls, 279, 285, 287, 289, 302, 374, 390
engulfment, 307
entry, defined, 307
entry permit, 308
entry supervisor, 308
environmental effect, defined, 209
epicenter, earthquake, 5
equipment grounds, 394
erosion, 35
Estonian deposits, 45
ethane (C_2H_6), 27, 30
ethanol, 26, 134
ethylbenzene, 30, 144, 147
evaporites, 36, 52–53
exemption, National Primary Drinking Water Regulations, 189
exit routes, requirements for, 268–270
Exxon Corporation, 48
eye protection, 286, 383–385

F

facies, sedimentary rock, 55
fall protection, 271–272
measures, 276
policies, 276–278
falls
causes of, 272–273
elevated, 272, 275
from stairs, 272, 275
types of, 272
fanning plume, 71, 75–77, 217
Fayetteville Shale
fracturing fluid, 118, 121
gas content, 77
horizontal vs. vertical wells in, 158
original gas-in-place estimate, 77
seasonal changes in river flow, and, 123
technically recoverable resources, 77
water needs per well, 121
Federal Water Pollution Control Act, 182
feldspar, 38, 40
filtration, 193
fire extinguishers, 265, 342–343, 345, 379, 380, 389
types of, 345
fire, necessary components of, 343
fire protection, 342–346, 377
fire watch, 377, 378–379, 380, 389
first aid, 341–342
Fischer assay, 41–42, 46, 51
fish, toxins in, 187
fit-testing, respirator, 293, 296–298
fixed guards, 365
flammable liquids, 282
flaring, 32
flashpoint, 400, 401
floor openings, 263
flow line, 10, 102
flowback, 11
Flower Mound, Texas, 159
flowmeters, 99
foaming agents, 132, 134, 143
focus, earthquake, 5
forced expiratory volume–1 second (FEV_1), 293, 294
forced vital capacity (FVC), 293, 294
forklifts, 354–361
safe work practices for, 355–359
sample operation exam, 359–360
formations, defined, 10
formation water, 71, 107, 113
fossil fuels, 3, 15, 23, 25, 26, 40, 53, 62, 209, 210, 222, 224, 234
fossiliferous limestone, 53
fossils, 42, 51, 55
fracking; *see* hydraulic fracking
fracture stimulation, 108, 109, 110, 112, 116–117, 157

fracturing fluids, 11, 107, 108, 110, 114, 116, 117–121
  constituents of, 141–148
  migration of, 138
  proprietary, 148–149
  types of, 131–141
friction reducers, 116, 117
fume, 293
fumigation, 217
fuses, 391, 396

G

gas cap, 61, 100
gas content
  Antrim Shale, 84–85
  Fayetteville Shale, 77
  Haynesville Shale, 79
  Marcellus Shale, 80
  New Albany Shale, 86
  Woodford Shale, 83
gas drive, 100
gas plus loss, 41
gas welding and cutting, 388
gasoline, natural, 30
gates, 368
Gaussian dispersion models, 220
gellants, 132–134
General Duty Clause, OSH Act, 253
geologic time scale, 70, 72–73
geopressurized zones, 63
geothermal energy, 26
geothermal maturity, 43
Giardia, 198
goggles, 286, 376, 383–385
goose-neck, 101
graded bedding, of sedimentary rock, 55
greenhouse gases, 3, 23, 208
Green River formation, 41, 43, 45, 46–50
ground-fault circuit interrupters (GFCIs), 395
Ground Water Protection Council (GWPC), 176
grounding, 392–395
  types of, 393
groundwater, 9, 10, 94, 96, 99, 106, 108, 112, 174, 191, 250
  as shale gas fracking water supply, 126, 135
  as source of drinking water, 16
  defined, 11
  non-permitted toxic releases into, 252
  protection of, 95, 106, 107, 161–163, 189, 192, 245, 282, 346
  radioactive minerals in, 198–199
grout, 97
guar, 132–133
guardrails, 263
guards, *see* safeguarding, machine
gypsum, 38

H

Haber process, 29
habitat, 11
habitat conservation plan (HCP), 248
halite (NaCl), 52
Halliburton Loophole, 3, 139
hand tools, safe work practices, 375–376
hard hats, 286
harness, safety, 316–317, 326–327
Haynesville Shale, 61, 71, 74, 77–79
  gas content, 79
  original gas-in-place estimate, 79
  technically recoverable resources, 79
  water needs per well, 121
Hazard Communication Standard, 148–149, 309, 397–402
hazardous air pollutants (HAPs), 144, 242
Hazardous and Solid Waste Amendments, 244
hazardous atmosphere, 308
hazardous material regulations, 281–284
hazardous materials, 219, 266, 267, 268, 285, 305, 327, 397–402; *see also* Material Safety Data Sheets (MSDSs)
  fire watch requirement, 378
  forklifts, and, 359
hazardous waste, 206, 244–247, 281–284
  categories of, 283
  disposal of, 245, 250, 266
  management of, 245
  permitting system, 245
  release of, 266
  requirements, exemption from, 245–246, 250, 252
  tracking system, 245

Hazardous Waste Operations and Emergency Response (HAZWOPER), 266
hearing conservation programs, 278–279
hearing protection, 280–281, 287
safe work practices, 281
helmets, 286, 294, 383–384
hematites, 55
holdback device, 367
horizontal drilling, 11, 64, 75, 76, 80, 82, 100, 113, 154, 157, 158, 160
horizontal wells, 75, 106, 107, 113, 114, 117, 118, 156–163
hot tap, 333
hot work permit, 309
procedure, 377–379
housekeeping, 263, 274, 275, 305
humic coal and carbonaceous shale, 44
Hutton's classification of oil shale, 44
hydraulic fracturing, 3, 64, 75, 76, 80, 91–126
Clean Air Act, and, 241–244
defined, 11
design, 112–113
environmental considerations, 153–163
process, 113–117
water supply, 121–126
hydrocarbons, 3, 27, 42, 62, 99, 221, 222, 224, 226, 227, 232
hydrochloric acid (HCl), 114, 121, 135, 252
hydroelectric energy, 26
hydrogen chloride, 148
hydrogen fluoride, 147–148
hydrogen sulfide (H_2S), 29, 30–32, 259–262, 318, 399
hydrologic cycle, 188
hydroretorting, 51
hydrostatic pressure, 11
hypocenter, earthquake, 5
Hytort process, 42

I

icing, vapor-plume-induced, 241
ignitability, 283
immediately dangerous to life or health (IDLH), 290, 293, 300, 301, 309
incineration, 32
inerting, 309
inertinite, 44, 45
infill drilling, 74
ingress, means of, 315–316
injection water, 107–108
injection well, 11, 107–108, 192, 201–207
Class II, 107
underground, *see* underground injection wells
Youngstown, Ohio, 9
inorganic chemical sediments, 52
inorganic chemicals, 194, 196
intensity, earthquake, 6, 7
interlocks, 366, 391–392
intermediate casing, 161, 162
internal combustion engines, 242–244
internal structure of Earth, 7–8
Interstate Oil and Gas Compact Commission (IOGCC), 177
inversion layer, 215–217
iso-butane, 30
isolation, permit space, 309
isopropyl alcohol, 143

J

jetted wells, 96

K

kelly drive, 103
kelly hose, 101
kerogen, 38; *see also* oil shale
analysis, 69, 70
kukersite, 44, 45
oil shale, 40

L

lacustrine algae, 42, 44, 45
ladders, 264, 315–316
lamalginite, 44, 45
lamosite, 44, 45
land quality, 244–249
lapse rate, 214–215
lead, 193, 196, 221, 222, 228
in fracturing fluids, 148
lease, 11

life-cycle assessment, 176
lighting, confined space, 314–315
lignite, 53
limestone, 16, 36, 38, 52, 74, 75, 78, 79, 82, 83, 85, 113, 119, 135, 234
linear gels, 132–133
line breaking, 310
lithification, 36
Local Emergency Planning Committees (LEPCs), 268
local regulations, 177–179
lockout/tagout, 309, 331–342, 375
  sample program, 334–341
lofting/looping, plume, 217
Lorain, Ohio, 2
low-permeability rock, 2, 14
lower explosive limit (LEL), 301, 308
lower flammable limit (LFL), 308, 343

M

machine guards, 361–375
  checklist, 371–373
  safe work practices, 363, 371
  training in use of, 370–373
  types of, 364–366
machine hazard warnings, 373–374
magnesium limestone ($CaMg(CO_3)_2$), 52
magnitude, earthquake, 6
maintenance roads, 159
mantle, of Earth, 7–8
Marble Falls Limestone, 74
marcasite, 40
Marcellus Shale, 1, 61, 71, 79–81, 91, 100, 110
  drilling rig, 243
  gas content, 80
  horizontal drilling in, 157
  hydraulic fracturing substages, 114
  original gas-in-place estimate, 81
  produced water, management of, 184
  technically recoverable resources, 81
  water requirements, 121, 122
marine oil shale, 44
marinite, 44, 45
Material Safety Data Sheets (MSDSs), 148–149, 251, 309, 327, 399, 401
maximum achievable control technology (MACT), 233, 242
maximum contaminant level (MCL), 189, 191, 192, 193, 194, 196
maximum contaminant level goal (MCLG), 189, 193, 194, 196
maximum residual disinfectant level (MRDL), 189
maximum residual disinfectant level goal (MRDLG), 189
Mcf (thousand cubic feet), 11, 12, 63
means of retrieval, 316–317
mechanical hazards, 364–365
medical emergencies, 265
Mercalli intensity scale, 7
metal ions, 133
methane (CH_4), 27, 28, 210, 224, 318, 398–399
  coal bed, 10, 62, 63
  hydrates, 63
methanol, 143, 148
microorganisms, 190, 194, 198
microseismic fracture mapping, 112, 113
mineral rights, 154–156
mist, 293
mitigation banking, 184
mixing, 217
  turbulent, 215
MMcf (million cubic feet), 11, 63
modified Fischer assay, 41–42, 46
monitoring, contaminant, 192
monkey board, 103
mountaintop mining, 17
mud cracks, 55
mud pits, 102, 104
mud pump, 101
mudstones, 36

N

nahcolite, 40, 47, 48, 49–50
naphthalene, 144
National Ambient Air Quality Standards (NAAQS), 193, 222, 230–231, 236–238, 242
National Drinking Water Standards, 193
National Emission Standards for Hazardous Air Pollutants (NESHAPs), 242
National Environmental Policy Act, 166, 175

National Estuary Program, 180
National Pollutant Discharge Elimination System (NPDES), 182, 183, 184, 185, 190, 235
National Primary Drinking Water Regulations, 189, 191, 193, 194–199
National Priorities List (NPL), 250
National Secondary Drinking Water Regulations, 193, 199
natural gas; *see also* unconventional natural gas resources
  calculations, 63
  composition of, 27–28
  conventional, 61
  equivalent to barrel of crude oil, 12
  industrial use of, 21
  measurement of, 63
  processing, 29–32
  reserves, 64
    U.S., 22, 138
  sweetening, 30
  technically recoverable, quantity of, 19–20
  U.S. consumption of, 21
  versatility of, 22–25
  vs. coal, 23
natural gas liquids (NGLs), 30
Natural Gas Policy Act (NGPA), 80, 81
naturally occurring radioactive materials (NORM), 12, 140–141
navigable waters, 190, 208
negative lapse rate, 214
nephelometric turbidity units (NTUs), 198
New Albany Shale formation, 71, 85–86, 108
  gas content, 86
  original gas-in-place estimate, 86
  technically recoverable resources, 86
new source performance standards (NSPS), 185
New Source Review (NSR), 237–238, 239
nitrate, 196
nitric acid (HNO_3), 224, 233
nitrogen dioxide (NO_2), 222, 223–224
nitrogen oxides (NO_x), 23, 209, 210, 221, 223–224, 226, 231, 232, 233
  thermal vs. fuel, 223
noise levels, limits on, 177, 278–281
nonattainment areas, 230–231, 237, 241
non-permit confined space, 310
non-point-source pollution, 188
nonspecific source wastes, 283
North Vernon Limestone, 85
numerical dispersion models, 220

O

occupational noise exposure, 279–280
Occupational Safety and Health Administration (OSHA), 141, 148, 166, 167, 179
  audits, 166, 167, 253, 256, 257, 262, 264, 288, 311, 321, 325, 328–331
  citations, 256–257, 259, 285, 341
  noise hazard requirements, 278–281
  oil and gas drilling standards, 252–262
  standards for shale gas operations, 262–402
Occupational Safety and Health (OSH) Act, 166, 169, 174, 252–257
Occupational Safety and Health Standards for General Industry, 166
oil-equivalent gas (OEG), 12
Oil Pollution Act (OPA), 179, 207–208, 246
oil shale, 38, 39–40
  byproducts, 46, 49
  classification of, 43–45
  deposits, U.S., 46–52
  geothermal maturity, 43
  grade, 41–42
  heating value of, 40, 41
  products made from, 46
  resources, evaluation of, 45–46
  vs. coal, 39
oil spills, 173, 207, 208
on-the-job injuries, 168
organic chemicals, 194–196
organic matter, origin of, 42
organic sediments, 52
original gas-in-place, 71
  Antrim Shale, 85
  Barnett Shale, 75
  Fayetteville Shale, 77
  Haynesville Shale, 79

Marcellus Shale, 80–81
New Albany Shale, 86
Woodford Shale, 83
Osage Limestone, 82
overcurrent devices, 391
overspeed switch, 392
oxidants, 226
oxidation–reduction reaction, 47
oxidizer, 402
oxyacetylene equipment, 388
oxygen deficiency, 293
oxygen-deficient/-enriched atmosphere, 310
ozone, 210, 211, 221, 222, 226–227, 230–231, 235–236, 237
cancer, and, 235
layer, 209, 224, 226, 235–236
precursors, 238

P

particle size, 222
particulate ash, 23
particulate matter, 12, 209, 210, 219, 221, 222, 227–228, 230, 231, 237
categories of, 228
Pasquill–Gifford dispersion model, 221
pay zone, 107
performance standards, 182, 185, 253
permeability, 12, 106, 110, 158
permissible exposure limit (PEL), 293, 301, 309, 399
permits, 174, 176–177, 178, 206, 208, 230, 231, 235, 245, 247–248, 310
air quality, 237–238, 244
Bureau of Land Management, 249
confined space, 260, 261, 290, 305, 306–307, 308, 309, 310, 311, 312–313, 314, 316, 317–320, 322–325, 328–330
construction, 94
hot work, 309, 323, 377–379, 381
NPDES, 182–187, 190, 235
operation, 94
wetland, 183
personal protective equipment, 285–304, 314, 374, 383–385, 399
classifications of, 286–287
hazard assessment, 288
OSHA requirements, 287
training requirements, 288–289
pesticide residues, 194
petrochemicals, 29
petroleum, conventional, 61
photochemical smog, 226–227
physical absorption, 30
physical dispersion models, 220–221
Piceance Creek Basin, 48
pipe rack, 103
Pitkin Limestone, 75
plume, 221, 240, 241
behavior, 217–218
rise formulas, 218
point source, 183, 188, 218
defined, 190
pollution, 170–174, 176, 182
air, 10, 209–244
defined, 170
effects of, 172–174
water, 182–189
porosity, defined, 12
Portland cement, 38
positive lapse rate, 214
potassium chloride (KCl), 132, 135
power plants, natural gas and, 21
presence/absence, of coliform bacteria, 198
presence-sensing devices, 366, 367
Prevention of Significant Deterioration (PSD), 220, 237, 238, 240
primacy
defined, 15
jurisdictional, 175, 184, 206–207
Process Safety Management (PSM), 267
produced water, 10, 15, 106, 160, 184, 210
prohibited condition, confined entry, 310
propane, 62
propane (C_3H_8), 27, 30
proppant, 15, 112, 114, 116, 117, 118, 119, 133, 134, 135, 143
propping agents, 15
proprietary chemicals, 148–149
Prototype Oil Shale Leasing Program, 48
protozoa, 198
proved reserves, 15, 19, 70

public disclosure requirements, 140, 142
Public Health Service Act of 1912, 188
public lands, oil and gas operations on, 249
publicly owned treatment works (POTW), 190
public relations, 268
public water systems, 181, 190, 192
pullback devices, 367
P-waves, 6
pyrite, 40

Q

quartz, 38, 40

R

radiation, 140–141
  background, 140
  inversions, 216–217
  solar, 211, 227, 235
  ultraviolet, 209, 226, 235
radiofrequency presence-sensing device, 367
radionuclides, 194, 198–199
radium, 140, 141, 199
radon, 141, 199
Range Resources Corporation, 80
reactance grounds, 393
reactivity, 283
Reasonable and Prudent Practices for Stabilization (RAPPS), 187
reasonably available control measures (RACMs), 231
receptor, pollutant, 171
recharge zone, 190
reciprocating internal combustion engine (RICE), 242–244
reclamation, 15, 159, 176
recordkeeping, 168–169, 242, 253, 321–322
recoverable resources, 40–41, 70; *see also* technically recoverable resources
Red Wash field, 43
reference dose (RfD), 194
refining, 29–34
reflectance, vitrinite, 43
regulated substances, 191
regulations, 165–402
  air quality, 208–244
  environmental, 170–172
  land quality, 244–250
  OSHA, 252–402
  structure of, 174–179
  water quality, 179–201
renewable energy, 25–27
reportable quantity (RQ), 251, 252
rescue, confined space entry, 310, 316, 318, 325–327
  equipment, 316–317
reserves, proved, 15
reservoir fluids, 106
resistance grounds, 393
resonant grounds, 393
Resource Conservation and Recovery Act (RCRA), 206, 244–247, 266, 284
Resource Management Plans (RMPs), 249
respirators, *see* respiratory protection
respiratory protection, 287, 289–305
  sample evaluation checklist, 302–304
  sample written program, 290–301
restraint device, 367
retention pits, 109–110
retrieval line, 316–317, 326, 327
retrieval system, confined space, 310
reverse osmosis (RO), 125
Richter scale, 6
rigging, 346–361
  inspections, 352–353
  proof testing, 349, 352
  safety program, 347–349
  safe work practices, 353–355
  training, 351
"right to know" station binder, 402
ripple marks, sedimentary rock, 55
riser, 99
Risk Management Planning (RMP), 267
Rock-Eval, 42
rock gypsum, 52
rock salt, 52–53
Rockford Limestone, 85
room-and-pillar mine, 48
ropes, 350–354
rotary bits, 101

S

saccharin, 293
Safe Drinking Water Act (SDWA), 3, 16, 139, 140, 142, 143, 144, 166, 174, 179, 180, 187–207, 246
  activities required, 192–193
  amendments, 180, 192, 199–200
  -regulated chemicals, 146–147
safeguarding, machine, 361–375, 380
safety controls, 367–368
safety devices, 366–368
safety harness, 316–317, 326–327, 383
safety policies, 276–278
safety programs, 169, 285, 323, 363, 370–371
  fall protection, 271–272, 276
  fire, 343–344
  machine guarding, 363, 370
  rigging, 346–349, 353
  scaffold, 276
  welding, 380–383, 386
sample taps, water well, 100
sandstone, 36, 37, 38, 75, 78, 106
sanitary seal, water well, 98
sanitation requirements, 305
screens, well, 98–99
sedimentary rocks, 61, 69, 70
  characteristics of, 54–55
  chemical, 52–54
  facies, 56
  types of, 36–52
sedimentation, 35–36
Seep Ridge project, 48
seismic waves, 5–6
  types of, 6
seismograph, 5, 6
self-adjusting guards, 366
self-contained breathing apparatus (SCBA), 289, 292, 294, 295, 296, 297, 300–301, 304, 314, 317
setback, 15
severed estate, 15
severed minerals, 154
shaker, shale, 101
shale development, beneficiaries of, 2
shale gas, 1–3, 19–32, 153
  defined, 15
  drilling development technology, 100
  drilling for, vs. water, 93
  geology, 35–56
  hydraulic fracturing, 91–126, 135, 153–163
  importance of, 19–22
  regulations affecting development of, 165–402
  operations on public lands, 249
  OSHA standards applicable to, 262–402
  plays, 61–66
    active, in United States, 71–86
  sources, 69–87
shale shaker, 101
shalenanza, 1–2, 64, 66
shales, 36, 37, 38, 39
  classification of, 38
shallow water wells, 95, 96
Shell Oil Company, 48
shields, confined space, 315
shock, electrical, 381, 387, 389, 390, 392, 393, 394
shorite, 47
SIC Industry Group 138, 257–262
significant hazard to public health, 191
silica sand, 15
silicates, 40, 47
silicon dioxide, 143
siltstone, 37
sink, pollutant, 171
slickwater, 15, 111, 114, 116, 117, 118
slings, rigging, 350–354
slip resistance, 274
slips, 272, 273–274
slope winds, 216
Smackover Limestone, 78
smog, 223, 226–227, 230
smoke, 293
soda ash, 50
sodium bicarbonate, 49, 50
sodium carbonate, 49, 50
soft coal, 54
soil compaction, 3
soil quality, *see* land quality
solar energy, 26
solar radiation, 211, 227, 235
solid bed absorbents, 30
solid grounds, 393
Solid Waste Disposal Act, 246

sound barriers, 161
source, of pollution, 171
spark ignition engines, 243–244
specific source wastes, 283
Spill Prevention, Control and Countermeasure (SPCC), 208
spirometric evaluations, 294, 298–299
split estates, 15, 154–156
spudding in, 106
Squaw Bay Limestone, 83
stage, fracturing, 114
stair treads, 263
stairway railings, 263, 275
Standard Industrial Classifications (SIC), 257
Standards of Performance for Stationary Spark Ignition Internal Combustion Engines, 242
standpipe, 101
stannic oxychloride, 293
State Implementation Plans (SIPs), 236–237
state regulations, 175–177
static electricity, 395–396
stationary internal combustion engines, 242–243
stationary reciprocating internal combustion engines, 243
statistical dispersion models, 220
stimulation, 15
stipulation, 15
stormwater discharges, 185–187, 190
strata, 36
stratification, of sedimentary rocks, 54
stratified atmospheres in confined spaces, testing, 328
streamflow source zone, 191
STRONGER (State Review of Oil and Natural Gas Environmental Regulation), 176–177, 247
subadiabatic rate, 214
sub-bituminous coal, 53
subsidence inversion, 217
substages, fracturing, 114–116
suction line, 101
sulfate particles (SO_4), 222
sulfates, authigenic, 47
sulfur dioxide (SO_2), 15, 23, 32, 209, 221, 222–223, 233, 234
sulfur trioxide (SO_3), 222
sulfuric acid (H_2SO_4), 222, 223, 233
sulfurous smog, 223
sundry notice, 176
superadiabatic rate, 214
Superfund Amendments and Reauthorization Act (SARA), 233, 250–251
surface disturbance, 157–158
surface rights, 154, 156
surface water, 97, 109, 139, 171, 190, 191, 207
  contamination of, 3, 104
  filtration, 193
  radionuclides in, 198–199
  use of for hydraulic fracturing, 122–126
  wastewater discharge into, 182–185, 188, 252
surface waves, 6
S-waves, 6
sweetening, natural gas, 30
switching devices, 391–392
synthetic fiber ropes, 350–351, 353–354
synthetic organic chemical (SOCs), 194–196
system ground, 392

T

tasmanite, 44, 45
Tcf (trillion cubic feet), 12, 15, 63
technically recoverable resources, 15, 19, 71
  Antrim Shale, 85
  Barnett Shale, 75
  Fayetteville Shale, 77
  Haynesville Shale, 79
  Marcellus Shale, 81
  New Albany Shale, 86
  Woodford Shale, 83
telalginite, 44, 45
terrestrial oil shale, 44
testing, confined space, 310, 318, 327–328
texture, sedimentary rock, 54–55
thermal cutout, 392
thermal maturity, 43, 69
thermogenic gas, 16
therms, 63

thixotrophy, 16
threatened and endangered species, 16
tight gas, 16, 21, 62, 110
tight sand, 16, 62
tillites, 38
toluene, 30, 144, 147
topography, 215–216, 217, 221
torbanite, 44, 45
torch-cutting safety, 388–389
Tosco II process, 42
total dissolved solids (TDS), 16
total organic carbon (TOC), 69
total reduced sulfur (TRS), 172
toxic and hazardous substances regulations, 397–402
toxic chemicals, 131, 144–148, 194, 221
toxic gases, 209, 289, 313, 318, 327, 329, 386
toxic pollutants, 179, 182, 185, 191, 209, 210, 221–228, 233, 242–243
Toxic Release Inventory (TRI), 251–252
toxicity, 283–284
toxins, in fish, 187
training, 169, 253, 256
  confined space entry, 320–323
  personal protective equipment, 288–289, 299
transformation, 210–219
transport, of pollutants, 218–219
traveling block, 101
treatment technique (TT), 191
tricone rotary bit, 101
trihalomethanes (THMs), 194–196
tripping, 104, 106
trips, 272, 274
trona, 50
turbidity, 191, 194, 198
turbulence, 212–213, 226
two-hand safety controls, 368

U

ultraviolet radiation, 209, 226, 235
unconventional natural gas resources, 21, 61–66, 71, 153; *see also* natural gas
  categories of, 62–63
  distribution of, 64–66
Underground Injection Control (UIC) program, 16, 176, 201–207
underground injection wells, 201–207
  classes of, 201, 206
underground source of drinking water (USDW), 16, 201, 206
  contamination of, 107–108
Unocal Oil Company, 48
unregulated monitored substances, 191
upper flammable limit (UFL), 343
uranium, 50, 199
U.S. Bureau of Mines, 48
USEPA, 3, 15, 16, 139, 146, 147, 153, 229, 230, 231, 237, 245, 246–247, 250; *see also* Clean Air Act, Clean Water Act, National Ambient Air Quality Standards, Oil Pollution Act, Safe Drinking Water Act, Superfund Amendments and Reauthorization Act (SARA)
  candidate contaminant lists, 147
  hazardous waste categories, 283
  permits, 235, 237
  regulations and standards, 174–208, 222, 231, 232, 233, 235, 241–246
  Risk Management Planning directive, 267
  Toxic Release Inventory, 251–252
U.S. Steel, 2
Utica Shale formation, 1

V

valley winds, 216
vapor, 294
variance, National Primary Drinking Water Regulations, 191
ventilating equipment, 314
ventilation, 385–386
vents, well casing, 99
vertical wells, 82, 94, 100, 114, 117, 157, 158, 160
viruses, 198
visibility, 239–240
  impairment analysis, 240
vitrinite, 43, 44, 45
volatile organic chemicals (VOCs), 193, 194, 196, 209, 210, 221, 224, 231, 237, 238
volcanoes, 5
voltage to ground, 392–393
vulnerability zones, 220

W

warning signs, 373–374
washout, 211
wastewater, 8, 9, 139, 140, 182, 185, 190, 198, 206, 283, 398–399
water
  drilling for, vs. shale gas, 93
  drive, 99, 100
  pollution, 182
    point source, 188
  quality
    defined, 16
    regulations, 179–181, 184, 185
  storage pits, 110
  supply, for hydraulic fracturing, 121–126
  surface, *see* surface water
  table, 11, 106
  waste, *see* wastewater
  wells, 94–97, 101
    components of, 97–100
    gas well setback from, 109
    process for developing, 94
    site requirements for, 95
    types of, 95–97
Water Quality Act (WQA), 185
waterborne diseases, 198
watershed, 16
weather, air pollution and, 211–212
welding, 376–389
  safety, 379–389
  safe work practices, 386–388
  supervisor responsibilities, 382
  ventilation, 385–386
well casing, 95, 97, 98, 99, 114, 116
  construction, 106–109
  vent, 99
well completion, 109
well gas, 29
well head, 29, 30, 32, 81, 95, 97, 99, 102, 103, 109, 116, 193
Well Head Protection Programs, 193
well pad, 97, 154–156
well screen, 98–99
wellbore, 101, 102, 103, 104, 106, 110, 112, 114, 116, 117, 162
wet adiabatic lapse rate, 215
wet deposition, 240
wet gas, 3, 20
wetlands, 183–184, 191, 247
  mitigation, 183
whipstock, 16
wildlife, impact on, 158–159
Williston Basin, 107–108
wind energy, 26
wire rope, 351
Woodford Shale formation, 82–83
  gas content, 83
  original gas-in-place estimate, 83
  technically recoverable resources, 83
workover, 16
workplace evaluation, for confined entry, 311–312

X

xylene, 30, 144, 147, 194

Y

Youngstown, Ohio, 1, 9

Z

zone of saturation, 11